Emerging Consequences of Biotechnology

Biodiversity Loss and IPR Issues

Other Books by the Same Author:

What I Require from Life? Popular Essays of JBS Haldane, Oxford University Press, 2008.

Malaria: Genetic and Evolutionary Aspects, Springer, 2006.

Infectious Disease and Host-Pathogen Evolution, Cambridge University Press, 2004.

Biological Wealth and Other Essays, World Scientific, 2002.

Biological and Social Issues in Biotechnology Sharing, Ashgate, 1998.

Science and Society, University Press of America, 1998.

Haldane's Daedalus Revisited, Oxford University Press, 1995.

If I Am to be Remembered: The Life and Work of Julian Huxley, World Scientific, 1993.

The History and Development of Human Genetics, World Scientific, 1992.

Selected Genetic Papers of JBS Haldane, Garland Publishing, 1990.

The Foundations of Human Genetics, C.C. Thomas, 1989.

Cleft Lip and Palate: Aspects of Reproductive Biology, C.C. Thomas, 1986.

Haldane: The Life and Work of JBS Haldane with Special Reference to India, Aberdeen University Press, 1985.

Haldane and Modern Biology, Johns Hopkins University Press, 1968.

Emerging Consequences of Biotechnology

Biodiversity Loss and IPR Issues

Krishna Dronamraju
Foundation for Genetic Research, USA

World Scientific
NEW JERSEY • LONDON • SINGAPORE • BEIJING • SHANGHAI • HONG KONG • TAIPEI • CHENNAI

Published by

World Scientific Publishing Co. Pte. Ltd.

5 Toh Tuck Link, Singapore 596224

USA office: 27 Warren Street, Suite 401-402, Hackensack, NJ 07601

UK office: 57 Shelton Street, Covent Garden, London WC2H 9HE

British Library Cataloguing-in-Publication Data
A catalogue record for this book is available from the British Library.

EMERGING CONSEQUENCES OF BIOTECHNOLOGY
Biodiversity Loss and IPR Issues

ISBN-13 978-981-277-500-9
ISBN-10 981-277-500-5

Typeset by Stallion Press
Email: enquiries@stallionpress.com

Printed by Fuisland Offset Printing (S) Pte Ltd. Singapore

Dedicated to the memory of Franz and Jeane Wambaugh

and

their daughter Michele who gave her unwavering and loving support for this project

Foreword

M.S. Swaminathan

Chairman, M.S. Swaminathan Research Foundation
Chennai, India

The elucidation of the double-helix structure of the deoxyribose nucleic acid (DNA) molecule in 1953 by Drs James Watson, Francis Crick, Maurice Wilkins, and Rosalind Franklin marked the beginning of what is now known as "new genetics". Research during the last 54 years in the fields of molecular genetics and recombinant DNA technology has opened up new opportunities in agriculture, medicine, industry, and environmental protection. The ability to move genes across sexual barriers has led to heightened interest in the conservation as well as sustainable and equitable use of biodiversity, since biodiversity is the feedstock for plant, animal, and microbial breeding enterprises.

Considerable advances have been made during the last 25 years in taking advantage of new genetics in the areas of medical research, production of vaccines, sero-diagnostics, and pharmaceuticals for human and farm animal health care. The production of novel bioremediation agents — for example, the development of a new *Pseudomonas* strain for clearing oil spills in oceans, rivers, and lakes by Dr Anand Chakrabarty — is also receiving priority attention because of increasing environmental and water pollution. There has also been substantial progress in agriculture, particularly in the area of crop improvement, through the use of molecular-marker–assisted breeding, functional genomics, and recombinant DNA technology. A wide range of crop varieties containing novel genetic combinations are now being cultivated in the USA, Canada, China, Argentina, and several other countries. A strain of cotton containing the *Bacillus thuringiensis* gene (Bt Cotton) that has resistance to bollworms is now under cultivation in India based on both official and unofficial (illegal) releases.

There is little doubt that new genetics has opened up unexpected opportunities for enhancing the productivity, profitability, sustainability, and stability of major cropping systems. It has also created scope for developing crop varieties tolerant/resistant to biotic and abiotic stresses through an appropriate blend of Mendelian and molecular breeding techniques. It has led to the possibility of undertaking anticipatory breeding to meet potential changes in temperature, precipitation, and sea level as a result of global warming. There are new opportunities for fostering prebreeding and farmer-participatory breeding methods in order to continue the merits of genetic efficiency with genetic diversity.

While the benefits are clear, there are also many risks when we enter the territory of the unknown and unexplored. Such risks include potential harm to the environment and to human and animal health. There are also equity and ownership issues in relation to biotechnological processes and products. The following issues are the major areas of concern to the public and policy makers:

1. What is inherently wrong with the technology?

   Is the science itself (e.g. the use of selectable marker genes conferring antibiotic or herbicide resistance) safe?

2. Who controls the technology? Will it be largely in the private sector?

   If the technology is largely in the hands of the private sector, the overriding motive behind the choice of research problems will be private profit and not necessarily public good. If this happens, "orphans will remain orphans" with reference to the choice of research priorities. Crops being cultivated in rain-fed, marginal, and fragile environments, which are crying for scientific attention, may continue to remain neglected.

3. Who will have access to the products of biotechnology?

   If the products arising from recombinant DNA technology are all covered by intellectual property rights, then the technology will result in social exclusion and will lead to a further enlargement of the rich–poor divide in villages.

4. What are the major biosafety issues?

   There are serious concerns about the short- and long-term impacts of genetically modified organisms (GMOs) on the environment, biodiversity, and human and animal health.

Thus, there is a need for transparent and truthful risk-benefit analysis in relation to GMOs, on a case-by-case basis. In the coming decades, farm women and men in population-rich but land-hungry countries like India and China will have to produce more food and other agricultural commodities to meet home needs and to take advantage of export opportunities, under conditions of diminishing per capita availability of arable land and irrigation water and expanding abiotic and biotic stresses. The enlargement of the gene pool with which breeders work will be necessary to meet these challenges. Recombinant DNA technology provides breeders with a powerful tool to enlarge the genetic base of crop varieties and to pyramid genes for a wider range of economically important traits.

The safe and responsible use of biotechnology will enlarge our capacity to meet the challenges ahead, including those caused by climate change. At the international level, the Cartagena Protocol on Biosafety provides a framework for risk assessment and aversion. At the national level, there is a need for a regulatory mechanism that inspires public, political, and professional confidence.

Six Areas for International Collaboration

The following six key areas in global cooperation can be identified for immediate attention:

1. Sustaining and expanding Africa's Green Revolution

   The Green Revolution in Africa is an idea which has been a long time in coming. The growth rate of African agriculture was 3.9% during 2004, compared to the global average of less than 2%. The work done by the Earth Institute of Colombia University in Malawi and in numerous Millennium Villages in Africa has shown that a doubling in maize production is possible, if fertilizers, seeds, and treadle pumps can be made available to small farmers at affordable prices. The support given to resource-poor farmers for adopting yield-enhancing and environmentally benign technologies should be referred to as "technology adoption support" rather than as a subsidy. In areas affected by HIV/AIDS, support for nutrition should be given, in addition to making relevant drugs available. Finally, opportunities for assured and remunerative marketing are essential to sustain farmers' interest in productivity improvement. The numerous agencies working in Africa, such as the Food and Agriculture Organization of the United States (FAO), the Bill & Melinda Gates Foundation, the Rockefeller Foundation, and the Earth Institute could join together to form an "African Green Revolution Symphony" in order to sustain and enlarge Africa's Green Revolution.

2. Harmonizing energy and food security

It would be useful to organize a scientific consultation on land use policies for biofuel production in order to develop approaches to make food and fuel security mutually reinforcing. The current steep rise in the price of maize in the global market, as a result of the increasing use of maize for ethanol production, has serious implications for the food security of the poor in Africa and Latin America.

It is important to accord priority to biotechnological approaches to energy production. Also, biomass utilization for energy generation (e.g. pyrolysis and gassification of biomass) deserves greater attention. Land use policies in every country should be based on a careful consideration of the needs for food and energy security in a holistic manner. Markets that use grain for energy generation should take into consideration the overriding importance of food security of the nearly one billion members of the human family who are currently undernourished.

3. Saving native breeds of livestock

African cattle breeds like Boran and N'Dama have trypanotolerance traits. Similarly, it is likely that some indigenous poultry breeds may have resistance to the avian influenza virus. The dreaded H5N1 strain of the avian influenza virus is leading to the indiscriminate killing of native breeds of poultry. The time has come for a scientifically designed international effort to conserve and evaluate the native breeds of livestock for resistance to important transboundary-disease–causing organisms like the H5N1 strain of the avian influenza virus.

The Svalbard International Seed Vault established by Scandinavian countries under permafrost conditions near the North Pole, with a holding capacity of three million seed samples for the preservation and posterity of plant genetic resources, is a good example of valuable international collaboration. A similar initiative for both conserving and evaluating genetic variability in livestock is needed urgently. Priority in the international effort can be given to diseases of transboundary importance.

4. Building the capacity for sustainable development

Investment in capacity building confers multiple benefits — on individuals, organizations, and the environment. Capacity building efforts should cover both grassroot workers and national and international policy makers. An action-education model of capacity building that integrates academic coursework and field experience will be essential for developing a cader of professionals well versed in the art and science of sustainable development and in bridging the growing gap between scientific know-how and field-level do-how. The Earth University in Costa Rica is a good example of designing institutions to encourage transformational agents who combine scientific humanism and humanistic science in a symbiotic manner. A hub-and-spokes model of organization will help such institutions to spread their pedagogic methods for sustainable development speedily around the world.

5. Bridging the divides

The world is witnessing many divides, such as economic, technological, digital, genetic, and gender divides. How can we bridge such divides and ensure social inclusion for access to technologies of importance to human food and health security? Patents and intellectual property rights should not come in the way of all members of the human family deriving benefit from the products of the human brain, particularly in areas relevant to achieving the UN Millennium Development Goals (UN MDGs). UN MDGs represent a global common minimum program for sustainable human security and well-being.

To achieve the goal of social inclusion for access to relevant technologies, it will be useful if the InterAcademy Council, the International Council for Science (ICSU), and the Academy of Sciences for the Developing World (TWAS) would jointly sponsor the establishment of an "International Patent Bank for Sustainable Human Security", to which scientists can assign their patents that are relevant for safeguarding food security and human and animal health as well as for mitigating the adverse impact of global warming and sea-level rise. The science academies can then help to ensure that access to relevant technologies is not denied to those who are unable to pay for them.

6. Building international networks

Global and regional action research networks, which can help to develop location-specific methodologies for conserving basic life support systems, will help buy time in technology development and dissemination. Many such networks already exist, particularly under the auspices of the Consultative Group on International Agricultural Research (CGIAR), ICSU, TWAS, and other organizations. It would be especially useful if interdisciplinary networks are organized to develop field-level action plans for the sustainable management of tropical rainforests and coral reefs. These habitats are rich sources of biodiversity, and a well-planned global effort for saving them will be valuable.

The present book by Professor Krishna R. Dronamraju is a timely contribution to the ongoing debate relating to the role of frontier technologies in stimulating and sustaining an "ever-green revolution" in agriculture, leading to the enhancement of productivity in perpetuity without associated ecological harm.

Acknowledgments

My foremost thanks go to M.S. Swaminathan for the Foreword as well as much advice and consultation. I am much appreciative of the photographs which were all taken by Michele Wambaugh. Others I have consulted from time to time included Peter Raven, Elof Carlson, David Hopwood, James F. Crow, William J. Schull, and Victor McKusick. Many years ago, it was my mentor, J.B.S. Haldane, who taught me the value of knowing one's ecological surroundings on biological, ethical, and moral grounds: "Any self-respecting biologist ought to be familiar with the local plant and animal species."

Contents

Introduction

"Healthy discontent is the prelude to progress."

— Mahatma Gandhi (1929)

The beneficial role of biotechnology in improving crop yields is well established. However, several recent studies have also revealed that there are significant risks associated with genetically modified (GM) crops — risks that may adversely affect our health and environment. Long-term studies are urgently needed to produce results which may be able to resolve these questions, yet such data have not been forthcoming. If this book raises too many questions, it is because they are timely. Biotechnology is at a critical juncture where several developing countries are poised to make large investments in GM crop research and expand their cultivation nationwide. The need for risk evaluation has never been greater than it is today. Underneath this critical approach is the basic optimism that biotechnology will be able to provide the means to feed the world's hungry, provided that the potential risks are first evaluated and dealt with. It would be unwise to proceed with the planting of GM crops on a large scale until the risks to health and environment are fully evaluated.

As a geneticist and biotechnologist, I have been keenly interested in the rapid development of biotechnology since 1980, when Ananda Chakrabarty successfully patented a live human-made microorganism, which was approved by the U.S. Supreme Court in a 5–4 decision in the case of Chakrabarty v. Diamond. It involved a GM bacterium, *Psudomonas aeruginosa*, which was designed to break down four of the main components of crude oil. This was an important step because it marked a new beginning for patenting man-made living organisms: it was based on the premise that the patent legislation, which was earlier enacted by the U.S. Congress, does not distinguish between living and nonliving matter. Until then, microorganisms were considered to be products of nature and thus not patentable.

The Chakrabarty patent provided the judicial framework for the U.S. Patent and Trademark Office to determine later that plants and animals are patentable subject matter under the U.S. code. It is a well-known fact that the Chakrabarty patent provided great economic stimulus to the patenting of microorganisms and cells, as well as a great incentive for the growth of the biotechnology industry.

Biotechnology and Biodiversity

Since the exciting and hopeful prospect of abundance that was initially expected of biotechnology, we have come a long way. Biotechnology has indeed produced many benefits. Food production has increased manifold, although population growth has outstripped achievements in agriculture. It has become clear during the last several years that the "green revolution", pioneered by Prof. M.S. Swaminathan, that was so successful in carrying India forward during the last 30 or 40 years is no longer adequate. This is due to several complex factors which have been discussed by others (e.g. Swaminathan 1999). What is needed is an "evergreen revolution", or perhaps several such revolutions, occurring periodically to meet the demands of an ever-growing population. The area of agricultural land is shrinking rapidly, due to the explosive growth of human and animal populations followed by habitat destruction.

Causes for Declining Agrobiodiversity

The principal underlying causes for declining agrobiodiversity include the rapid expansion of industrial and Green Revolution agriculture, intensive livestock production, and industrial fisheries and aquaculture (some production systems using GM varieties and breeds) that cultivate relatively few crop varieties in monocultures, rear a limited number of domestic animal breeds, or fish for or cultivate few aquatic species. Variety replacement is the main cause of losses. The replacement of local varieties or landraces by "improved" and/or exotic varieties and species is reported to be the major cause of genetic erosion around the world.

Globalization of the food system and marketing, as well as the extension of industrial patenting and other intellectual property systems to living organisms, has led to the widespread cultivation and rearing of fewer varieties and breeds for a more uniform and less diverse, but more competitive, global market. The consequences are marginalization of small-scale, diverse food production systems that conserve farmers' varieties of crops and breeds of domestic animals, which form the genetic pool for food and agriculture in the future; reduced integration of livestock in arable production, which reduces the diversity of uses for which livestock are needed; and reduced use of "nurture" fisheries techniques, which conserve and develop aquatic biodiversity.

Genetic erosion refers to the loss of genetic diversity, including the loss of individual genes and gene complexes (particular combinations of genes), such as those manifested in locally adapted landraces. The main cause of genetic erosion in crops, as reported by almost all countries, is the replacement of local varieties by "improved" or exotic varieties and species. As old varieties in farmers' fields are replaced by newer ones, genetic erosion frequently occurs because the genes and gene complexes found in the diverse farmers' varieties are not contained in the modern varieties. In addition, the sheer number of varieties is often reduced when commercial varieties are introduced into traditional farming systems. There have been few systematic studies of the genetic erosion of crop genetic diversity that have provided quantifiable estimates of the actual rate of genotypic or allelic losses.

Benefits of Agricultural Biodiversity

Prime Minister Manmohan Singh of India pointed out that in the past, the community food tradition assured that a wide range of food crops rich in protein, iron, micronutrients, and vitamins was available to the people; however, commercial agriculture has narrowed the range of food crops available.

Agricultural biodiversity can help in developing decentralized community food security systems that benefit local communities. They are also beneficial for long-term security through the establishment of gene banks, seed banks, and grain banks, which can be managed by local people. The diversity of crops could also reduce pesticide use. Furthermore, tropical fruits, sweet potato (with beta-carotene), and other vegetable crops can fight vitamin A deficiency in children.

Agricultural biodiversity provides the important raw material for improving the quality of crops, livestock, and fish. It can also create opportunities for entrepreneurship by generating employment and additional income from a whole range of value-added foods, medicines, nutraceuticals, biofuel, and other sources. On a global scale, nearly 2.5 billion people depend directly on wild and traditionally cultivated plant species to meet their daily needs.

Toxic Effects

Unfortunately, there are other consequences of biotechnology that are clearly undesirable. The picture that has emerged from recent studies is unfavorable, even alarming, if the results are confirmed in long-term studies. While we are even more convinced today that biotechnology will continue to be needed for feeding the world's billions in the future, there is evidence which indicates that a cautious approach is very much warranted. The method of achieving greater food production by utilizing GM crops, as we have done so far, appears to have some significant associated risks to our health and environment. A few examples will suffice.

An analysis of results obtained when rats were fed GM corn (MON 523 produced by Monsanto Corp.) was recently published by Prof. Gilles-Eric Seralini (2007) from the University of Caen, France. The study, completed at CRIIGEN (Caen, France), examined the raw data on MON 863 feeding experiments on rats, initially suppressed by Monsanto but later obtained by others in 2005 after a court action in Germany. Using more sophisticated analytical methods than those employed by Monsanto, the new study uncovered an increase of up to 40% in blood triglycerides in female rats, and a more-than-30% decrease in urine phosphorus and sodium in male rats, specifically linked to the GM diet. The reasons for these changes are unclear, but they may provide clues to the deaths of many animals which consumed *Bacillus thuringiensis* (Bt) feed in other animal experiments. However, these data should be confirmed further using large numbers of experimental animals.

Similar observations were reported by the Russian scientist Irena Ermakova, who fed GM soy to female rats.[a] Other studies have shown the deleterious effects of GM pollen on Monarch butterflies and caterpillars as well as other insects.

Hellmich *et al.* (2001) conducted laboratory tests to establish the relative toxicity of Bt toxins and pollen from Bt corn in monarch larvae. They found that first instars are sensitive to Cry1Ab and Cry1Ac proteins, and that pollen contaminants can dramatically influence larval survival and weight gain and produce spurious results.

The biologist Michelle Marvier and her colleagues (2007) at Santa Clara University in California, USA, examined the ecological

[a] Irina Ermakova's research was critically examined in the following paper: Marshall, A. (2007) GM soybeans and health safety — a controversy reexamined. *Nat Biotechnol* **25**: 981–987.

consequences of transgenic Bt in a meta-analysis of 42 field experiments. They indicated that nontarget invertebrates are generally more abundant in Bt cotton and Bt maize fields than in nontransgenic fields managed with insecticides; however, in comparison with insecticide-free control fields, certain nontarget taxa are less abundant in Bt fields. The central goal of this study was to quantitatively investigate whether changes in invertebrate abundance were statistically significant. The failure to find significant differences in previous studies was generally viewed as a signal of environmental safety, but they were often based on small samples. Whether statistically significant differences in abundance truly indicate ecologically significant changes is not clear. The study revealed, however, that Bt crop acreage has less insect biodiversity than untreated fields. It is unclear whether the reduced abundance of various groups (coleopterans, hemipterans, and hymenopterans) is due to direct toxicity or is a response to reduced availability of prey in Bt crops.

1

Impact of GM Crops on Biodiversity and the Environment

The environmental impact of genetically modified (GM) crops seemed almost benign when studies first began several years ago. However, after numerous studies by various public and private groups in several countries, the situation looks increasingly dire. While some reports have indicated a mildly adverse impact of GM crops, several recent studies have produced results that are quite alarming in indicating toxicity and a real danger to the environment and ecology. These discussions are summarized in this chapter, with an appraisal of their implications for biodiversity.

A GM crop plant that is toxic to insect pests could have a direct harmful effect on nontarget insects; it could also have an indirect effect by reducing the food source for other wildlife, such as birds. GM crops that are tolerant to herbicides could lead to a reduction in weeds which may harbor beneficial insects, and could also indirectly impact on their predators, i.e. the bird populations.

An insecticide could create insects which become resistant to those chemicals when used on pest-tolerant GM crops repeatedly. This would naturally increase the number of insect pests, thus creating an imbalance in nature by altering the predator/prey ratio.

But, there are other effects as well. Monarch butterfly larvae fed only on leaves covered in *Bacillus thuringiensis* (Bt) pollen grew more slowly and showed higher death rates. Aphids fed on GM potatoes producing a different toxin were also reported to have a harmful effect on ladybirds. Over 10 million birds are reported to have disappeared from the countryside in the U.K. in recent years.

Summary of Exposure of Animals and Human Beings to GMOs

Species	GM species	Transgene trait	Effect
Rat	Soya	Roundup Ready	Stunting, death, sterility
Humans	Cotton	Cry1Ac/Cry1Ab	Allergy symptoms
Sheep	Cotton	Cry1Ac/Cry1Ab	Death, liver toxicity
Cows	Cotton	Cry1Ac/Cry1Ab	Death, liver toxicity
Goats	Cotton	Cry1Ac/Cry1Ab	Death, liver toxicity
Mice	Pea	Alpha-amylase inhibitor	Lung inflammation, general food sensitivity
Mice	Soya	Roundup Ready	Liver, pancreas, and testis affected
Humans	Maize	Cry1Ab	Illnesses, death
Rats	Maize	Cry3Bb	Liver and kidney toxicity
Cows	Maize	Cry1Ab/Cry1Ac	Death, illnesses
Rats	Potato	Snowdrop lectin	Damage in every organ system, stomach lining twice as thick as controls
Mice	Potato	Cry1A	Gut lining thickened
Rats	Tomato	Delay ripening	Holes in the stomach
Chickens	Maize	Glufosinate tolerance	Deaths

(Ho 2007)

An Avalanche of Bans and Rulings Strikes GM Crops Worldwide

Other rulings and bans on genetically modified organisms (GMOs):

- GM alfalfa ban in the USA was made permanent.

- Nine towns in Massachusetts, USA, voted against GM food and crops.
- Santa Cruz County, California, USA, imposed a moratorium on GM crops.
- The Ecuador government imposed a ban on GM food aid.
- Bolivia outlawed GM crops and went organic.
- Mexico banned planting of GM corn.
- South Australia extended GM ban, and Western Australia was against GM trials.
- Romania banned GM soya from January 2007.
- GM seeds were banned in Greece.
- Germany imposed much stricter regulations on GM maize.
- Hungary passed the "strictest" GMO crop law.
- Poland imposed a ban on GM animal feed and planting of GM seeds.
- India imposed a ban on further GM field trials and introduced stringent new conditions on trials approved.
- The Netherlands will either return or burn US shipment of GM maize that lacks safety clearance.
- Cyprus intends to declare itself GM-free, agricultural minister Photis Photious announced on June 6, 2007.

Thirty Years of GMOs Are More than Enough (Ho 2007)

- No increase in yields; on the contrary, GM soya decreased yield by up to 20% compared with non-GM soya, and up to 100% failure of Bt cotton was found in India.
- No reduction in pesticide use; on the contrary, GM crops increased pesticide use by 50 million pounds from 1996 to 2003 in the United States.
- GM crops harm wildlife, as revealed by the UK's farm-scale evaluations.
- Bt-resistant pests and Roundup-tolerant superweeds render the two major GM crop traits practically useless.
- Vast areas of forests, pampas, and cerrados were lost to GM soya in Latin America (15 million hectares in Argentina alone); this may worsen with the demand for biofuels.

- Epidemic of suicides was found in the cotton belt of India involving 100 000 farmers between 1993 and 2003; a further 16 000 farmers a year have died since.
- GM food and feed were linked to deaths and sicknesses in the field and in lab tests.
- Roundup herbicide is lethal to frogs and toxic to human placental and embryonic cells, yet is used in more than 80% of all GM crops planted in the world.
- Transgene contamination is unavoidable; scientists have found GM pollination of non-GM crops and wild relatives 21 km away.

Potential Hazards of GMOs

- Synthetic genes and gene products new to evolution could be toxic and/or immunogenic for humans and other animals.
- Genetic modification is uncontrollable and unreliable, as it mutates and scrambles genomes, generating deformities and toxic or immunogenic products. These problems are multiplied by the instability of transgenic DNA.
- Viruses in the host genome that cause diseases may be activated by genetic modification.
- Spread of antibiotic resistance genes to pathogens by horizontal gene transfer makes infections untreatable.
- Genetic modification greatly facilitates and enhances horizontal gene transfer and recombination, a main route to creating disease agents.
- Transgenic DNA is designed to invade genomes, and its strong synthetic promoters may trigger cancer by activating oncogenes.
- Herbicide-tolerant GM crops accumulate herbicide and herbicide residues highly toxic to humans and animals as well as plants.

DuPont in India

DuPont Co. will open its first plant biotechnology center outside the United States at a new research center in Hyderabad, India. The biotech center will be located in the $22.5 million DuPont

(*Continued*)

(Continued)

Knowledge Center, which is expected to break ground this month and open in early 2008. DuPont is leasing a facility near the site, and has already hired 20 new crop genetics scientists for the center. Another 80 scientists will be hired by the end of the year.

"The center will allow us to access tremendous scientific talent in the region in support of DuPont's efforts to create products that address the food, feed, fuel, and materials challenges of the 21st century," said Balvinder S. Kalsi, president and chief executive of DuPont India.

Dupont opens first major plant — Biotechnology Research Center outside United States. Dupont Press Release, New Delhi, March 14, 2007.

Strong Suspicions of Toxicity in One GMO Corn

For the first time in the world, an independent study on the health risks of a GM maize authorized for consumption shows signs of hepatorenal toxicity. It is a countervaluation performed by the Committee for Independent Research and Information on Genetic Engineering (CRIIGEN), France, of a regulatory study by the Monsanto Company on rats fed with a GM maize (MON 863) over a 3-month period. The raw data were used to obtain the commercial release of this GM maize at an international level. These revelations are certainly sufficient to require an immediate ban of GM maize MON 863 and all of its hybrids from human or animal consumption, as well as new and more carefully conducted feeding studies.

A report in the French Newspaper *Le Monde* (March 14, 2007) claimed that the consumption of MON 863, a transgenic corn invented by Monsanto, disturbs numerous biological parameters in rats to a greater or lesser extent: weight of the kidneys, weight of the liver, the level of reticulocytes (new red blood cells), the level of triglycerides, etc. Urinary chemistry is also changed, with reductions in excreted sodium and phosphorus going as high as 35%. The effects vary with the sex of the animals: female rats exhibit an increase in blood fat and sugar levels, and an increase in body weight — all associated with greater hepatic sensitivity.

The following headline stories appeared:

Revealed: Why Monsanto Suppressed GM Maize Feeding Study

Press Notice

GM Free Cymru

March 13, 2007

Independent analysis of data uncovers evidence of scientific fraud

It has been revealed today, at a Paris Press Conference (1), that the Monsanto GM maize referred to as MON863 caused serious damage to the liver and kidneys of rats which consumed it during feeding trials. This is the first time in the world that a study on the health risks of a GM maize authorized for consumption shows signs of hepatorenal toxicity (2). The study is published today in the peer-reviewed journal *Archives of Environmental Contamination and Toxicology*.

The study, completed at CRIIGEN (Caen, France), contains an examination of the raw data on MON863 feeding experiments initially suppressed by Monsanto but later obtained in 2005 after a Court action in Germany. Prior to that court action, Monsanto had refused public access to the data on the spurious grounds of "commercial confidentiality", although it had been widely leaked that the feeding studies showed statistically significant negative health effects on animals fed with the GM maize (3).

The GM maize in question produces a new insecticide called "modified Cry3Bb1" which has the capacity to kill the coleopteran insect *Diabrotica virgifera*. The plant also contains a gene coding for antibiotic resistance. (There are many other commercial GM varieties which produce new insecticides, and many others which are herbicide-tolerant or herbicide-resistant. Almost all of these new varieties have been heavily criticized by independent scientists on the grounds that their safety has never been fully established.) In America the variety is classified as a pesticide since every cell is toxic to insects. In spite

(*Continued*)

(Continued)

of widespread concern and protests from the scientific community and consumer organizations, MON863 was given formal approval by the EC on 8th August 2005.

The new study (4) involved a new and rigorous statistical analysis of all the raw data in the 1130 page document, concentrating on the blood and urine analyses of the test animals. The French researchers claim that the Monsanto statistics were not detailed enough and that their protocols were questionable. Real damage to test animals was therefore masked by the analytical methods chosen — and there can be little doubt that Monsanto knew this. Upon detailed analysis, the French team uncovered an increase of up to 40% in blood triglycerides in females, and a more than 30% decrease in urine phosphorus and sodium in males, specifically linked to the GM diet. The reasons for these changes are unclear, but they may provide clues to the deaths of many animals which have consumed Bt feed in other animal experiments (5).

Professor Seralini said: "These revelations are profoundly disturbing from a health point of view. They are certainly sufficient to require new and more carefully conducted feeding studies and an immediate ban from human or animal consumption of GM maize MON 863 and all its hybrids. This maize cannot now be considered safe to eat. We are now calling urgently for a moratorium on other approved GMOs while the efficacy of current health testing methods is reassessed."

Speaking for GM Free Cymru, Dr Brian John said: "Now we know why Monsanto wanted so desperately to keep this animal feeding study out of the public domain. There is scientific fraud here, and this must now be apparent to all of us, including the regulatory bodies. Goodness knows how many other studies showing real harm to animals fed on GM crops and foods have simply been hidden away from independent scrutiny. We support Professor Seralini's call for an immediate moratorium on ALL GM varieties, approved or unapproved, while the regulators put into place the robust and independent health

(Continued)

(Continued)

testing methods that we have been calling for since 2001. There can now be no further doubt that GM crops and foods are damaging to health."

Notes

1. From the Independent Committee for Research and Information on Genetic Modification: Corinne LEPAGE, Présidente du CRIIGEN (Comité de Recherche et d'Information Indépendantes sur le Génie Génétique) vous convie à une conférence de presse le mardi 13 Mars 2007 à 17h30 sur le thème : « OGM : de nouvelles révélations » En présence du Professeur Gilles-Eric SERALINI, des Docteurs Dominique CELLIER et Joël SPIROUX de VENDOMOIS du Conseil Scientifique du CRIIGEN (www.criigen.org) Au Pain Quotidien 18 rue des Archives 75004 Paris Métro: Hôtel de Ville Merci de confirmer votre présence auprès de Céline ALONZO par téléphone au 06 03 53 19 07 ou par mail calonzo@free.fr

   Un cas grave : un maïs OGM autorisé est impropre à la consommation Pour la première fois au monde, une étude des risques sur la santé d'un maïs transgénique autorisé à la consommation montre des signes de toxicité hépatique et rénale. Des exigences et recommandations seront formulées.
2. The article, entitled "New analysis of a rat feeding study with a genetically modified corn reveals signs of hepatorenal toxicity" is by Gilles-Eric Séralini, Dominique Cellier, and Joël Spiroux de Vendomois. It is published on line (www.springerlink.com/content/1432-0703) by the American journal *Archives of Environmental Contamination and Toxicology*. It will be printed in April.
3. The peer review of the MON863 feeding study by Dr Arpad Pusztai was subject to a gagging order imposed by Monsanto as a condition for the report to be examined. Dr Pusztai also revealed statistically significant differences between the "GM-fed" group of rats and

(Continued)

(Continued)

the control group. See these items: http://www.gmfreecymru.org/news/Press_Notice31May2005.htm; http://www.gmfreecymru.org/pivotal_papers/monsanto.htm

4. From CRIIGEN press release: Animal feed made from MON863 maize was given to rats in a laboratory over a period of 13 weeks. The associated tests on the GM-fed group and control groups were the longest and most detailed ones involving mammals which have consumed this plant, and they were used in support of its authorization throughout the world. These tests were controversial from the outset in France, and in 2003 they provoked a disagreement between experts, in particular in the French CGB (Commission du Génie Biomoléculaire). CRIIGEN (the Committee for Independent Research and Information on Genetic Engineering) was concerned about possible scientific fraud, and asked the GM regulatory authorities for sight of the raw data. These data were kept confidential until Greenpeace Germany won a Court verdict against Monsanto; this forced the company to make public the blood and urine analyses of rats fed with MON863 during the 3 month feeding trials.

   The data are contained within more than 1130 pages of tables of numbers and calculations. A group from CRIIGEN comprising Prof. Gilles-Eric Séralini (researcher on pesticides and governmental expert on GMOs, University of Caen), Dr. Dominique Cellier (biostatistician, University of Rouen), and Dr. Joël Spiroux de Vendomois (physician and specialist on environmental health), has now performed a re-evaluation of these data. The work has been done quite independently of Monsanto or any other GMO producer.

   The effects of the GM maize on animal weight variations were not studied by the Monsanto scientists. In 2006 the company published certain studies based on the feeding trials, but the scientists did not analyse animal weight or urine data. The statistics were not detailed enough and their protocols were questionable. Upon detailed analysis, the data are now shown to reveal an increase of up to 40% in blood triglycerides in females, and a

(Continued)

(Continued)

more than 30% decrease in urine phosphorus and sodium in males, specifically linked to the GM diet. However, these effects were not picked up by the regulatory authorities including EFSA, and they did not request any repeat or prolongation of these experiments.

5. http://www.gmfreecymru.org/news/Press_Notice16Feb2007.htm
   http://www.gmfreecymru.org/pivotal_papers/ermakova.htm
   http://www.gmfreecymru.org/pivotal_papers/toxic.htm

Additional Notes

The authors of this work used data drawn from an experiment sponsored by Monsanto, which bore on the study of 400 rats for 90 days. The statistical treatment applied to these data by the experts of the agrochemical firm was published in August 2005, by *Food and Chemical Toxicology*. That work brought to light significant variations in biological parameters between animals fed MON 863 and those fed with its isogene — the same plant variety without the genetic modification.

Monsanto researchers, for their part, had concluded that those disparities were within the frame of the natural variability of the measured parameters. The effects produced by the GMO were therefore not considered pathological. As for the "natural variability," it had been established by measuring the same series of data on rats fed with other varieties of non-GMO corn, with different nutritional values from MON 863 and its isogene.

The raw experimental data — over a thousand pages — were kept confidential by the agrochemical firm until Greenpeace obtained a court order for its publication in spring 2005 from the Appeals Court of Munster (Germany).

Criigen was thus able to examine the data in detail and to apply a new statistical treatment to them. According to Mr. Seralini, that, notably, consisted of extracting from the raw data the most significant effects specifically imputable to GMO absorption.

(*Continued*)

(Continued)

"Of the 58 parameters measured by Monsanto," the researcher details, "all those that were altered concern kidney or liver functioning." He continued, "Furthermore, Monsanto had deemed that, because the males and the females responded differently, there was no reason for worry." He added, "Yet, the liver, for example, is an organ that reacts differently as a function of sex." In the same way, the fact that the measured biological response was not always in exact correlation with the dose of GMO received was interpreted by the company's experts as proof that the transgenic corn being tested was not the cause. Mr. Seralini contests that principle: "When the disturbances are hormonal, for example, the impact may not be proportional to the dose."

Toxicologist Gerard Pascal, a member, like Mr. Seralini, of the Committee on Bio-molecular Engineering, deems certain that Criigen's conclusions are erroneous. "I reject the analysis of the animals' weight curves, conducted without taking their feeding into account," says Mr. Pascal. "But I agree that the biological responses may vary between males and females and with the principle that the effects of a GMO corn must be compared with its isogene only and not take into account effects produced by other corn varieties."

According to Mr. Pascal, the lack of direct correlation between the GMO doses received and the impacts observed on the hepatic parameters disqualifies the conclusions about liver toxicity. Significant differences with respect to "kidney weight" and "urinary sodium, phosphorus, and potassium" suggest a renal impact. "However," Mr. Pascal recalls, "at my request, the CGB pressed for investigations of the kidneys and had not found any definitive evidence of toxicity" (December 15th, 2004, Le Monde). "The variations in the levels of reticulocytes and eosinophiles (white blood cells) remain," adds M. Pascal. "I don't know how to interpret that, but those are parameters that move around a lot in experiments." As far as Mr. Pascal is concerned, the information developed by Criigen is not of a nature to call into question the favorable opinions delivered with respect to MON 863. "All that is nothing but a personal interpretation," adds the toxicologist.

A Serious Concern: Authorized GM Maize Is Unfit for Consumption

CRIIGEN

Press release

March 14, 2007

The case of Bt GM maize MON 863

Abstract

It is a countervaluation performed by CRIIGEN (France), of a regulatory study by the Monsanto Company, on rats fed with a GM maize (MON 863) over a three-month period. The raw data were used to obtain the commercial release of this GM maize at an international level. The symptoms discovered in re-analyzing the data are consistent, and are evidenced in comparison to control rats of the same genetic origin, the same age, and caged in strictly similar conditions. They have eaten a diet of equilibrated chemical composition, assessed as equivalent to controls, but without the Bt toxin which is the insecticide produced by the GM maize itself. On average, females show a gain of weight, a significant increase of sugar and fat in the blood, an increase of liver weight relative to body weight, and disruption of renal function. Inversely, the males lose weight, they are more sensitive at the renal level, the kidneys also lose weight in comparison to the body, and ions analyses are modified in urine. This may have a relationship with the diagnosed nephropathies. This latter phenomenon may be naturally developed with age in this rat strain, but in this case the rats were young, reaching only five months by the end of the experiment. Markers of hepatic function are also reached. We can notice that toxic products such as pesticides regularly provoke different effects according to the sex, like during a cancer initiation. It is not possible for such short tests to identify the precise beginning of a particular disease. However, the detoxification organs are reacting.

The body weight variations of these animals were not statistically evaluated by Monsanto, who published a study on this subject in

(*Continued*)

(Continued)

2006. Monsanto's paper also omitted the urine chemistry analyses. The statistics were not detailed enough and their protocols were questionable.

1. We raise concern about the reasons for which the authorities did not require an independent study of the statistical analyses performed by Monsanto, which would have exposed these problems.
2. We question why the authorities did not require the renewal and the prolongation of these experiments, controversial since 2003.
3. And we question whether the authorities did not ask for the sexual hormones measurements, that may be disrupted because of the different effects based on gender.

The raw data of Monsanto that allowed this countervaluation were obtained via Court action. These data were considered as confidential not only by the Company, but also by European States and the European Community. The data thus concern the MON 863 maize producing a new insecticide called "modified Cry3Bb1" supposedly there to kill Chrysomelidae (coleopteran insect, *Diabrotica virgifera*). This insect is a particularly devastating pest to the maize. It was also recently introduced by plane several times in Europe. This recently authorized GM maize also contains a gene coding for antibiotic resistance. Monsanto's tests prove quite insufficient, although they are at the same time the most detailed, and the longest ones, ever performed over the world on mammals, after consumption of this plant; and these are typical of actual regulatory tests for GMOs (lasting only 90 days maximum on rats).

Because it produces a new internal insecticide, this GMO belongs to the second most important category of cultivated and commercialized GMOs throughout the world. The other GMOs absorb an herbicide without dying. Thus, most of GMOs are pesticide-plants.

(*Continued*)

(Continued)

For the record, these tests were controversial from the outset in France, and in 2003 they provoked a disagreement between experts, in particular in the French CGB (Commission du Génie Biomoléculaire). CRIIGEN (the Committee for Independent Research and Information on Genetic Engineering) was concerned about possible scientific weakness, and asked the GM regulatory authorities for sight of the raw data. These data were kept confidential until Greenpeace Germany won a Court verdict against Monsanto; this forced the company to make public the blood and urine analyses of the rats under experiment. The raw data are contained within more than 1130 pages of tables of numbers and calculations. A group from CRIIGEN comprising Prof. Gilles-Eric Séralini (researcher on pesticides and governmental expert on GMOs, University of Caen), Dr. Dominique Cellier (biostatistician, University of Rouen), and Dr. Joël Spiroux de Vendomois (physician and specialist on environmental health), have concluded a study and re-evaluation of these data. The work has been done independently of Monsanto or any other GMO producer.

These revelations are certainly sufficient to require an immediate ban of GM maize MON 863 and all its hybrids from human or animal consumption, as well as new and more carefully conducted feeding studies. This maize cannot now be considered safe to eat. We are calling urgently for a moratorium on other approved GMOs while the efficacy of current health testing methods is reassessed.

(1) The article, entitled "New analysis of a rat feeding study with a genetically modified maize reveals signs of hepatorenal toxicity" is by Gilles-Eric Séralini, Dominique Cellier, and Joël Spiroux de Vendomois. It is published on line (http://dx.doi.org/10.1007/s00244-006-0149-5 (you may need to copy and paste the URL into your browser)) by the American journal *Archives of Environmental Contamination and Toxicology*. It will be printed in May. The editor is Dr. Doerge from the Food and Drug Administration (FDA).

The effects of the GM maize on animal weight variations were not studied by the Monsanto scientists. In 2006, the company published certain studies based on the feeding trials, but the scientists did not analyze animal weight or urine data. The statistics were not detailed enough and their protocols were questionable. Upon detailed analysis, the data are now shown to reveal an increase of up to 40% in blood triglycerides in females and a more-than-30% decrease in urine phosphorus and sodium in males, specifically linked to the GM diet. However, these effects were not picked up by the regulatory authorities including EFSA, and they did not request any repeat or prolongation of these experiments.

EFSA Rejects Concerns over Monsanto Maize

On June 29, 2007, The European Food Safety Authority's (EFSA) GMO panel has no safety concerns after reviewing data from French scientists suggesting toxicity concerns in rats fed the MON863 variety of GM maize from Monsanto. It received European approval for use in animal feed in 2005 and for human consumption in 2006. The corn is also authorised in Australia, Canada, China, Japan, Mexico, the Philippines and the USA. In the US, no re-evaluation of the data was announced by FDA.

After re-evaluating the safety data relating to MON863, EFSA however have come to a different set of conclusions, stating: "EFSA considers that the paper does not present a sound scientific justification in order to question the safety of MON 863 maize.

"Observed statistically significant differences reported by Monsanto, Séralini *et al.*, and EFSA, were considered not to be biologically relevant. In the absence of any indications that the observed differences are indicative of adverse effects, the GMO Panel does not consider that this paper raises new issues with respect to the safety of MON 863 maize.

(*Continued*)

(Continued)

"Therefore, the GMO Panel sees no reason to revise its previous Opinions that the MON 863 maize would not have an adverse effect in the context of its proposed use," said the authority.

EFSA rejects concerns over Monsanto maize. www.foodnavigator.com, June 29, 2007/Europe, IWC Paris 2008.

FDA

The Food and Drug Administration (FDA) of the United States does not require any safety tests on genetically modified foods. If Monsanto or other biotech companies declare their foods safe, the agency has no further questions. The rationale for this hands-off position is a sentence in the FDA's 1992 policy that states, "The agency is not aware of any information showing that foods derived by these new methods differ from other foods in any meaningful or uniform way." The statement, it turns out, was deceptive. Documents made public from a lawsuit years later revealed that the FDA's own experts agreed that GM foods are different and might lead to hard-to-detect allergens, toxins, new diseases or nutritional problems. They had urged their superiors to require long-term safety studies, but were ignored. The person in charge of FDA policy was, conveniently, Monsanto's former attorney (and later their vice president). One FDA microbiologist described the GM food policy as "just a political document" without scientific basis, and warned that industry would "not do the tests that they would normally do" since the FDA didn't require any. He was correct.

There have been less than 20 published, peer-reviewed animal feeding safety studies and no human clinical trials — in spite of the fact that millions of people eat GM soy, corn, cotton, or canola daily. There are no adequate tests on "biochemistry, immunology, tissue pathology, gut function, liver function and kidney function," and animal

(*Continued*)

(Continued)

feeding studies are too short to adequately test for cancer, reproductive problems, or effects in the next generation.

Kessler, D. (1992) FDA proposed statement of policy clarifying the regulation of food derived from genetically modified plants. Washington, DC:FDA Reports, March 20, 1992. (Statement of policy: foods derived from new plant varieties. *Federal Register* **57**(104): 22991, May 29, 1992.)

American Academy of Environmental Medicine Urges NIH to Follow Up Study

Dr. Jim Willoughby, President of the Academy, presented the Russian study of Dr. Irene Ermakova which showed greater mortality in rats which were fed on Monsanto's GM corn, at the annual conference of the American Academy of Environmental Medicine (AAEM) in Tucson on October 27, 2005. In response, the AAEM board passed a resolution asking the US National Institutes of Health (NIH) to sponsor an immediate, independent follow-up of the study. Dr. Willoughby said, "Genetically modified soy, corn, canola, and cottonseed oil are being consumed daily by a significant proportion of our population. We need rigorous, independent and long-term studies to evaluate if these foods put the population at risk."

However, there are other complications. In 2003, a French laboratory analyzed the inserted genes in five GM varieties, including Roundup Ready soybeans. In each case, the genetic sequence was different than that which had been described by the biotech companies years earlier. Had all the companies made a mistake? That's unlikely. Rather, the inserted genes probably rearranged over time. A Brussels lab confirmed that the genetic sequences were different than what was originally listed. But the sequences discovered in Brussels didn't all match those found by the French. This suggests

(Continued)

that the inserted genes are unstable and can change in different ways. It also means that they are creating new proteins — ones that were never intended or tested. The Roundup Ready soybeans used in the Russian test may therefore be quite different from the Roundup Ready soybeans used in follow-up studies.

Genetically Altered Foods Prone to Side Effects

Jeffrey Smith, who is the executive director of the Institute for Responsible Technology in Iowa, presented the AAEM conference with results from other published studies as well. Animals fed GM food developed potentially precancerous cell growth, stunted organs, damaged immune systems, problems in blood cell and liver cell development, lesions in the stomach, kidneys, and livers, and higher death rates. Also, nearly 25 farmers claim that varieties of GM corn caused their pigs to become sterile.

There has only been a single human feeding study, which, according to Smith, verified that the gene inserted into GM soy transferred into the DNA of intestinal bacteria. "This means that even if you stop eating GM soy, you may still have the foreign protein being produced inside you, possibly for the long term."

According to Jeffrey Smith, the process of gene insertion can turn genes off, permanently turn them on, change the expression of hundreds of other genes, create mutations, and introduce new allergenic proteins. "Even the FDA's own scientists warned of possible toxins, allergens, new diseases and nutritional problems," says Smith, who refers to agency memos made public from a lawsuit. "Government scientists had urged their superiors to require long term safety tests but were ignored by the person in charge of policy — who was the former attorney for biotech giant Monsanto and later their vice president." FDA policy states that the manufacturers can decide if their own GM foods are safe, without required studies.

Smith, J. M. (2005) American Academy of Environmental Medicine. Ermakova study on rats and GM soy. GM Free Cymru. October 31, 2005. www.gmfreecymru.org

Ecological Impacts of GM Cotton on Soil Biodiversity

Below ground production of Bt by GM cotton and Bt cotton impacts on soil biological processes

The research programme directed by Vadakattu Gupta and Stephanie Watson for CSIRO in 2004 (Gupta and Watson 2004) focused on the impacts of genetically modified (GM) cotton crops (*Bacillus thuringiensis* cotton; Bt cotton for short) on soil biodiversity and ecosystem function in Australia. Their results are described here.

The experimental work was based upon the need to establish the risk of production and release of Bt toxin by below ground plant parts of cotton and its potential persistence in the soil where Bt cotton crops are grown in Australia. In addition, the potential impacts of new gene products may affect key soil biological processes essential for a number of ecosystem functions.

They measured the levels of Bt toxin production in different plant parts of cotton, especially below ground parts, and also evaluated the mechanisms through which the Bt toxin enters the soil environment. In controlled environments (glasshouse and growth chamber) and field experiments, Bt cotton varieties expressed Bt genes and produced measurable amounts of Bt toxin in different parts of the cotton root system (tap, secondary and fine roots, and root hairs). The levels of Bt toxin in roots were similar to those observed in leaves, whereas the levels of Bt toxin in stems were the lowest. For example, Bt toxin levels in the leaves of cotton variety Sicot 289i ranged from 2,900 to 20,300 ppb; and in the roots, from 4,900 to 18,700 ppb.

It was found that as the plants grew older, the levels of Bt toxin in roots of 8-week old Bt cotton (Sicot 289i) were higher (4,900 and 7,000 ppb dry weight in taproot and fine roots, respectively) than that in leaves (2,900 ppb dry weight). In most situations, Bt toxin levels in the fine roots were higher than other parts of the root system and plant-related reductions in this part of the root system were smaller compared to other plant parts. This higher level of Bt toxin below ground was attributed to the continued growth of new root systems through the later stages of the cotton season.

The results show that Bt toxin was produced in every major part of Bt cotton plants (leaves, stems, and roots), that root Bt toxin production was comparable (or higher in the later stages of cotton plant) to that in cotton leaves, and that the above observations held true for all three soil types.

They also found that the roots of Bt cotton varieties release Bt toxin, both *in vitro* (solution culture) and by soil-grown plants, through presumably passive release from the roots or as cell lysates, and the levels of release (cell-free) of Bt toxin from roots were significantly increased (>6-fold) following any damage to the root system (e.g. fine roots). The non-Bt cotton cultivars, as expected, released no detectable Bt toxin. They found Bt toxin release from plants that were 2 to 12 weeks old, and found no evidence for the presence of Bt toxin from roots of non-Bt cotton varieties.

Root hairs and sloughed epidermal cells contribute a significant amount of root material in the rhizosphere of actively growing plants. They found that the sloughed epidermal cells and fine-root hair fragments from Bt cotton (Sicot 289i) plants contained large concentrations of Bt toxin (e.g. 1317 ppb/g wet weight), whereas non-Bt control (Sicot 189) cells/fine-root hairs showed no Bt toxin. Results suggested that Bt toxin has the potential to enter the soil system throughout the Bt cotton growing season, through both a root release process and root turnover. Levels of Bt toxin entering the soil system could therefore be significantly higher than previously suggested on the basis of contributions of Bt toxin to soil from above-ground cotton material only.

Unlike the Bt toxin from leaves and other above-ground plant parts, which may enter soil only after defoliation (leaves) and cotton harvest (stems), roots with Bt toxin are in constant contact with the soil system (including soil biota) and Bt toxin levels in fine roots were found to be as high as that in younger leaves. In view of the results reported above (large concentrations of Bt toxin in Bt cotton roots and demonstrated root release), more detailed investigations on the environmental fate of the root-derived Bt toxin, binding to soil components and buildup, and movement beyond the rhizosphere and root zone are warranted. Results from our initial work found detectable levels of Bt toxin in the rhizosphere of Bt cotton varieties by using both immunological tests and insect bioassays.

Microbial growth indicators measured in this study (decomposition rates, substrate induced respiration, and respiration quotients) suggest that microbial population growth on Bt cotton leaf litter might be different than for non-Bt varieties. Microscopic examination revealed an apparent increase in fungi and fungal spores on the Bt cotton residues compared to the non-Bt residues. Experiments did not indicate whether these changes were likely to be detrimental, neutral, or beneficial in an agricultural situation.

Their research showed that different plant parts of Bt cotton (leaves, stubble, and roots) contain large concentrations of Bt toxin and therefore have the potential to be a reservoir of Bt toxin in agricultural fields of Australia. Large concentrations of Bt toxin (above soil background) in decomposing Bt cotton leaf residues, even after the decomposition of >40% of leaf residue, indicate that Bt toxin from dead leaves is not easily degraded by soil microorganisms, which one would expect for such a protein substance. If more Bt toxin enters the soil environment than is degraded by microbes, eaten by insect larvae, or inactivated by sunlight, there is potential for the toxin to accumulate if it is bound and protected by soil particles (clays, minerals, and humic acids). Could accumulation of active Bt toxin constitute a hazard to non-target organisms and impact the biodiversity and functionality of the organisms inhabiting the soil?

The Bt Toxin Does Not Simply Disappear

The accumulation of Bt toxin in the soil was reported by Christophe Tebbe and Susanne Baumgarte at the Federal Agricultural Research Centre (FAL), Institute of Agroecology, in Braunschweig.

(Continued)

(Continued)

No Harmful Ecological Effects So Far

Even if the Bt values increase from one year to the next, they are evidently too small to harm the soil life. "In our estimation, the effect on the diversity of microorganisms is on the low side if anything," says Christoph Tebbe. "We know this from our studies of rhizospheres." Of greater interest is the impact of the high Bt toxin concentrations in the crop residues. "Many minute arthropods and worms which are involved in mineralising these crop residues are dependent upon them. This affects entire terrestrial food chains." Wolfgang Büchs and his team from the Federal Biological Research Centre (BBA) in Braunschweig are studying sciarid fly larvae, for example, which live on rotting plant material.

Accumulation — Even over Several Years?

The critical question, however, is whether Bt toxin gradually accumulates in the soil when Bt maize is grown over a period of several years. It is also conceivable that the Bt toxin gradually breaks down and levels off.

But the areas from which the FAL group took their soil samples will not be planted with Bt maize next year. The study is coming to an end — and the question of whether Bt values would have continued to rise in 2004 remains unanswered. "We can see a trend", says Christoph Tebbe, "but after only two years of study we cannot conclusively say how Bt toxin contents in fields of Bt maize change." It is also not clear at present what effect crop rotation or soil preparation has on the persistence of Bt toxin in the soil. Further studies in the coming years will be able to shed more light on this matter now that the sensitive detection system is available.

Bt maize and soil — hot on the trail of transgenic molecules. *GMO Safety*, November 21, 2007.

In the second cultivation year, all measured Bt values were significantly higher than those of 2002 at both sites. The increase in toxin content was five or seven times the previous year's level, depending on the site. Even in soil samples taken in April 2003, i.e. before the next planting season, something could still be detected. Although the measured values in the soil increased from one year to the next, they are still very low and, by way of comparison, are equivalent to just one thousandth of the Bt concentrations typically measured in the roots of Bt maize.

GMO Area to Surge in Europe

In defiance of several adverse reports, Monsanto's representative in Paris, Sybille de La Hamaide, issued a statement that Europe will increase its genetically modified crop area by 50,000–100,000 hectares a year over the next decade, from 100,000 ha in 2007 (Reuters, 26 June 2007). "It will be slow but within 10 years GMOs will have reached the point of no return," Jean-Michel Duhamel, Monsanto's director for southern Europe, told reporters in Paris. "The technology will not impose itself to consumers but consumers will better understand the usefulness of GMO technology as farmers increasingly adopt it," he added.

de la Hamaider, S. (2007) Europe GMO area to surge over 10 years: Monsanto. June 25, 2007. www.reuters.com

May 24, 2007

Center for Food Safety Publishes Guide to Avoiding GE Foods

Andrew Kimbrell, of the Center for Food Safety, has published *Your Right to Know: Genetic Engineering and the Secret Changes in Your Food.*

- Today, an estimated 52% of all corn, 87% of all soy, 55% of all canola, and 79% of all cotton grown in the US are genetically engineered.
- Genetically engineered plants can cause serious allergies, increase our resistance to antibiotics, depress our immune systems and remove the nutrition from our food.

(*Continued*)

(Continued)

- The FDA requires no labeling or human health testing of GE foods.
- While US companies are labeling GE food exports for the more than 50 countries that require it, they are not for Americans.

(see Kimbrell and Newman, 2007)

Genetically modified crops can impact on the environment both directly as well as indirectly. So far, the available evidence on their impact does not indicate systematic adverse effects nor is it adequate to understand the long-term consequences of the impact of GM crops. Furthermore, from a scientific point of view, research should be conducted to understand the impact in diverse environments, ecosystems, and soils in tropical, semitropical, and temperate countries. Indeed, thus far no extensive research data are available from the developing countries, where much of the world's biodiversity is located. Matching control data are also required to reach a meaningful conclusion.

Kerala, India

The Government of the State of Kerala and many environmental activists in India oppose field trials of genetically modified crops, because of their fears for endangering biodiversity. A press report in *The Hindu* (July 22, 2007) summarized the situation. The following are some readers' concerns on both sides of this issue.

A growing number of plant foods have been developed in recent years by inserting a new gene into a crop to get advantageous characteristics. Genetically modified crops improve production, are resistant to pests, and have a better shelf life and nutrient value. It is a pity that our environmentalists and scientists are at loggerheads on this issue. There are concerns about endangering biodiversity but environmentalists should see what promise these crops hold.

(Continued)

(Continued)

Need for more studies

The Government and many environmental activists have opposed field trials of genetically modified crops in the State while some experts are of the view that we should go ahead with the profitable use of biotechnology to attain food security.

Hitherto technological developments have brought about revolutionary changes in our lives and biotechnology is not an exception. As such, carrying out field trials of genetically modified organisms or crops will not endanger biodiversity though it is being opposed by some groups. But long-term effects have to be studied well before these are introduced on a large scale.

Threat to food safety

There are apprehensions all round about the widespread cultivation and use of genetically modified crops. The European Union and Japan have made their rules regarding import of GM seeds stringent. In Andhra Pradesh where Bt cotton is cultivated, farmers and livestock developed allergies and other health problems. If this is the case with cotton, use of these seeds in food crops will be more disastrous. It will pose a threat to our food security and biodiversity.

Not a solution

The idea of genetically modified crops has been under fire from its very inception as it involves certain contentious issues. Its application is advocated by those who have patented the technology. However, one has to view a technology not only from its commercial proposition but also from the social impact it will have. The social compatibility of the knowledge and its use has to be ensured before the application is advocated; otherwise, it can be counterproductive.

Our own scientists have evolved a good number of hybrid seeds. It is also a fact that we have lost hundreds of local varieties which were resistant to pests and diseases. What is required is to allow farmers

(*Continued*)

(Continued)

to maintain genetic diversity in agriculture crops. The answer to the problems of Indian agriculture is not genetically modified crops but ensuring remunerative price for the produce.

Must for future needs

India could attain self-sufficiency in wheat production after the introduction of high-yielding, disease-resistant varieties of wheat developed by Norman Borlaug by using techniques in classical genetics. The contributions of Prof. M.S. Swaminathan, the father of Green Revolution in India, in introducing and propagating genetically modified Mexican Dwarf variety of wheat in India is acknowledged the world over. This has been a phenomenon that has not led to any environmental problems affecting the biodiversity. . . . The future of the food situation in the country looks grim unless a second Green Revolution is brought about by application of molecular genetics. This can be done only with a political will and a clear mandate of the public. It is the responsibility of environmentalists and social activists to point out with precise scientific basis how the introduction of genetically modified plant varieties can adversely affect the biodiversity of the concerned region.

Exercise caution

Our existing biodiversity has been evolved through millions of years of complicated and subtle processes. Hence biodiversity is a more valuable asset for a nation and the world at large than the shorter goal of increasing productivity and production. Any attempt at introducing genetically modified crops should therefore be done with utmost caution and sensitivity.

Address concerns

Genetically modified food crops are grown with success in developed and developing countries resulting in high quantitative and qualitative productivity. . . . Countries where genetically modified crops are

(*Continued*)

(Continued)

grown have not reported significant health damage or environmental harm and biodiversity. Moreover farmers are using less pesticides or using less toxic ones, reducing harm to water supplies and workers' health and allowing return of beneficial insects to the fields. Some of the concerns relating to "gene flow" and pest resistance have been addressed. Therefore the trials with GM crops should be encouraged and the benefits reaped in the best interest of the people.

Kerala: GM-free status sought for State. *The Hindu Business Line.* Chennai, India. July 13, 2007.

Wynand J. van der Walt

Wynand van der Walt explores the risk of GMO's, the risks of not adopting new technology and also providing opinion on fears of super pests, super weeds, impact of GMO's on biodiversity and more.

Natural ecosystems were disturbed on the day some 11000 years ago that the first planet dwellers decided to dig holes in the ground and plant seeds and tubers. Today the planet has to cope with over 6 billion people and urbanization, industrialization, abject poverty, while agricultural practices continue to impact on the environment. Assessing impact on the environment is much more than just a tunnel vision approach on hypothetical risks of GM crops. South Africa's existing problems relate to the listed 117 major invader species and another 84 new invaders. We should all agree that approval of GM and new conventional varieties with potential risk should be subject to prior assessment for adverse impact on the environment. Identifying problems proactively is substantially better than corrective action after the problem has surfaced.

(*Continued*)

(Continued)

It seems that there is inadequate capacity at local government level and this is precisely where the environmental management legislation could fail. Delegation and fragmentation of responsibilities to provinces and local authorities run the risk of inadequate capacity, funding or awareness. Obvious examples include bugweed (*Solanum*) lining the road shoulders around Pretoria, liberally blended with a vast array of farm weeds, or pink pom-pom (*Campuloclinium*) now having invaded thousands of hectares all over Gauteng, including nature reserves, or billions of new black wattle plants having taken the place of trees chopped down. The primary factor for roadside invasion is mowing of grass only once per year when weeds have set seed, serving as a massive redistribution system. Baling of grass cuttings for hay will transfer the roadside problem to the farm. This picture conveys the message that local government continues to add to the problem while central government is forking out billions in control measures. In contrast, the GMO Act regulates at national level and consensus decision making involves six government departments which includes environmental legislation, prior to release of new GM varieties.

Genetic changes in plants, animals and micro-organisms have taken place with increasing impact and are as old as agricultural practices itself, as farmers domesticated and selected biodiversity. South Africa grows some 100 000 hectares of triticale, a wheat–rye cross, and consumers can enjoy plumcots, pluots and apriums from apricot–plum crosses, all coming from non-genetic modification techniques. Vegetative propagation through budding, grafting, tubers, rootstocks and tissue culture are major industries.

Genetic modification (GM) application only started when scientists began to understand the periodic jump of genes from their positions on one chromosome to other positions in the genome (transposons), micro-organisms being able to incorporate foreign naked DNA (transformation) and vectors carrying genes from one species to another (transduction). The first applications of GM technology were in human health (insulin in 1982), food processing (chymosin enzyme), industrial production, and, only by the early 1990s, GM crop plants.

U.K. GM Crop Farm-Scale Evaluations: Background Papers

The Farm-Scale Evaluations of Genetically Modified Herbicide Tolerant Crops Rationale and Chronology

A Paper by the Biotechnology Safety Unit, Department of the Environment, Transport and the Regions

1. Summary

This paper sets out the rationale and chronology relating to the decisions to set up the farm scale evaluations of certain herbicide tolerant genetically modified crops (GMHT crops) and the subsequent developments.

- The Government announced the farm-scale evaluations (FSE) in 1998 as part of a set of initiatives to strengthen the process for making decisions on whether or not to allow commercial cultivation of certain GMHT crops grown and managed with their associated herbicide regimes.
- In the event that cultivation were to be permitted the results will also inform decisions on what conditions or restrictions should be applied.
- The evaluations will assess the impact on farmland wildlife of the management of the GMHT crops with their companion herbicide as compared with equivalent plantings of non-GM crops.
- The crops involved, rape, maize and beet, were all on the verge of entering commercial agriculture in the EU. Under the voluntary agreement with industry, they will not now be grown, other than in the evaluations, until the programme is complete.
- The European regulatory authorities and their scientific advisors were content with the safety of the GMHT crops themselves, but questions remained about the impact of the new herbicide regimes on the abundance and diversity of farmland wildlife. The

(*Continued*)

(Continued)

FSE were therefore set up to address this specific remaining area of uncertainty.

- Background papers on the regulatory regime and the evaluations prepared for the AEBC by DETR in August 2000, are:
  - *The legal framework for decision making on the release and marketing of GMOs in the EU*
  - *Risk assessment for releases and marketing of GMOs in the European Union*
  - *The history of the farm scale evaluations and*
  - *The science of the farm scale evaluations.*

2. Regulatory Background

In the early 1990s the European Community set up a comprehensive system for the assessment and control of GMOs. Under Directive 90/220 no product comprising or containing GMOs can be placed on the market until it had been shown that measures have been taken to avoid adverse affects on human health and the environment. In addition any GM product to be used as or in food had to be approved under the EU Novel Foods Regulation.

In addition GM crops have to satisfy the same requirements as conventional varieties for addition to the National List of Seeds or the European Common Catalogue. This requires a series of tests to demonstrate distinctiveness, uniformity and stability. Any use of pesticides on the crops also has to be approved.

In 1998, several types of GM crop were working their way through the regulatory process and could have received all the necessary approvals for commercial cultivation by spring 1999.

3. Concerns

At that time concerns were raised about:

- The environmental impact assessment required under 90/220.
- The safety of GM crops in the food and feed chain.

(Continued)

(Continued)

- Aspects not evaluated as part of the regulatory process, in particular the change in the pattern of use of herbicides on these crops which could lead to adverse affects on farmland wildlife.
- The acceptabilty of GM crops and GM food generally; strong feeling that the move to commercialisation was happening too fast.

Although the environmental and food and feed aspects had been considered by the Advisory Committee on Releases to the Environment (ACRE) and the Advisory Committee on Novel Foods and Processes (ACNFP), many critics felt that there was too much uncertainty in the assessment and that a more precautionary approach should be taken. Others raised concerns which they felt had not been examined by the committees. In 1996 ACRE had itself raised the need to consider the environmental impact of widespread cultivation of GM herbicide tolerant crops.

In early 1998 English Nature and the other statutory nature conservation agencies called for a moratorium on the introduction into commercial agriculture of GM crops modified for insect resistance or herbicide tolerance until more was known about the impact of their cropping systems on farmland biodiversity. Many other organisations joined the call for a moratorium. Many went further calling for a halt to import of GM foodstuffs and all outdoor testing of GM crops as well.

During 1998 ministers and officials in DETR and MAFF had a series of meetings with English Nature and NGOs from both sides of the debate. In October 1998 amid mounting pressure, DETR officials consulted the leading organisations campaigning for a moratorium and separately the industry body SCIMAC. The meetings focussed on the legality and terms of a possible moratorium and the further research and information thought necessary.

4. Introduction of the Farm Scale Evaluations

At the same time the House of Lords Agriculture Select Committee was conducting an enquiry into GMOs and had taken evidence from a wide

(*Continued*)

(Continued)

range of organisations and individuals over the summer. Taking this evidence into account together with the other discussions noted above the Government drew up a series of measures to strengthen and improve the assessment of GM crops and the decision making process. Michael Meacher and Jeff Rooker used their appearance before the committee on 21 October to announce the package of measures.

The main elements were:

- An agreement with SCIMAC for a programme of managed development of GMHT crops to limit their introduction whilst ecological monitoring was carried out and a three year pause on the introduction of insect resistant GM crops;
- The farm scale evaluations to assess the effects of the agricultural management of field scale releases of GMHT crops on farmland wildlife as compared with comparable plantings of conventional crops;
- Consideration of the establishment of a stakeholder forum to discuss and advise on environmental issues raised by biotechnology to work alongside ACRE (This led to the establishment of the AEBC);
- The setting up of a new Ministerial Group on Biotechnology (MISC6);
- UK action to ensure that the amendment of directive 90/220 had well defined and broad requirements for environmental risk assessment and monitoring;
- A scientific review of pesticides used on GM crops comparing the likely impact on biodiversity of current and possible future practice;
- A reassessment of herbicides to be used on GMHT crops including their effect on non-target species and a requirement for new approvals for the use of the relevant herbicides on GMHT crops;
- Consideration of the introduction of long term monitoring capable of picking up any unexpected effects.

(*Continued*)

(Continued)

The Ministers made it clear that these measures made a moratorium unnecessary and confirmed the Government's view that:

- The approval under directive 90/220 for cultivation of a GM crop could only be revoked if there was new evidence of harm. If such evidence came to light, action to impose a ban in the UK could be taken using powers in article 16 of the directive.
- The National Seed List Trials are a series of objective tests, so if the GM variety passes the tests there are no grounds for refusal.

The following paragraphs describe how the relevant decisions associated with this announcement were taken forward.

5. Agreement with SCIMAC to Limit Commercialisation of GM Crops

No approvals for products consisting of or containing GMOs have been issued in the European Union since August 1998, when the Europe-wide approval for cultivation of GMHT maize was issued. There is now a backlog of 14 products. In light of this and the continuing concern in the UK, the government made a new agreement with SCIMAC in November 1999. The terms of the agreement include:

- Renewal of the voluntary agreement on the conduct of the farm scale evaluations through until the end of the evaluations following harvest of the crops planted in 2002;
- No unrestricted cultivation of GM crops in the UK until the FSE are complete;
- None of the produce from GM crop plantings in the UK will be used in a way that is of direct commercial benefit to the consent holders during the FSE period.

At the same time ministers agreed to include GMHT sugar beet and fodder beet in the evaluations on the same terms as the rape and maize.

(*Continued*)

(Continued)

6. Decisions on Setting up the Farm Scale Evaluations

After the announcement in October 1998, DETR scientists had discussions with other government departments, members of ACRE and wildlife and research advisors and then drew up the specification for the ecological studies.

- The hypothesis to be tested was *that there are no significant differences between the biodiversity associated with the management of the particular GMHT crop and the comparable non-GM crop at the farm scale.*
- The secondary objective was to contribute to an assessment of the wider question of whether the commercial use of GMHT crops will change the management of farming systems and the agricultural landscape.

Fifteen leading research organisations were invited to tender for the work which involved the design and implementation of the monitoring programme, specification of the methodologies to be used and the level of statistical significance which could be obtained. Officials in DETR agreed the practical arrangement with SCIMAC, who were to provide the GM seeds and arrange for suitable farmers to grow and manage both the GM crops and the conventional crops used in the trials. The ecological studies are funded by DETR with small contributions from MAFF and the Scottish Executive.

During the tendering period, DETR wrote to NGOs and other interested parties inviting comments on the specification for the research. The comments received informed the tender review.

Tenders were received from eight organisations. A tender review panel comprising, Professor John Lawton (then director of the Centre for Population Ecology at Imperial College) and scientists from English Nature, DETR, MAFF, the Scottish Office and the British Society of Plant Breeders considered the various proposals. Ministers announced the decision on the appointment of the successful

(*Continued*)

(Continued)

research consortium on 15 April 1999. Once those involved in carrying out the ecological research had been decided, ministers appointed an independent Scientific Steering Committee (SSC) to oversee the research programme and advise on the outcome. The membership, terms of reference and minutes of meetings of the Scientific Steering Committee are published and available on their web site. The research consortium submits half yearly reports to the Committee; these are also published and available on the web site.

The first year of the evaluations was a pilot phase with a small number of fields of each GMHT crop sown in 1999. The full programme started in 2000 and is due to run for 3 years. The SSC will consider the results from the spring-sown crops when they become available in autumn 2002 and the results from the autumn sown crops in autumn 2003. The SSC will supervise the publication of the results. It will then be for ministers, taking the advice of ACRE and others on the interpretation of the outcome, to decide how to go forward.

7. Decisions on the Risks to the Environment from the Farm Scale Evaluations

Safety of the GM plants. The developers of the four GMHT crops involved in the farm scale evaluations have submitted applications for their approval for EU wide cultivation under Part C of directive 90/220. These dossiers either have been approved (in the case of maize) or are in the late stages of approval by member states (see section on regulatory approval). ACRE had considered these dossiers at various stages of their development and advised ministers; their advice is public.

The Aventis GMHT maize was granted Europe wide approval for cultivation in August 1998 so no specific approval for the FSE was needed. For the GMHT oil seed rape and the sugar and fodder beet to be grown in the evaluations specific to Part B, research approvals under directive 90/220 were needed. ACRE and the statutory nature conservation agencies considered the applications and in particular

(Continued)

(Continued)

the impact on the environment and advised that there were no grounds relating to safety for human health or the environment for not granting consent for the trials. The relevant regulatory authorities in England, Scotland and Wales have granted consents; these and the applications are available on the public register held at DETR. In advising on these applications ACRE considered the various concerns raised about the safety of the GM plants and voiced by scientists, pressure groups and the public, such as promoters, cross-pollination, horizontal gene transfer and effects on bees.

Safety of the herbicide use. Following advice from the Advisory Committee on Pesticides, ministers have given specific approvals under the pesticides legislation for use of the broad-spectrum herbicides in the farm scale evaluations. The decision on full commercial approval for this use of the herbicides awaits the outcome of the FSE.

Agronomic safety. Through 1998 officials at MAFF had been leading discussions with the industry body SCIMAC on development of a code of practice for the supply and agronomic management of GMHT seeds and crops. The code was published in June 1999 and endorsed by ministers. The voluntary code, which is binding on participants, covers the arrangements that would be necessary to ensure integrity of the supply chain for both GM and non-GM crops should GM crops enter commercial production. The code also includes measures to avoid agronomic problems such as herbicide tolerant volunteers. Ministers and SCIMAC agreed that where relevant the GMHT crops in the FSE will be grown in compliance with the SCIMAC code. This includes separation distances between the GM crops and nearby conventional or organic crops.

Food and feed safety. The rape and beet do not have approval for use in food or feed and the consents require that at harvest they are disposed of by ploughing in or removal to land fill. The maize has Europe wide approval for use in food and feed, however the agreement

(*Continued*)

(Continued)

with SCIMAC requires that at harvest the plants are disposed of by ploughing in or removal to land fill. ACRE and ACNFP have advised that any cross-pollination with neighbouring crops during the trials or volunteers arising in subsequent years do not pose a risk to food or feed safety.

8. Decisions on Strengthening the Regulatory Process

The European Union is in the final stages of agreeing a revised Directive 90/220 to reflect current best practice in member states and to introduce new measures to strengthen the regulatory process. This includes new requirements for risk assessment and monitoring. The scope of the risk assessment will now include possible impacts of the specific techniques used for the management of GMOs where these are different from those used for non-GMOs. Therefore in future aspects such as changes in the patterns of rotation will be considered in the assessment. European Environment Ministers agreed in December 1998 to adopt these new procedures straight away for new applications, using the powers of the existing directive and without waiting for implementation of the new directive.

In October 1998 Ministers asked ACRE for advice on how the management of the GM crop can be taken into account in the approvals process. An ACRE sub-group under the chairmanship of Sir John Beringer was set up in February 1999 with the publication of ACRE's report on commercialisation of GMHT crops and has been considering these wider environmental issues. The group consulted on a draft guidance note in September 2000 and is currently considering the responses. This ACRE sub-group and the Environment Panel of the Advisory Committee on Pesticides have jointly been considering how the environmental impact of the changed pattern of use of herbicides on GMHT and other crops should be assessed using the powers in both the GMO and pesticides legislation.

(Continued)

(Continued)

9. The GM Crops, Information and Status

Before a GM crop plant may be placed on the market, ie sold to and grown by farmers commercially, it requires separate approvals:

- under part C of Directive 90/220 to place a GMO product on the market, to ensure the crop plant will not cause harm to human health and the environment;
- under seeds legislation to market the seeds and add them to the National List;
- under pesticides legislation for a new use of the herbicide on a GM crop and
- under novel foods regulations for use of the produce in foods.

In the summer of 1998 no GM crops had all the necessary approvals for commercial cultivation. However several had or were close to getting Part C product approval under directive 90/220. As discussed below it was possible that GMHT rape and maize could have obtained all the necessary approvals in time for sowing in the spring 1999 with beet following in spring 2000.

It is not possible to prevent cultivation of crops which have EU wide approval under Directive 90/220 unless new information is available on the risks to the environment or human health which justifies taking action under Article 16 to impose a temporary local ban.

Oil seed rape

Genetically modified herbicide tolerant oil seed rape (GMHT rape) has been grown in trials in Britain and many European Member States since 1988. It is in wide spread cultivation in North America. There are three types of GMHT rape relevant to the FSE.

In 1994 the UK considered a type of GMHT rape known as 'MS1RF1' tolerant to the broad spectrum herbicide 'Liberty', glufosinate ammonium, from the company PGS (now Aventis) in an application for

(*Continued*)

(Continued)

a Part C approval under Directive 90/220 for seed production. After due consideration and agreement by Member States, the UK issued consent in 1996. A further application under Part C was made through the French Authorities for general cultivation and animal feed. Approval was given by Member States in 1997 but the French Authorities did not issue the consent because of concerns in France. UK National List seed trials for a variety of this rape were complete and the oil is approved for food use. Therefore this GMHT rape could have had full approval for commercial cultivation at any time.

In October 1998, two more types of GMHT rape from AgrEvo (Aventis) were in the final stages of the Part C process and National List trials. Both are tolerant to 'Liberty'. The GMHT rape used in the evaluations is known as MS8RF3, reference C/BE/96/01.

In September 1998 DETR commissioned Prof Alan Grey and his team to review the application for Part C approval for cultivation for GMHT rape. He was asked to identify any new information that had become available on the environmental risks since ACRE had considered the dossier in 1996. Prof Grey was not a member of ACRE when they considered the original application. ACRE considered Prof Grey's review in January 1999 and advised that having taken the new information into account their original advice was unchanged.

Maize

GM insect tolerant maize is grown extensively in the US. For many years research trials have been carried out in other EU member states, notably France, Italy and Spain. Both insect resistant and herbicide tolerant varieties have been developed. Two so-called Bt maize types, giving resistance to the European corn borer have Part C approval for commercial cultivation in the EU. There is no expectation that these particular crops would be grown in the UK as the corn borer is not prevalent.

In 1995 AgrEvo (now Aventis) applied through the French Competent Authorities for EU-wide product (Part C) approval for

(*Continued*)

(Continued)

import, cultivation and animal feed for T25 maize tolerant to the herbicide glufosinate ammonium 'Liberty'. After due consideration by member states consent was granted by the French in August 1998. UK National Seed List trials on a variety of this maize known as Chardon LL were due to be completed in 1999 and novel food approval had been given. Therefore, again there was the prospect of imminent commercial cultivation.

Beet

An application from Monsanto to market GM fodder beet tolerant to the herbicide 'Round-up', glyphosate, was made through Danish Authorities in 1997. Both GMHT sugar beet and fodder beet have been extensively trialed in the UK. A decision on an EU product approval (Part C) for commercial cultivation was expected in early 1999.

10. Extent of Governmental, Commercial and Public Consultation

The circumstances surrounding the setting up of the farm scale evaluations and the consultations undertaken have been described in sections 3–6 above and in the background paper on the history of the FSE. Since then DETR held a seminar for representatives from NGOs in July 1999. DETR, the research consortium, English Nature, SCIMAC and the Scientific Steering Committee made presentations and then answered questions. DETR has published a report of the meeting including the Q&A session.

DETR has published information about the evaluations in a leaflet and through the DETR web site. To coincide with the spring sowings in 2000 DETR organised 12 public meetings in the main trial areas. Representatives from DETR, the research consortium and SCIMAC gave information about the evaluations with an alternative view presented by Genewatch or Friends of the Earth, followed by a question

(*Continued*)

(Continued)

and answer session. DETR also sent information to each parish council where a trial was being held. The Scottish Executive also held a public meeting.

In the autumn 2000, the publicity concentrated on early notification of parish councils. Mr Meacher invited the chairman of each council to a meeting with him in London to be briefed on the evaluations. DETR also offered to send a representative to a parish meeting should the council wish to organise one. Three out of the 23 councils arranged such meetings.

DETR hosted a meeting for the farmers involved in the evaluations in October 2000.

11. The Future

The farm scale evaluations of the spring-sown crops will be complete in the autumn of 2002. The report of the research will be published in 2003 and open to scrutiny. The Government will need to evaluate the findings in the context of public views on acceptability of GM crops. The remit of the AEBC sub-group includes adding value to the future decision-making process. The sub-group might wish to consider providing advice on how the Government might manage the process as regards decisions on possible future commercialisation of GM crops.

GM Farm Scale Trials. BBC News, London. March 9, 2004.

The U.K. Farm Scale Trials

In 2000, in response to public concerns, the U.K. Government introduced a 3-year program of farm-scale genetically modified (GM) herbicide-tolerant crops. The Government and the biotechnology industry reached an agreement not to grow such crops commercially until the results of the trials were evaluated and published in 2004.

The purpose of the trials was to establish whether there is any significant impact on farmland biodiversity from growing GM herbicide-tolerant crops, compared to the equivalent non-GM crops. Four GM crops were involved in the trials:

1. Forage (fodder) maize tolerant to the herbicide glufosinate;
2. Spring oil seed rape tolerant to glufosinate;
3. Winter oil seed rape tolerant to glufosinate; and
4. Sugar and fodder beet tolerant to glyphosate.

Each trial site was divided into two parts. One half was planted with the non-GM crop grown with conventional weed control, and the other half with the GM crop and its associated herbicide. The effect of different herbicides on the abundance of plants and insects was monitored. No other issues were considered.

Over the next three years, about 70 field trials per crop were conducted in several locations. This reflected the range of farming practices and geographical distribution.

The trials have been criticized for their limited scope, especially for lacking any ethical or socioeconomic considerations. Another criticism was that the industry played an active role in selecting the locations; there was no consultation with neighboring farmers or beekeepers, and not enough advance time was allowed for public notices to facilitate real consultation. Furthermore, separation distances between the trials and neighboring crops were too small to prevent cross-fertilization and to protect the economic interests of organic and conventional farmers.

Gene Transfer

Transfer of genes between plants within varieties and species has been a natural occurrence, both in crop plants as well as wild species. However, similar transfer between plants involving GM varieties and non-GM plants is viewed with much concern and apprehension by some. In principle, unintentional transfer of DNA would make it difficult to control certain selected characteristics in wild weeds, for instance, the genetically engineered ability to withstand herbicide applications. It is feared that this could lead to the creation of "superweeds", which would require stronger herbicides. They could become invasive, threatening agricultural biodiversity.

The following concerns are evident:

1. What direct effects could genetically modified plants have on the environment?
   - Gene flow
   - Nontarget species
2. What indirect effects could genetically modified plants have on the environment?
   - Agricultural practices
   - Pesticide use
   - Herbicide use
   - Pest and weed resistance
   - Difficult agricultural conditions
3. How should these environmental effects be assessed?

Gene Flow

The case of transgenic contamination of maize in Mexico illustrates the problem of gene flow. Despite a moratorium on the environmental release of GM maize in Mexico since 1998, transgenic DNA had in fact made its way into Mexican maize landraces. In November 2001, Quist and Chapela (2001) published their research findings, causing much concern.

A report released by the North American Commission for Environmental Cooperation (CEC) in November 2004, traced the arrival of the GMOs in Oaxaca to imports of GM maize from the U.S. Small-scale farmers in Mexico planted the seeds which were originally

(*Continued*)

(Continued)

meant for consumption only. The report suggested that, from a scientific point of view, transgenic maize does not threaten genetic diversity more than other methods of modern agriculture such as hybridisation. The report also stressed that maize holds cultural, symbolic and spiritual values for many Mexicans, especially the campesinos (or small-holder farmers) who regard GM maize as a direct threat to their cultural identity, personal safety and biodiversity.

The CEC report recommended that the GM maize planting moratorium should be continued and strengthened by minimising the import of living transgenic maize grain from countries that grow transgenic maize commercially, such as the U.S. and Canada. Later, in 2005, another report showed no evidence of GMOs in more than 150,000 seeds taken from 870 plants in Oaxaca in 2003 and 2004, and calmed the fears, at least temporarily (Ortiz-Garcia *et al.* (2005)).

Maize and biodiversity: The effects of transgenic maize in Mexico. Commission for Environmental Cooperation (CEC), Secretariat Report, Key Findings and Recommendations. Montreal, Canada, 2004.

Chemical Contamination

Accidental spread of chemical compounds to soils, ecosystems and other plants is feared when the gene flow involves genes modified to produce pharmaceutical or chemical compounds. For example, in 2002, seeds from plants genetically modified to generate an animal vaccine germinated in a field and were mixed inadvertently with soybeans that were subsequently grown on the land. The soybean crop was destroyed as the toxic effect of the vaccine, if any, on human health and the environment was unknown.

Impact on health and environment was not considered when a Texas-based biotech company, ProdiGene, developed the GM maize. Further research is in progress to assess the risk of contamination to other crops and wild plants.

(*Continued*)

(Continued)

Gene Flow to Wild Relatives

Most ecological scientists agree that gene flow is not an environmental problem unless it leads to undesirable consequences. In the short term, the spread of transgenic herbicide resistance via gene flow may create logistical and/or economic problems for growers. Over the long term, transgenes that confer resistance to pests and environmental stress and/or lead to greater seed production have the greatest likelihood of aiding weeds or harming non-target species. However, these outcomes seem unlikely for most currently grown transgenic crops. Many transgenic traits are likely to be innocuous from an environmental standpoint, and some could lead to more sustainable agricultural practices. To document various risks and benefits, there is a great need for academic researchers and others to become more involved in studying transgenic crops. Similarly, it is crucial that molecular biologists, crop breeders and industry improve their understanding of ecological and evolutionary questions about the safety of new generations of transgenic crops.

The presence of wild and weedy relatives varies among countries and regions. Major crops are grouped by their ability to disperse pollen and the occurrence of weedy relatives in the continental United States. This can be useful in identifying cases where gene flow from a transgenic crop to a wild relative is likely. For crops where no wild or weedy relatives are grown nearby — as with soybean, cotton and maize shown here in green — gene flow to the wild would not occur. Rice, sorghum and wheat have wild relatives in the United States and a relatively low tendency to outcross, which could allow transgenes to disperse into wild populations. The crops that have a high tendency to outcross and have wild relatives in the United States are shown in red. There is a high potential for gene flow between these crops and their wild relatives, so care should be taken in growing transgenic varieties that might confer a competitive advantage on their hybrids.

(Continued)

(Continued)

Non-target Species

GM crops can also have direct impacts on non-target species that consume them or their pollen. Crops which use *Bacillus thuringiensis*, a soil bacterium that kills many of the worm-like insects that destroy crops, is a case in point. While Bt saves the crop from pests that destroy the crop, it could also hurt other harmless worm-like insects that are found in the fields. There is also the possibility that insects will become immune to the Bt toxin since such resistance would provide them with an evolutionary advantage in the presence of widespread Bt use. This could have adverse longterm effects on the invasiveness of these insects in the environment and on farms, because use of Bt — including through sprays and non-GM methods — is one of the most effective, cheapest and least environmentally harmful ways to tackle the spread of pests. This problem has not emerged thus far — possibly owing to the requirement in many countries to have small areas of non-Bt plants ("refuges") near any Bt fields to minimise evolutionary advantages any Bt-resistant insects would have (IFATPC, 2004).

Impact on Monarch butterflies

Collaborative research effort by scientists in several states and in Canada has produced information to develop a formal risk assessment of the impact of *Bt* corn on monarch butterfly (*Danaus plexippus*) populations. Information was sought on the acute toxic effects of *Bt* corn pollen and the degree to which monarch larvae would be exposed to toxic amounts of *Bt* pollen on its host plant, the common milkweed, *Asclepias syriaca*, found in and around cornfields. Expression of Cry proteins, the active toxicant found in *Bt* corn tissues, differed among hybrids, and especially so in the concentrations found in pollen of different events. In most commercial hybrids, *Bt* expression in pollen is low, and laboratory and field studies show

(*Continued*)

(Continued)

no acute toxic effects at any pollen density that would be encountered in the field. Other factors mitigating exposure of larvae include the variable and limited overlap between pollen shed and larval activity periods, the fact that only a portion of the monarch population utilizes milkweed stands in and near cornfields, and the current adoption rate of *Bt* corn at 19% of North American corn-growing areas. This 2-year study suggests that the impact of *Bt* corn pollen from current commercial hybrids on monarch butterfly populations is negligible (Sears *et al.* 2001).

Laboratory Studies

Hellmich *et al.* (2001) conducted laboratory tests to establish the relative toxicity of *Bacillus thuringiensis* (*Bt*) toxins and pollen from *Bt* corn to monarch larvae. Toxins tested included Cry1Ab, Cry1Ac, Cry9C, and Cry1F. Three methods were used: (i) purified toxins incorporated into artificial diet, (ii) pollen collected from *Bt* corn hybrids applied directly to milkweed leaf discs, and (iii) *Bt* pollen contaminated with corn tassel material applied directly to milkweed leaf discs. Bioassays of purified *Bt* toxins indicate that Cry9C and Cry1F proteins are relatively nontoxic to monarch first instars, whereas first instars are sensitive to Cry1Ab and Cry1Ac proteins. Older instars were 12 to 23 times less susceptible to Cry1Ab toxin compared with first instars. Pollen bioassays suggest that pollen contaminants, an artifact of pollen processing, can dramatically influence larval survival and weight gains and produce spurious results. The only transgenic corn pollen that consistently affected monarch larvae was from Cry1Ab event 176 hybrids, currently <2% corn planted and for which re-registration has not been applied. Results from the other types of *Bt* corn suggest that pollen from the Cry1Ab (events Bt11 and Mon810) and Cry1F, and experimental Cry9C hybrids, will have no acute effects on monarch butterfly larvae in field settings.

(Bickham *et al.* 2000)

(Continued)

Belgium Rejects GM Oilseed Rape

05/02/2004 — The possible cultivation of GM crops within the European Union suffered a setback this week after the Belgian government decided to reject an application to grow GM oilseed rape. Officials said that the decision was based on the belief that the crop would be more damaging to wildlife than conventionally-grown varieties.

The decision has dealt a serious blow to the biotechnology industry and in particular to Bayer Cropscience, the manufacturer of the GM variety. Two other Bayer-produced GM rape varieties are awaiting approval by German regulators.

The Belgian ruling will now be reviewed by other EU Member States, though it is expected that they will back the decision. As officials point out, the ruling is backed by the conclusions of one of the world's largest scientific studies into the impact of GM crops on the environment. The UK's farm-scale trials suggested that growing GM spring-sown rape and sugar beet were more damaging to wildlife such as insects, though GM maize was less harmful.

Anti-GM pressure groups such as the UK-based Friends of the Earth (FoE) believe that governments should not be rushing into the cultivation of oilseed rape. There have been cases in North America where this has led to the contamination of non-GM farmland.

"The fact is that oil seed rape is a very small seed, and can easily be blown around," said FoE campaigner Pete Riley. "The pollen can spread several kilometres. This means that genetically modified seeds are easily blown onto non-GM farmland, and these 'volunteer' crops, as they are called, are finding their way into our food chain whether we like it or not."

But to complicate matters further, the European Commission has warned that it might challenge member-state bans on biotech crops in the European Court of Justice. Bans have also been imposed by Austria, France, Germany, Greece, Italy, Luxembourg and the UK.

While farmers can group together voluntarily into biotech-free zones, the commission said in a seven-page document on food

(*Continued*)

(Continued)

biotechnology that it would act against any national or regional laws preventing coexistence of biotech and conventional crops.

Bayer now has the right to appeal to the Belgian courts to have the decision overturned. Belgian ministers have already given the go-ahead for the GM rape variety to be imported from outside the EU and processed, since this would have no environmental impact.

The biotechnology industry will be looking closely at this case. It would no doubt regard a final decision to prevent it from marketing spring-sown rape for cultivation in the EU as a serious setback.

However it regards the future of winter-sown GM variants as far more important to their commercial success since these would be used over much greater areas across Europe.

www.foodqualitynews.com

Austria Bans Monsanto's GM Oilseed Rape

Immediate release: Monday 23 January 2006

Friends of the Earth has today welcomed the decision by the Austrian Government to ban Monsanto's genetically modified (GM) oilseed rape. This brings the total number of European bans on GM foods or crops to twelve (1). The decision by the current EU presidency follows November's referendum in Switzerland which put in place a five year moratorium on growing GM crops and comes ahead of the WTO GM dispute ruling, which will include whether countries are allowed to impose such bans. Friends of the Earth is calling on the UK Government to follow suit.

The Austrian decision to ban Monsanto's oilseed rape, GT73, is based on the risk of genetic contamination and the inadequate risk assessment carried out before the European Commission authorised it for import in August 2005. This authorisation came

(Continued)

(Continued)

despite a majority of EU Environment Ministers blocking its approval in December 2004 for environmental and health reasons.

Friends of the Earth wrote to the Government following the approval of GT73 oilseed rape, urging it to impose a national ban in the UK to protect health and the environment, but it refused.

Friends of the Earth's GM Campaigner, Clare Oxborrow said:

"The Austrian decision to ban this GM oilseed rape is a clear sign of the growing frustration with the EU's undemocratic decisions to approve GM food. Opposition to GM crops and food is growing — in the UK over eighteen million people live in GM free areas. It is time the UK Government accepted that there are genuine concerns surrounding this GM oilseed rape, and banned it too."

GMOs are now banned in seven European countries, while the number of EU regions banning GMOs is also growing: 172 Regions in the European Union and 4500 local authorities and other zones have now declared themselves GMO free and are calling for the right of Regions to decide whether or not to grow GMOs. This includes 60 local authorities in the UK. In June 2005, the EU Commission was defeated by Member States when it tried to force them to drop national GMO bans.

Note

(1) GMO bans in the European Union:

Germany (2000)

Syngenta's Bt176 maize (banned 31/03/2000) — Reason: effects on non-target insects + transfer of antibiotic resistance genes to humans and animals + insects could develop resistance to the Bt.

Bayer's oilseed rape Topas 19/2 (banned 16/11/1998) — Reason: impact of genetic escape and spread of herbicide tolerance.

Bayer's oilseed rape MS1×Rf1 (banned 16/11/1998) — Reason: impact of genetic escape and spread of herbicide tolerance.

(Continued)

(Continued)

France

Bayer's oilseed rape Topas 19/2 (banned 16/11/1998) — Reason: impact of genetic escape and spread of herbicide tolerance.

Bayer's oilseed rape MS1×Rf1 (banned 16/11/1998) — Reason: impact of genetic escape and spread of herbicide tolerance.

Austria

Syngenta's Bt176 maize (banned 13/02/1997) — Reason: effects on non-target insects such as butterflies + transfer of antibiotic resistance genes to humans and animals.

Bayer's T25 maize (banned 28/4/2000) — Reason: protection of sensitive areas, lack of monitoring plan and concerns about the herbicide used.

Monsanto's MON810 maize (banned 10/06/1999) — Reason: effects on non-target insects.

Hungary

Monsanto's maize MON810 seeds (banned 20/01/2005)

Luxembourg

Syngenta's Bt176 maize (banned 07/02/1997) — Reason: Transfer of antibiotic resistance genes to humans and animals.

Greece

Bayer's oilseed rape Topas 19/2 (banned 08/09/1998) — Reason: impact of genetic escape.

Monsanto's maize MON810 seeds (Commission ruled to overturn ban earlier this month).

(Continued)

Poland

Monsanto's maize MON810 seeds

Friends of the Earth Europe. Press Release. Brussels, Belgium. January 23, 2006.

EU on Line to Prohibit GM Oilseed Rape Crops

Greens hail an environmental victory for biodiversity as Belgium rejects Bayer application and urges all member states to follow suit.

Genetically modified oilseed rape crops may be barred throughout the EU for many years after Belgium's rejection yesterday of a union-wide application by the German company Bayer CropScience.

The Belgian government will call on other EU countries to follow suit. They are likely to do so, although, in theory, the application can be reopened by any country.

Despite heavy lobbying by biotech companies, Belgian ministers followed the advice of their bio-safety advisers, which drew heavily on several years of evidence from GM crop trials in Britain.

This showed, broadly, that herbicide-tolerant GM oilseed rape reduces biodiversity.

The British scientists found that bees and butterflies were less abundant in the GM oilseed rape crops than in non-GM crops, because of the lack of weeds and wild plants. There were also substantially fewer weed seeds present.

Weed seeds are an important source of food for small mammals and birds, particularly during the winter.

A spokesman for Bayer CropScience Belgium, Henk Joos, said: "We have serious concerns about the way the Belgian government handled this. We believe the decision was highly influenced by Belgian politics.

"The experts raised some concerns but indicated that with proper controls it would be possible to cultivate this crop without impacting on the environment."

(*Continued*)

(Continued)

Although the decision applies to only one variety of herbicide-tolerant GM rape, European green groups said it set an important precedent which would make it hard to justify the commercial production of any similar rape crop in Europe.

Karen Simal, Greenpeace Belgium's GM campaigner said: "This is a slap in the face of the biotech industry and a victory for the environment.

"The Belgian government has acknowledged that growing GM oilseed rape is harmful to the environment."

The decision was softened, however, by the Belgian government's decision to let GM oilseed rape be imported and processed.

A health ministry spokesman, Karim Ibourki, said the imported rape would be used for fuel, not for human or animal consumption.

Arian Bebb, GM campaigner for Friends of the Earth Europe, said: "GM oilseed rape will harm the environment and contaminate non-GM agriculture, whether it is grown in the EU or elsewhere in the world.

"It is inconsistent to ban the cultivation yet allow it for import."

European countries will have to rule on many applications from GM companies this year.

Bayer CropScience has two applications for similar GM oilseed rape varieties before the German government, and Belgium and Denmark must soon rule on whether to allow GM sugar and fodder beet to be grown.

Meanwhile there is a host of applications before EU countries from various GM companies to grow different varieties of GM maize.

"The Belgian decision is very significant," said Sue Mayer of the British group Genewatch.

"Bayer should withdraw all its remaining applications to grow GM oilseed rape in Europe as the tide of scientific and public opinion is clearly against them.

"This is another body blow to the biotech industry."

Vidal, J. EU on line to prohibit GM oilseed rape crops. *The Guardian*. February 3, 2004.

Biotech Industry withdraws GM foods, March 20, 2007

Helen Holder, GMO campaigner for Friends of the Earth Europe stated:

> "There is no market for GM food and crops, and companies are even withdrawing them from the market. European citizens want GM-free food and EU leaders need to take the necessary steps to make this happen."
>
> "These genetically modified foods should have never been allowed to be grown in the first place, as no one knows the long term effects to both people and the environment."
>
> "It is an absolute disgrace that European taxpayers' money was spent defending a trade dispute about products that biotech companies were about to withdraw. The biotech industry should be forced to pay the EU compensation for the time and money they have wasted."

Bans by EU Member States on three of these five GM crops were central to the transatlantic trade dispute in the World Trade Organisation (WTO) which ended in 2006. The WTO ruled that countries did have the right to prohibit GM crops but that the bans in the EU had not followed WTO procedures.

Friends of the Earth Europe. Press Release. Brussels, Belgium. March 20, 2007.

Impact of Agriculture on Biodiversity

Modern agricultural practices are known to have caused a decline in biodiversity. Its impact can be found in many species, particularly insects and birds. The degree of decline depends on several factors, including the amount and kinds of pesticides and insecticides used, the cropping patterns employed, and the frequency of tillage. It also depends on the nature of transgenics used as well as the nature of the surrounding ecosystems located near the fields of cultivation.

Meta-analysis of Bt Cotton and Maize on Non-target Insects

There has been much concern and debate about the possible adverse effects of Bt cotton and corn on non-target insects, at least since 1996. However, appropriate quantitative studies are lacking.

Biologist Michelle Marvier and her colleagues at Santa Clara University in California (Marvier *et al.* 2007) approached this problem from a different perspective. To examine the ecological consequences of transgenic *Bacillus thuringiensis* (Bt), the authors constructed a searchable database for non-target effects of Bt crops. A meta-analysis of 42 field experiments indicates that nontarget invertebrates are generally more abundant in Bt cotton and Bt maize fields than in nontransgenic fields managed with insecticides. However, in comparison with insecticide-free control fields, certain nontarget taxa are less abundant in Bt fields.

Crops modified to produce insecticides against pests are relatively kind to other insects, an analysis of 42 field experiments suggests. Fields of transgenic cotton and corn contain more non-target insects than those of traditional crops sprayed with insecticides, the study shows. But both have fewer such insects than traditional fields that aren't sprayed at all.

The finding eases worries that crops engineered to produce an insecticidal toxin made by the *Bacillus thuringiensis* (Bt) bacterium might kill more insects than intended, thus harming wildlife. The toxin is intended to target specific groups of plant pests, such as corn borers and cotton bollworms.

This is such a controversial issue, says ecologist Michelle Marvier of Santa Clara University, California. "There's a lot of public fear, in part because there's not a lot of transparency in the testing process."

Marvier and her colleagues used the US Freedom of Information Act to obtain the results of field trials submitted to the Environmental Protection Agency as part of the approval process for the engineered crops.

Fig. 1. Women cutting sugarcane near Visakhapatnam, Andhra Pradesh, India. Courtesy of Michele Wambaugh.

The field studies, she found, tended to use sample sizes that were too small to reveal small but statistically meaningful differences. So the researchers combined the data from field studies that measured invertebrate populations near Bt crops, in the hope of getting a large enough sample to spot small differences.

The central goal of this study is to investigate quantitatively whether changes in invertebrate abundance are statistically significant. Failure to find significant differences in previous studies was generally viewed as a signal of environmental safety. Whether statistically significant differences in abundance truly indicate ecologically significant changes is not clear.

Marvier *et al.* concluded that their analyses provided some support to claim that GM plants can reduce environmentally undesirable aspects of agriculture, particularly the nontarget impacts of insecticides. However, their study involved only one type of genetic modification, but also involved various control experiments. The vast majority of Bt maize acreage is concerned with varieties that are specially suited for silage or processed foods for which insecticide use has been limited. On the other hand, insecticides are far more frequently used in cotton production, over 71% of cotton acreage surveyed in 2005.

Fig. 2. The mysterious disappearance of bees in recent years has been attributed to various causes, including the possible toxicity of GM pollen, but was finally traced to a virus infection. To help unravel the mystery, Jeffery Pettis at the Bee Research Laboratory of the U.S. Department of Agriculture in Meltsville, Maryland, and his colleagues collected genetic samples from 51 colonies across the U.S., 30 of which had been devastated. By comparing the gene fragments they found against the recently published honeybee genome sequence (http://www.newscientist.com/channel/life/mg19225754.000-honeybees-have-their-genome-sequenced.html), Pettis and his colleagues quickly spotted the presence of foreign DNA belonging to the Israeli acute paralysis virus. However, the role of GM pollen toxicity is not completely ruled out in some cases. Courtesy of Michele Wambaugh.

June 07, 2007

Getting the Bugs Out of Genetically Modified Crops

Are crops genetically altered to resist insects really better for the environment?

In 1985 scientists inserted genetic information into tobacco plants that enabled them to produce a crystal that was toxic to butterflies, moths and other insect pests. Derived from the bacterium *Bacillus thuringiensis*, the Bt toxin has since been engineered into crops from corn to cotton, because it is lethal to pests yet seemingly

(*Continued*)

(Continued)

harmless to other insects and animals, including people. Now a new review of 42 field experiments indicates that fields planted with Bt crops have more insects and other critters than those treated with broad-spectrum insecticides. But it also exposes holes in the available research, such as the impact of genetically modified crops on neighboring ecosystems.

Biologist Michelle Marvier of Santa Clara University in California and her colleagues found that Bt corn and cotton that were not sprayed with pesticides had more fauna than treated traditional crops. "What the study really tells us is that conventional insecticides kill non-target insects," says entomologist Bruce Tabashnik of the University of Arizona (U.A.) in Tucson, who was not involved in the study. The question becomes, he adds: "Do Bt crops reduce insecticide use?"

The study also reveals, however, that Bt crop acreage has less insect biodiversity than untreated fields. "It is unclear whether the reduced abundance of these [insect] groups (coleopterans, hemipterans and hymenopterans) is due to direct toxicity or is a response to reduced availability of prey in Bt crops," Marvier's team reports today in *Science*.

U.A. entomologist Yves Carrière, who was also not involved in the study, notes that farm practices will ultimately determine the value and impact of such genetically modified plants. "If broad-spectrum insecticides are commonly used and Bt crops reduce such use, then Bt crops could have positive impacts," he says. "If insecticides are rarely used, then Bt crops do not bring advantages, and it is still unclear whether they may bring significant disadvantages."

Previous studies have indicated that Bt crops could lead to increased use of narrowly targeted pesticides. But they also show that they have reduced use of the most damaging broad-spectrum insecticides, which could be good news considering that an estimated 71 percent of U.S. cotton fields are treated with them, according to U.S. Department of Agriculture data from 2005.

"It's not just [that] the percentage of the acreage treated is higher, it's that the use per acre is more intense," Tabashnik notes.

(*Continued*)

(Continued)

"If you've got Bt cotton, you've got a bigger potential to reduce insecticide use."

Tabashnik also says that studies to determine the impact of genetically modified crops on creatures great and small in surrounding nonfarm environments are now needed. "How do transgenics affect wildlife in native habitats in the U.S.?" he asks. "That is the next frontier in this environmental assessment."

(Biello 2007)

Impact of Growing GM Crops on Biodiversity

Crops modified to tolerate pests or herbicides could potentially influence biodiversity

A crop plant modified to be toxic to insect pests could have a direct harmful effect on non-target insects if they eat the plant. It could also have an indirect effect by reducing the insects that are a food source for other wildlife, such as farmland birds.

GM crops that are tolerant to herbicides could also lead to a reduction in weed populations that act as refuges for beneficial insects, and/or those that are eaten by birds.

Repeated use of the same insecticide on pest-tolerant GM crops could also cause insects that become resistant to the chemical. This could affect the populations of pest insects and lead to an imbalance in their predator/prey relationships.

What is the evidence that this is likely to happen?

Several preliminary laboratory based studies suggest an adverse impact on biodiversity. Monarch butterfly larvae fed only on leaves covered in pollen from Bt corn grew more slowly and suffered higher death rates. Similar results were reported for pink bollworm fed on cotton producing the Bt toxin. Aphids fed on GM potatoes producing

(*Continued*)

(Continued)

a different toxin were also reported to have a harmful effect on lady-birds feeding on the aphids.

The results of many of these experiments have been questioned by international scientists and UK Government advisors

Further experiments on Monarch butterflies to test the likely levels of exposure of the larvae to the toxin strongly suggest that Bt maize has a "negligible" impact on biodiversity in the field. Many feeding studies have also been widely criticised for lack of scientific rigour and use of appropriate controls. The UK Government, however, requires people wishing to release GM crops in trials to give more prominence to risk assessments of the effect on GM pollen on non-target organisms.

Any impact of GM crops would operate against a biodiversity landscape already affected by post-War intensive agriculture

In the past 20 years in the UK, over 10 million breeding individuals of 10 species of farmland birds have disappeared from the countryside. Experimental changes in farming practices to encourage wildlife have shown a positive impact on bird populations.

Changes in crop husbandry could be incorporated into adoption of GM crops to encourage biodiversity.

Understanding the ecology of the area is crucial to ensuring this can happen. Recent research suggests changes in timing of herbicide applications to GM sugar beet can encourage weed plants without compromising crop yield.

It is a big challenge to design meaningful experiments to measure an impact on biodiversity in the field

Laboratory studies to test this often indicate a "worst case scenario" or "snap-shot" of the situation. To test effects over time and on a suitable scale, field tests have to be carried out.

(*Continued*)

(Continued)

The UK has been monitoring the impact of GM herbicide-tolerant sugar beet, fodder beet, maize and oil seed rape in Farm Scale Evaluations for the last 4 years. The numbers of several indicator species (such as earthworms, snails, slugs, ground beetles, bees and butterflies) have been measured throughout the trials.

Soybeans resistant to Dicamba

Behrens *et al.* (2007) from the University of Nebraska described the development of soybean and other broadleaf plant species resistant to dicamba, a widely used, inexpensive, and environmentally safe herbicide. The dicamba resistance technology will augment current herbicide resistance technologies and extend their effective lifetime.

The new plants are resistant to a compound called dicamba, and could offer farmers an alternative in areas where glyphosate-resistant weeds have become a problem.

Dicamba, which kills broadleaf weeds but spares grasses, has been used for decades to protect fields planted with corn, a member of the grass family. The researchers have now created transgenic soya beans, tomatoes and other broad-leaved crops that are resistant to this herbicide — a development that will expand the range of dicamba's uses.

Dicamba lasts only a few months in soil, and rarely contaminates water. The chemical itself is stable, but it is quickly devoured by hungry hordes of microbes living in the soil.

Genetic evidence

Don Weeks and his collaborators at the University of Nebraska in Lincoln isolated a gene from *Pseudomonas maltophilia* that is responsible for the breakdown of dicamba. They then transferred this gene into tobacco, soya beans, tomatoes and the model plant

(*Continued*)

(Continued)

Arabidopsis thaliana. In every case, the plants became resistant to dicamba, the researchers report in this week's *Science*.[1]

Monsanto, the makers of the 'Roundup Ready' line of glyphosate-resistant crops in St Louis, Missouri, has already licensed the dicamba technology. The company says it hopes to make dicamba-resistant soya beans available commercially in three to seven years, with cotton to follow after that. Monsanto does not market dicamba itself.

Some 90% of soya-bean crops in the United States and 60% of its cotton are genetically engineered to resist glyphosate. Although critics argue that transgenic technology encourages use of herbicides, advocates say that the crops have helped many farms to stop tilling the earth before they plant — a practice that causes added soil erosion and water pollution. Instead, farmers can plant glyphosate-resistant crops directly after 'burning down' existing weeds with glyphosate.

Dicamba also has some problems. The compound's volatility means that it can kill off broad-leaved plants on fields and houses up to half a kilometre away, meaning that farms close to sensitive areas such as vineyards or private gardens may not be able to use dicamba-resistant crops.

Isolated dicamba-resistant weeds have sprung up, and they have not posed a threat to agriculture. But that's not to say that resistance won't become a problem with more widespread future use.

2

Biodiversity Loss

The problem of extinction of species has been recognized for centuries. In c. 1760, the great Swedish naturalist Carl Linnaeus wrote in his journal: "I do not know how the world could persist gracefully if but a single animal species were to vanish from it." I wonder what he would say today when so many species of both animals and plants have already disappeared and many others are facing extinction in the foreseeable future!

Several leading authorities are making dire predictions regarding the rapid loss of our planet's biodiversity, with examples of several species of plants and animals becoming extinct over time. The main point of contention is not whether extinction of species has occurred or will occur, but rather the rate at which it will occur over the next 25 or 50 years and the magnitude of factors which will influence its pace. Estimates of species extinction rates range from as low as 10% to as high as 75%, the latter based on 90% loss of tropical forests.

Earth is presently in the midst of a sixth great extinction, the Holocene. Unlike previous mass extinctions in the past (i.e. the Ordovician, the Devonian, the Permian, the Triassic, and the Cretaceous), the current extinction is resulting from human activities such as habitat destruction, soil and genetic erosion, and the introduction of alien species into established environments.

Extinction rates are presently 1000–10 000 times the historical background rate of about 1 species per million per year. They are expected to increase significantly in the future, especially as the effects of climate change intensify in the coming years.

How Many Species Are Threatened?

According to "Countdown 2010", hosted by The World Conservation Union (IUCN) Species Survival Commission, 16 119 species are threatened with extinction, but this may be a gross underestimate because fewer than 3% of the world's 1.9 million described species have been assessed by the Red List.

Class	No. of species threatened
All species	16 119
Animals	7725
Plants and lichens	8394
Birds	12% or 1 in 8
Mammals	23% or 1 in 4
Amphibians	32% or 1 in 3
Turtles and tortoises	About 42%
Conifers	25% or 1 in 4
Cycads	52%

IUCN Red List 2007

Life on Earth is disappearing fast and will continue to do so unless urgent action is taken. Such is the considered assessment of the 2007 IUCN Red List of Threatened Species. This respected index, produced by an international network of species experts and partner organizations, analyzes global factors contributing to extinction risk such as rate of decline, population size, area of geographic distribution, and degree of population and distribution fragmentation. The numbers are numbing: there are now 41 415 species on the IUCN Red List, and 16 306 of them are threatened with extinction. According to the newest data, one in four mammals, one in eight birds, one third of all amphibians, and 70% of the world's assessed plants are in jeopardy. Great apes are in decline, corals and seaweeds are fading, and even some plants are becoming extinct.

According to Julia Marton-Lefèvre, Director General of IUCN, "This year's IUCN Red List shows that the invaluable efforts made so far to protect species are not enough. The rate of biodiversity loss is increasing and we need to act now to significantly reduce it and stave off this global extinction crisis. This can be done, but only with a concerted effort by all levels of society" (*Science Daily*, September 13, 2007).

Some highlights from this year's IUCN Red List

These highlights from the 2007 IUCN Red List are merely a few examples of the rapid rate of biodiversity loss around the world. The disappearance of species has a direct impact on people's lives. Declining numbers of freshwater fish, for example, deprive rural poor communities not only of their major source of food, but of their livelihoods as well.

(a) The decline of the great apes

The Western Gorilla (*Gorilla gorilla*) has moved from Endangered to Critically Endangered, after the discovery that the main subspecies, the Western Lowland Gorilla (*Gorilla gorilla gorilla*), has been decimated by the commercial bushmeat trade and the Ebola virus. Their population has declined by more than 60% over the last 20–25 years, with about one third of the total population found in protected areas killed by the Ebola virus over the last 15 years.

The Sumatran Orangutan (*Pongo abelii*) remains in the Critically Endangered category and the Bornean Orangutan (*Pongo pygmaeus*) in the Endangered category. Both are threatened by habitat loss due to illegal and legal logging and forest clearance for palm oil plantations. In Borneo, the area planted with oil palms increased from 2,000 km^2 to 27,000 km^2 between 1984 and 2003, leaving just 86,000 km^2 of habitat available to the species throughout the island.

(*Continued*)

(Continued)

First appearance of corals on the IUCN Red List

Corals have been assessed and added to the IUCN Red List for the very first time. In addition, 74 seaweeds have been added to the IUCN Red List from the Galápagos Islands. Ten species are listed as Critically Endangered, with six of those highlighted as Possibly Extinct. The cold water species are threatened by climate change and the rise in sea temperature that characterizes El Niño.

Yangtze River Dolphin listed as Critically Endangered (Possibly Extinct)

After an intensive, but fruitless, search for the Yangtze River Dolphin, or Baiji (*Lipotes vexillifer*), last November and December, it has been listed as Critically Endangered (Possibly Extinct). The dolphin has not been placed in a higher category as further surveys are needed before it can be definitively classified as Extinct.

India and Nepal's crocodile, the Gharial (*Gavialis gangeticus*), is also facing threats from habitat degradation and has moved from Endangered to Critically Endangered. Its population has recently declined by 58%, from 436 breeding adults in 1997 to just 182 in 2006. Dams, irrigation projects, sand mining and artificial embankments have all encroached on its habitat, reducing its domain to 2% of its former range.

Vulture crisis

This year the total number of birds on the IUCN Red List is 9,956 with 1,217 listed as threatened. Vultures in Africa and Asia have declined, with five species reclassified on the IUCN Red List. In Asia, the Red-headed Vulture (*Sarcogyps calvus*) moved from Near Threatened to Critically Endangered while the Egyptian Vulture (*Neophron percnopterus*) moved from Least Concern to Endangered. The rapid

(*Continued*)

(Continued)

decline in the birds over the last eight years has been driven by the drug diclofenac, used to treat livestock.

In Africa, three species of vulture have been reclassified, including the White-headed Vulture (*Trigonoceps occipitalis*), which moved from Least Concern to Vulnerable, the White-backed Vulture (*Gyps africanus*) and Rüppell's Griffon (*Gyps rueppellii*), both moved from Least Concern to Near Threatened. The birds' decline has been due to a lack of food, with a reduction in wild grazing mammals, habitat loss and collision with power lines. They have also been poisoned by carcasses deliberately laced with insecticide. The bait is intended to kill livestock predators, such as hyenas, jackals and big cats, but it also kills vultures.

North American reptiles added to IUCN Red List

After a major assessment of Mexican and North American reptiles, 723 were added to the IUCN Red List, taking the total to 738 reptiles listed for this region. Of these, 90 are threatened with extinction. Two Mexican freshwater turtles, the Cuatro Cienegas Slider (*Trachemys taylori*) and the Ornate Slider (*Trachemys ornata*), are listed as Endangered and Vulnerable respectively. Both face threats from habitat loss. Mexico's Santa Catalina Island Rattlesnake (*Crotalus catalinensis*) has also been added to the list as Critically Endangered, after being persecuted by illegal collectors.

Plants in peril

There are now 12,043 plants on the IUCN Red List, with 8,447 listed as threatened. The Woolly-stalked Begonia (*Begonia eiromischa*) is the only species to have been declared extinct this year. This Malaysian herb is only known from collections made in 1886 and 1898 on Penang Island. Extensive searches of nearby forests have failed to reveal any specimens in the last 100 years.

(*Continued*)

(Continued)

The Wild Apricot (*Armeniaca vulgaris*), from central Asia, has been assessed and added to the IUCN Red List for the first time, classified as Endangered. The species is a direct ancestor of plants that are widely cultivated in many countries around the world, but its population is dwindling as it loses habitat to tourist developments and is exploited for wood, food and genetic material.

Banggai Cardinalfish heavily exploited by aquarium trade

Overfishing continues to put pressure on many fish species, as does demand from the aquarium trade. The Banggai Cardinalfish (*Pterapogon kauderni*), which is highly prized in the aquarium industry, is entering the IUCN Red List for the first time in the Endangered category. The fish, which is only found in the Banggai Archipelago, near Sulawesi, Indonesia, has been heavily exploited, with approximately 900,000 extracted every year. Conservationists are calling for the fish to be reared in captivity for the aquarium trade, so the wild populations can be left to recover.

These highlights from the 2007 IUCN Red List are merely a few examples of the rapid rate of biodiversity loss around the world. The disappearance of species has a direct impact on people's lives. Declining numbers of freshwater fish, for example, deprive rural poor communities not only of their major source of food, but of their livelihoods as well.

IUCN News Release, September 12, 2007. www.iucn.org/redlist/

Species Loss Is Our Loss

Conservation action is slowing down biodiversity loss in some cases, but there are still many species that need more attention from conservationists. This year, only one species has moved to a lower category of threat. The Mauritius Echo Parakeet (*Psittacula eques*), which was one of the world's rarest parrots 15 years ago, has moved from Critically Endangered to Endangered. The improvement is a result of successful conservation action, including close monitoring of nesting sites and supplementary feeding combined with a captive breeding and release program.

Human activities and climatic or biological factors, among other causes, have led to the universal phenomenon of ecosystem fragmentation and, in some cases, total destruction. The present rate of human impact on biodiversity is unprecedented and is increasing dramatically. Humans have utilized and transformed over half of the land's surface, and it is difficult to find any area that can be described as untouched by the human hand today.

Causes of Extinction

Three basic causes of extinction have been frequently mentioned in the literature: (1) conversion of land use from forest to agriculture to meet the nutritional demands of rapidly growing human and animal populations, especially in the tropical countries; (2) increasing hunting practices and illegal poaching for commercial exploitation of large and small animals as well as for the medicinal use of both animal and plant species; and (3) changing climatic factors such as drought and floods which make the habitat inhospitable for several existing species of animals and plants, but may offer a hospital niche for invasive species.

Contrary View

However, a contrary view was recently presented in an interesting paper by Joseph Wright from the Smithsonian Tropical Research Institute in Balboa, Panama, and Helene Muller-Landau from the University of Minnesota, who argued that species loss may be more moderate than the commonly cited figures (Wright and Muller-Landau 2006). Although some have regarded their work as seriously understating the tropical biodiversity crisis, it has raised some fundamental questions about future conservation priorities that may lead to more effective strategies for conserving biodiversity.

The authors' projections are based largely on a log-linear relationship between human population density and the percentage of original forest cover remaining. Wright and Muller-Landau's projections of future extinctions are based on a standard method of using a species–area curve, calculating the number of species in a given area that is then utilized to estimate species loss accompanying a habitat loss. It is a logarithmic relationship, predicting approximately a 10%–20% extinction of species for a 50% habitat loss.

The key points presented by Wright and Muller-Landau (2006) are as follows. The central thesis of their argument is that population density and forest cover are linked together.

1. With the projection of falling rural population densities by 2030, net deforestation will fall as agricultural areas are abandoned and secondary forest recovers in previously deforested areas.
2. As primary forest will continue to be cleared for timber, the net effect will be that primary forest will be replaced with degraded forest and recovering secondary forest, where recolonization by forest specialists will take place.
3. Consequently, the net result is little change in total forest cover between now and 2030, as primary forest is replaced with degraded forest and recovering secondary forest.
4. Using species–area curves to predict extinction, the authors forecast a 21%–24% extinction in Asia, 16%–35% in Africa, and a significantly lower projection in Latin America.

Wright and Muller-Landau (2006) predict that many species which are currently at risk from habitat loss will not become extinct, but will benefit from the projected abandonment of agricultural lands and subsequent regrowth of secondary forest. Their hypothesis is based on the following assumptions: (1) it is based on an observed correlation, i.e. human population density in the tropical nations and net forest cover, which could result from a variety of causes; (2) the relationship between forest cover and population density is variable; (3) the United Nations (UN) estimates for population growth and urbanization are correct; (4) their deforestation projections represent net forest cover including regenerating growth; and (5) their projections are valid for the tropical regions of Asia, Africa, and the American continents.

Although Wright and Muller-Landau (2006) agree that the secondary forest is not as biodiverse as the primary forest, tropical forests are known to have retreated to smaller areas before, e.g. during the Ice Age. Today's forest species have survived extensive hunting and land clearing pressure from large sustained indigenous populations in the Amazon, Congo, and New Guinea. They predict, as many other researchers have done, that most of the primary tropical forests today will be cleared for timber harvesting and then be replaced by secondary forests regenerating after previous clearing. Any remaining primary forest area will be restricted to protected areas, such as the sacred areas maintained by tribal communities, as I have described in my book *Biological Wealth and Other Essays* (Dronamraju 2002). These are more likely to survive in areas of low human population densities that are not valuable for agriculture or development, such as the Amazon basin, the Guianas, Papua New Guinea, and Gabon.

The most hopeful situation, according to Wright and Muller-Landau (2006), is in Latin America, where the extent of forest still remaining is relatively high. Surprisingly, they mention low projected population growth (in spite of lack of birth control practices in the Catholic populations) and intense urbanization. Consequently, they predict a substantial net increase in the forest area, occurring before 2030.

Population migration from the tropical forest regions to the urban centers, as hypothesized by Wright and Muller-Landau (2006), is only effective in the present context if it exceeds the birth rate significantly. But, no one seems to have considered this point so far. Are there specific quantitative data on birth and death rates as well as population movements and changes in forest cover area in the tropics that would support the Wright–Muller-Landau hypothesis? Until such supporting data are available, their hypothesis will remain a mere hypothesis.

Habitat Destruction

Peter Raven, distinguished expert on biodiversity, wrote: "The greatest threat to biodiversity is habitat destruction. . . . (O)ur forecast that half of all species becoming extinct or being on the way to becoming extinct by the end of this century is still a modest and quite reasonable projection" (Raven, personal communication, 2007).

On the hypothesis proposed by Wright and Muller-Landau (2006), Raven commented:

> The vast majority of people in the field pretty well discount what they said. Wright and Muller-Landau argue that population growth will slow down and cutting will be alleviated, but if you consider the fact that the global footprint of humans is rising then this is a tough argument to follow. Globalfootprint.org does a good job of explaining this. It estimates that we are currently using 120% of Earth's productivity, meaning that we are using more than what Earth can sustainably produce. This has climbed from about 70% in 1970, so it's risen rapidly. . . . Even in a scenario of stabilizing population. . . ., I think it is unrealistic to think that forest loss is going to slow down. (Peter Raven, personal communication, 2007)

Hotspots

The concept of biodiversity "hotspots" identified 25 areas with high numbers of endemic species where large-scale habitat loss has already occurred and further losses are likely in the future (Myers *et al.* 2000). Covering 12% of the earth's surface, these hotspots contain 44% of vascular plants and 35% of terrestrial vertebrates. Sixteen of these hotspots in the tropical regions have already lost about 90% of their forest cover (Brooks *et al.* 2002). A species–area curve would predict that this loss would eventually result in an extinction of 50% of the species endemic to these regions.

However, Wright and Muller-Landau (2006) argue that most tropical species are found outside the area covered by the hotspots, inhabiting one of the four great tropical forest regions (they have also emphasized that most species have large geographic distributions, which are expected to buffer them from extinction):

1. Indo-Malaya;
2. Mesoamerica;
3. The Amazon basin and Guiana Shield; and
4. The Congo Basin and humid Western Africa.

Others have disputed their claim about buffering against extinction. For instance, a colleague of Wright at the same institution, William Laurance, stated:

> However, these areas also sustain numerous local endemics. . . . Even within seemingly monotonous expanses of forest, current and historical barriers, such as rivers, mountains and past forest refugia, have created complex patterns of species endemism. . . . (E)ven the largest tropical forest tracts currently in existence contain many restricted endemics that are inherently vulnerable to habitat disruption. (Laurance *et al.* 2006)

Furthermore, Laurance *et al.* (2006) wrote:

> A related tenet of the Wright and Muller-Landau argument is that future species extensions will be buffered by expanding forest regeneration. . . . (H)owever, many vulnerable species do not use secondary forests. There is, moreover, little compelling evidence that most tropical nations will experience substantial forest recovery, because of escalating demands for arable land . . ., and because their rural populations are not so much expected to fall as to plateau or grow less rapidly than will urban populations.

It has also been argued that, contrary to the view of Wright and Muller-Landau (2006), even sparse or declining rural populations can still cause heavy and sustained deforestation (Sloan 2007). It is also known that the poorest countries of the world, which include many developing countries, are not likely to, and cannot afford to, make the transition from deforestation to afforestation (Ewers 2006).

In contrast to the view of Wright and Muller-Landau (2006), who argued that only those species that undergo the most dramatic and historically unprecedented population declines will become extinct, Brooks *et al.* (2002) believe that many species are already on the way to extinction because of major declines and fragmentation of their populations. Wright and Muller-Landau emphasized that surviving forests are not just shrinking, but are also being severely fragmented. By missing this point, Wright and Muller-Landau have seriously underestimated the extinctions. Several factors such as climatic changes, overhunting and emerging pathogens have led to habitat degradation, which in turn has hastened the rate of species loss (Laurance *et al.* 2006).

Laurance and Peres (2006) pointed out that Wright and Muller-Landau (2006) violated the precautionary principle that one should err on the side of caution in conservation matters. Numerous changes which are caused by biological, genetic, climatic, geological, demographic, sociological, and political factors are rapidly transforming the biodiversity landscape of the planet. Many species including several predators have already become extinct. It is very likely that several species have already disappeared even before they were discovered and described. Species losses have inflicted severe damage to environmental wealth. These losses have already resulted in irreversible and adverse consequences to future generations. Laurance cautioned that it is too early to be optimistic in this situation.

Population Size and Forests

Laurance *et al.* (2006), among others, questioned the validity of the assumptions underlying Wright and Muller-Landau's (2006) hypothesis. Wright and Muller-Landau assumed that the density of rural inhabitants, rather than total population density, is most strongly correlated with percent forest cover at a national level. Their justification is that rural slash-and-burn farmers cause most forest loss. However, the UN population projections used by Wright and Muller-Landau contain large uncertainties. Global population projections for the year 2100 vary enormously from 7 billion to 15 billion people. Certain assumptions, such as that dramatic population growth in Equatorial Africa will soon decline even though industrialization and urbanization are likely to remain weak, are questionable. At a regional level, projected population trends in developing countries that underlie Wright and Muller-Landau's hypothesis remain very much uncertain. Such assumptions seem incredulous to professional demographers. There is not even a consideration of the possibility that one can project a birth rate which will balance the outward migration to the urban areas, although such a possibility appears to be unlikely in view of the future estimates projected by the UN.

The Wright and Muller-Landau (2006) hypothesis appears to be too simplistic at best. Threats to forest cover come from many directions, including violent or drastic climatic factors, oil and gas exploration, major highway construction, commercial exploitation such as lumber extraction, and other major changes brought on by globalization and industrialization such as irrigation dams and major hydroelectric projects. These threats to forest cover are not eliminated simply by population movement to urban centers alone.

World deforestation rates and forest cover statistics, 2000–2005

The Food and Agriculture Organization of the United Nations (FAO) released its *2005 Global Forest Resources Assessment*, a regular report on the status of the world's forest resources. Overall, FAO concludes that net deforestation rates have fallen since the 1990–2000 period, but some 13 million hectares of the world's forests are still lost each year, including 6 million hectares of primary forests. Primary forests — forests with no visible signs of past or present human activities — are considered the most biologically diverse ecosystems on the planet.

Industrial logging, clearing and forest conversion for agriculture, fuelwood collection by rural poor, and forest fires — often purposely set by people — are considered the leading causes of deforestation.

Central America, South Asia, and Southeast Asia lead deforestation rates. The regions with the highest tropical deforestation rate were Central America — which lost 1.3% or 285,000 hectares of its forests each year — and tropical Asia.

The region of Bangladesh, Bhutan, Brunei, Cambodia, East Timor, India, Indonesia, Laos, Malaysia, Maldives, Myanmar, Nepal, Pakistan, Philippines, Singapore, Sri Lanka, Thailand, and Vietnam lost about 1% of its forests each year. According to FAO, Vietnam lost a staggering 51% of its primary forests between 2000 and 2005, while Cambodia lost 29% of its primary forests between 2000 and 2005 (Cambodia's figures were revised by the FAO after this article was published. Original data showed Cambodia's primary forest cover declining to 122,000 hectares in 2005 from 356,000 hectares

(*Continued*)

(Continued)

in 2000. The new FAO data says Cambodia's current primary forest cover stands at 322,000 hectares]. Illegal logging, combined with rapid development, is blamed for much of Cambodia's forest loss.

Plantations offsetting natural forest

Due to a significant increase in plantation forests, forest cover has generally been expanding in North America, Europe and China while diminishing in the tropics. Plantations help offset the loss of natural forests but essentially result in an overall decline in global biodiversity as single species plantations replace their biologically richer natural counterparts.

New U.S. Policy: A disaster

The United States has the seventh largest annual loss of primary forests in the world, according to FAO. In the 2000–2005 period, the United States lost an average of 831 square miles (215,200 hectares, 2,152 square kilometers or 531,771 acres) of such lands which are sometimes termed "old-growth forests."

Overall, when plantations are added to the picture, the US gained a net 614 square miles (159,000 hectares) of forest per year. The FAO report suggests America's primary forests are losing ground to modified natural, seminatural, and plantation forests. Earlier this year, the government revoked President Clinton's 2001 "Roadless Area Conservation Rule" that protected 58.5 million acres of undeveloped national forest, in effect opening more than 90,000 square miles of forests to road construction, logging and industrial development.

UN Figures Contested

- Some environmental groups have criticized the UN numbers as "misleading and inaccurate" saying that FAO is using industrial plantations to offset deforestation figures for natural forests while relying on flawed figures provided by governments that have varying

(Continued)

(Continued)

standards of forest monitoring. The London-based Rainforest Foundation notes that "the UN figure is based on a definition of forest as being an area with as little as 10% actual tree cover, which would therefore include areas that are actually savannah-like ecosystems and badly damaged forests." Further, says a press release from the organization, "areas of land that presently have no trees on them at all, but that are 'expected' to regenerate, are also counted as forests."

- Despite the criticism, industry experts say that FAO has the best figures available across virtually all countries in the world. According to Mila Alvarez, who tracks forest trends for World Resources Institute and Global Forest Watch (globalforestwatch.org), "The F.A.O. is doing the best it can given what the governments are providing."
- The World Resources Institute and other organizations are developing a way to use satellite imagery to analyze forest changes and to verify government estimates.

Deforestation Tables

All area figures are in hectares (with the kind permission of Rhett Butler, mongabay.com).

Worst deforestation rate of primary forests, 2000–2005. All countries.

1	Nigeria	55.7%
2	Viet Nam	54.5%
3	Cambodia	29.4%
4	Sri Lanka	15.2%
5	Malawi	14.9%
6	Indonesia	12.9%
7	North Korea	9.3%
8	Nepal	9.1%
9	Panama	6.7%
10	Guatemala	6.4%

(Continued)

(Continued)

Highest average annual deforestation of primary forests, 2000–2005, by area. All countries.

1	Brazil	–3,466,000
2	Indonesia	–1,447,800
3	Russian Federation	–532,200
4	Mexico	–395,000
5	Papua New Guinea	–250,200
6	Peru	–224,600
7	United States of America	–215,200
8	Bolivia	–135,200
9	Sudan	–117,807
10	Nigeria	–82,000

Highest average annual deforestation of primary forests, 2000–2005, by area. Tropical countries.

1	Brazil	–3,466,000
2	Indonesia	–1,447,800
3	Mexico	–395,000
4	Papua New Guinea	–250,200
5	Peru	–224,600
6	Bolivia	–135,200
7	Sudan	–117,807
8	Nigeria	–82,000
9	Cambodia	–66,800
10	Colombia	–56,160
11	Panama	–43,200
12	Malawi	–39,600
13	Guatemala	–26,834
14	Viet Nam	–20,400
15	Democratic People's Republic of Korea	–17,400
16	French Guiana	–12,000
17	Senegal	–11,000
18	Nepal	–7,000
19	Madagascar	–6,800
20	Sri Lanka	–6,000

(*Continued*)

(Continued)

Most primary forest cover, 2005. All countries.

1	Brazil	415,890
2	Russian Federation	255,470
3	Canada	165,424
4	United States of America	104,182
5	Peru	61,065
6	Colombia	53,062
7	Indonesia	48,702
8	Mexico	32,850
9	Bolivia	29,360
10	Papua New Guinea	25,211

Most primary forest cover, 2005. Tropical countries.

1	Brazil	415,890
2	Peru	61,065
3	Colombia	53,062
4	Indonesia	48,702
5	Mexico	32,850
6	Bolivia	29,360
7	Papua New Guinea	25,211
8	Suriname	14,214
9	Sudan	13,509
10	Madagascar	10,347
11	Guyana	9,314
12	French Guiana	7,701
13	Congo	7,464
14	Thailand	6,451
15	Ecuador	4,794

Most "tropical rainforest", 2005. These rankings are estimates.

1	Brazil
2	Congo, Dem. Rep.
3	Peru
4	Indonesia

(Continued)

(Continued)

5	Colombia
6	Papua New Guinea
7	Venezuela
8	Bolivia
9	Mexico
10	Suriname
11	Guyana
12	Madagascar
13	French Guiana
14	Congo
15	Ecuador
16	Thailand
17	Malaysia
18	Panama
19	Guatemala
20	Nicaragua
21	Honduras
22	Laos
23	Philippines
24	Côte d'Ivoire
25	Belize

Most number of native tree species, 2005. All countries.

1	Brazil	7,880
2	Colombia	5,000
3	Madagascar	5,000
4	Belize	4,000
5	Philippines	3,000
6	Bolivia	2,700
7	Malaysia	2,650
8	Zambia	2,621
9	Peru	2,500
10	China	2,500
11	Guinea-Bissau	2,243
12	Australia	2,100

(*Continued*)

(Continued)

13	Singapore	2,013
14	Brunei Darussalam	2,000
15	Myanmar	2,000
16	Zimbabwe	1,747
17	Mali	1,739
18	Lao People's Democratic Republic	1,457
19	Togo	1,451
20	Venezuela (Bolivarian Republic of)	1,360

The Democratic Republic of Congo should be on this list, but FAO does not have figures for this war-torn country.

Highest total forest cover as a percentage of total land cover, 2005. All countries.

1	Suriname	94.7
2	French Guiana	91.8
3	Micronesia (Federated States of)	90.6
4	American Samoa	89.4
5	Seychelles	88.9
6	Palau	87.6
7	Gabon	84.5
8	Pitcairn	83.3
9	Turks and Caicos Islands	80.0
10	Solomon Islands	77.6
11	Guyana	76.7
12	Finland	73.9
13	Guinea-Bissau	73.7
14	Belize	72.5
15	Northern Mariana Islands	72.4
16	Anguilla	71.4
17	Lao People's Democratic Republic	69.9
18	Japan	68.2
19	Bhutan	68.0
20	Sweden	66.9

Includes plantations, nonnatural and degraded forests.

(Continued)

(Continued)

Highest total forest cover as a percentage of total land cover, 2005. All tropical countries.

1	Suriname	94.7
2	French Guiana	91.8
3	Micronesia (Federated States of)	90.6
4	American Samoa	89.4
5	Seychelles	88.9
6	Palau	87.6
7	Gabon	84.5
8	Solomon Islands	77.6
9	Guyana	76.7
10	Guinea-Bissau	73.7
11	Belize	72.5
12	Northern Mariana Islands	72.4
13	Anguilla	71.4
14	Lao People's Democratic Republic	69.9
15	Bhutan	68.0
16	Cook Islands	66.5
17	Congo	65.8
18	Papua New Guinea	65.0
19	Malaysia	63.6
20	Dominica	61.3

Includes plantations, nonnatural and degraded forests.

Highest total forest cover as a percentage of total land cover, 2005. All tropical countries excluding small islands.

1	Suriname	94.7
2	French Guiana	91.8
3	Seychelles	88.9
4	Gabon	84.5
5	Guyana	76.7
6	Guinea-Bissau	73.7
7	Belize	72.5
8	Lao People's Democratic Republic	69.9
9	Bhutan	68.0

(*Continued*)

(Continued)

10	Congo	65.8
11	Papua New Guinea	65.0
12	Malaysia	63.6
13	Cambodia	59.2
14	Democratic Republic of the Congo	58.9
15	Colombia	58.5
16	Equatorial Guinea	58.2
17	Panama	57.7
18	Brazil	57.2
19	Zambia	57.1
20	Bolivia	54.2

Includes plantations, nonnatural and degraded forests.

Total forest cover, 2005. All countries.

1	Russian Federation	808,790,000
2	Brazil	477,698,000
3	Canada	310,134,000
4	United States of America	303,089,000
5	China	197,290,000
6	Australia	163,678,000
7	Democratic Republic of the Congo	133,610,000
8	Indonesia	88,495,000
9	Peru	68,742,000
10	India	67,701,000
11	Sudan	67,546,000
12	Mexico	64,238,000
13	Colombia	60,728,000
14	Angola	59,104,000
15	Bolivia	58,740,000
16	Venezuela (Bolivarian Republic of)	47,713,000
17	Zambia	42,452,000
18	United Republic of Tanzania	35,257,000
19	Argentina	33,021,000
20	Myanmar	32,222,000

Includes plantations, nonnatural and degraded forests.

(Continued)

(Continued)

Total forest cover, 2005. Tropical countries.

1	Brazil	477,698,000
2	Democratic Republic of the Congo	133,610,000
3	Indonesia	88,495,000
4	Peru	68,742,000
5	India	67,701,000
6	Sudan	67,546,000
7	Mexico	64,238,000
8	Colombia	60,728,000
9	Angola	59,104,000
10	Bolivia	58,740,000
11	Venezuela (Bolivarian Republic of)	47,713,000
12	Zambia	42,452,000
13	United Republic of Tanzania	35,257,000
14	Myanmar	32,222,000
15	Papua New Guinea	29,437,000
16	Central African Republic	22,755,000
17	Congo	22,471,000
18	Gabon	21,775,000
19	Cameroon	21,245,000
20	Malaysia	20,890,000

Deforestation figures for Brazil

Year	Deforestation (sq mi)	Deforestation (sq km)	Change (%)
1988	8,127	21,050	
1989	6,861	17,770	−16%
1990	5,301	13,730	−23%
1991	4,259	11,030	−20%
1992	5,323	13,786	25%
1993	5,751	14,896	8%

(*Continued*)

(Continued)

1994	5,751	14,896	0%
1995	11,220	29,059	95%
1996	7,012	18,161	–38%
1997	5,107	13,227	–27%
1998	6,712	17,383	31%
1999	6,664	17,259	–1%
2000	7,037	18,226	6%
2001	7,014	18,165	0%
2002	8,187	21,205	17%
2003	9,711	25,151	19%
2004	10,590	27,429	9%
2005	7,256	18,793	–31%
2006	5,421	14,040	–49%
2007	3,865	10,010	–47%

All figures derived from official National Institute of Space Research (INPE) data. Individual state figures.

(Continued)

(Continued)

Country Forest Data (sorted by region)

Country/ Area	Total area (sq km)	Total forest area 2005 (sq km)	Forest cover 2005 (% of land)	Other wooded land 2005 (sq km)	Total forest area 1990 (sq km)	Total forest area 2000 (sq km)	Annual change rate 1990–2000 (sq km/ yr)	Annual change rate 1990–2000 (%/yr)	Annual change rate 2000–2005 (sq km/ yr)	Annual change rate 2000–2005 (%/yr)	Primary forest 2005 (sq km)	Primary forest 2005 (% total forest)	Annual loss of primary forest 2000–2005 (sq km/ yr)	Annual loss of primary forest 2000–2005 (%/yr)	Production plantation 2005 (sq km)
Angola	1,246,700	591,040	47.4	N/A	609,760	59,728	−1,250	−0.2	−1,250	−0.2	0	0	0	N/A	1,310
Botswana	581,730	119,430	21.1	347,910	137,180	12,535	−1,180	−0.9	−1,180	−1	N/A	–	N/A	N/A	N/A
British Indian Ocean Territory	80	30	32.5	0	30	30	0	0	0	0	N/A	–	N/A	N/A	N/A
Comoros	1,860	50	2.9	N/A	120	80	N/A	−4	−10	−7.4	0	0	0	N/A	10
Kenya	580,370	35,220	6.2	349,200	37,080	35,820	−130	−0.3	−120	−0.3	7,040	20	−24	−0.3	2,020
Lesotho	30,350	80	0.3	310	50	70	N/A	3.4	N/A	2.7	10	7.5	0	0.0	70
Madagascar	587,040	128,380	22.1	170,540	136,920	130,230	−670	−0.5	−370	−0.3	103,470	80.6	−68	−0.1	2,340
Malawi	118,480	34,020	36.2	N/A	38,960	35,670	−330	−0.9	−330	−0.9	11,320	33.3	−396	−3.5	2,040
Mauritius	2,040	370	18.2	150	390	380	N/A	−0.3	N/A	−0.5	0	0	0	N/A	110
Mayotte	370	50	14.7	N/A	60	60	N/A	−0.4	N/A	−0.4	N/A	–	N/A	N/A	N/A
Mozambique	801,590	192,620	24.6	409,190	200,120	195,120	−500	−0.3	−500	−0.3	N/A	–	N/A	N/A	380
Namibia	824,290	76,610	9.3	84,730	87,620	80,330	−730	−0.9	−740	−0.9	N/A	–	N/A	N/A	N/A
Réunion	2,510	840	33.6	550	870	870	N/A	−0.1	−10	−0.7	550	65	−0.2	0.0	20
Seychelles	450	400	88.9	N/A	400	400	0	0	0	0	20	5	0	0.0	50

(Continued)

(Continued)

Country/ Area	Total area (sq km)	Total forest area 2005 (sq km)	Forest cover 2005 (% of land)	Other wooded land 2005 (sq km)	Total forest area 1990 (sq km)	Total forest area 2000 (sq km)	Annual change rate 1990–2000 (sq km/yr)	Annual change rate 1990–2000 (%/yr)	Annual change rate 2000–2005 (sq km/yr)	Annual change rate 2000–2005 (%/yr)	Primary forest 2005 (sq km)	Primary forest 2005 (% total forest)	Annual loss of primary forest 2000–2005 (sq km/yr)	Annual loss of primary forest 2000–2005 (%/yr)	Production plantation 2005 (sq km)
South Africa	1,219,090	92,030	7.6	214,090	92,030	92,030	0	0	0	0	N/A	–	N/A	N/A	14,260
Swaziland	17,360	5,410	31.5	2,890	4,720	5,180	50	0.9	50	0.9	N/A	–	N/A	N/A	1,140
Uganda	241,040	36,270	18.4	11,500	49,240	40,590	–860	–1.9	–860	–2.2	N/A	–	N/A	N/A	360
United Republic of Tanzania	945,090	352,570	39.9	47,560	414,410	373,180	–4,120	–1	–4,120	–1.1	N/A	–	N/A	N/A	1,500
Zambia	752,610	424,520	57.1	31,610	491,240	446,760	–4,450	–0.9	–4,450	–1	N/A	–	N/A	N/A	750
Zimbabwe	390,750	175,400	45.3	N/A	222,340	191,050	–3,130	–1.5	–3,130	–1.7	N/A	–	N/A	N/A	1,540
Total Eastern and Southern Africa	8,343,800	2,265,340	27.8	1,670,230	2,523,540	2,350,470	–17,310	–0.7	–17,020	–0.8	0		0	N/A	0

Human Activities and Ecosystem Damage

Human activity has been changing the diversity of life on Earth — our biodiversity — more than any other time in history. This includes biodiversity loss that harms the natural systems, known as ecosystems, which sustain all life on the planet. Biodiversity conservation is one of the most urgent tasks today.

The leading agricultural scientist of India, M.S. Swaminathan, aptly stated the problem as follows:

> Biodiversity conservation is increasingly being recognized as a fundamental component of sustainable development. Although policy-makers often associate biodiversity conservation only with wildlands conservation and the prevention of species extinction, the concept actually encompasses the entire range of life's variety from wild *ecosystems* to the genetic diversity of agricultural species. More specifically, biodiversity is the variety of the world's organisms, including their genetic diversity, and the assemblages they form. (Swaminathan and Jana 1992, Introduction)

The rural poor in developing countries and the tribal or indigenous communities are often hit hardest, because they are more directly dependent on the resources and services that ecosystems provide.

The Millennium Ecosystem Assessment (MEA), the result of five years' research by 1360 of the world's leading scientists, documented how the growing human population is depleting resources and degrading the ecological systems that provide the fundamentals of life — clean water, breathable air, productive soil, and a stable climate (*Ecosystems and Human Well-being*, 2005).

Ecosystems, especially the tropical rainforests that harbor vast biological riches, provide services that clean our air and water, and provide food, medicines, energy, and raw materials. They regenerate soils and pollinate crops, regulate the climate, control floods, and offer recreational opportunities and spiritual renewal.

Ecosystem services are valued at US$30 trillion — more than the combined domestic product of all nations. Degrading them causes economic harm, as well as human suffering. For example, the removal and degradation of mangroves and other coastal ecosystems for development meant the loss of natural buffers to the December tsunami in Asia, increasing the devastation.

Communities closest to an ecosystem are most affected by change and biodiversity loss, the MEA notes. Converting or clearing a forest for cash-crop agriculture or timber means the loss of ecosystem services such as wild sources of food, water for drinking and crop irrigation, firewood and building materials, along with the recycling of wastes into nutrients.

It also makes clear that the benefits that biodiversity provides have not been accurately considered in decision-making and resource management. For example, the costs of lost ecosystem services frequently exceed the benefit of habitat conversion. Often, economic calculations fail to properly account for ecosystem services or tend to privilege the gains of one group over the losses of the wider community.

Subsidies for agriculture or extractive industries distort the relative costs and benefits of ecosystem services. The end result is often the majority of local inhabitants disenfranchised by the changes.

Protecting biodiversity cannot be justified by economic rationale alone. Relying solely on the numbers will fail to halt biodiversity loss. Ultimately, more biodiversity will be conserved if ethical, equitable distribution, and spiritual concerns are taken into account.

Biodiversity conservation should be part of strategies and programs for meeting the overall goals. Conservation in the form of protected areas and habitat restoration is urgently required.

Does Biodiversity Increase with Global Warming?

Carlos Jaramillo, at the Smithsonian Tropical Research Institute (STRI), and colleagues seek explanations for the longest Central and South America pollen record, published in the March 31, 2006 issue of the journal *Science*. "Plant diversity seems to increase when tropical forests cover large areas. Shrinking ecosystems may experience biodiversity loss lasting for millions of years" (Jaramillo *et al.* 2006).

"Jaramillo's intriguing findings provide an evolutionary perspective on a modern crisis," William Laurance of STRI and the Biological Dynamics of Forest Fragments project in Brazil comments. "They suggest that the rapid contemporary loss and fragmentation of rainforests will lead to striking, long-term biodiversity declines" (Laurance and Peres 2006).

Jaramillo *et al.* (2006) used cores drilled through 5 km of rock in eastern Colombia and western Venezuela to get at the fossil pollen record in a sequence of samples representing 10 million to 82 million years before the present. Then, they correlated pollen diversity with global temperature estimates for the middle part of that sequence (20–65 mybp).

> We found that pollen diversity tracks global temperature through time over millions of years. Diversity increases as the planet warms and decreases as it cools. The mystery is that even when global temperatures vary enormously, average temperatures in the tropics don't change much, so why do we see global temperature patterns reflected in tropical plant diversity?

Jaramillo *et al.* (2006) proposed that changes in area drive speciation and extinction in the tropics:

> There is good correlation between area and number of species: more area implies more species. During global warming, tropical areas expand and diversity goes up, the opposite happens during global cooling. If this is the case, fragmentation of modern tropical forest could be equated to a global cooling period, because forested areas are shrinking dramatically, resulting in plummeting diversity in the forests that remain.

Since this is an idea of much interest with respect to any study of tropical biodiversity, I quote the following excerpt from Jaramillo *et al.* (2006):

> Several mechanisms have been proposed to explain the high levels of plant diversity in the Neotropics today, but little is known about diversification patterns of Neotropical floras through geological time. Here, we present the longest time series compiled for

> palynological plant diversity of the Neotropics (15 stratigraphic sections, 1530 samples, 1411 morphospecies, and 287,736 occurrences) from the Paleocene to the early Miocene (65 to 20 million years ago) in central Colombia and western Venezuela. The record shows a low-diversity Paleocene flora, a significantly more diverse early to middle Eocene flora exceeding Holocene levels, and a decline in diversity at the end of the Eocene and early Oligocene. A good correlation between diversity fluctuations and changes in global temperature was found, suggesting that tropical climate change may be directly driving the observed diversity pattern. Alternatively, the good correspondence may result from the control that climate exerts on the area available for tropical plants to grow.

There is a correspondence between the global temperature curve for the Cenozoic and the diversity pattern shown. The increasing temperature trend from the early Paleocene to the early Eocene thermal maximum is paralleled, although slightly offset, by an increase in floral diversity. The subsequent long drop in temperature between the late middle Eocene and the early Oligocene is also paralleled by a similar drop in diversity, with a larger drop in both temperature and diversity at the Eocene–Oligocene boundary. This correspondence between diversity patterns and global temperature suggests a causal relationship.

> During the early and middle Eocene, there was a major global warming event that allowed tropical lineages to expand well into the modern temperate areas. High-diversity forests existed in the early Eocene of northern Patagonia, which was located near the southern tip of the tropical belt during the Eocene. This increase in the area with tropical-like climate could be the main factor enhancing the increase in local diversity in the Neotropics during the Eocene. Larger regions can support more species, which enhance both regional and local diversity by reducing the risk of extinction and increasing niche opportunities. In contrast, a cooling event in the late Eocene to early Oligocene reduced tropical areas drastically and, thus, drove local extinction in the Neotropics. A recent analysis of biome size integrated over time and diversity also found a primary role for changes in biome area over time in determining current species richness. (Jaramillo *et al.* 2006)

In conclusion, Jaramillo *et al.* (2006) stated:

> The overall pattern shows that plant diversity in the Neotropics has fluctuated greatly through time, as it is sensitive to global temperature. Temperature or precipitation change in the tropics may explain the pattern. An alternative hypothesis involves the control of global climate change on the area available for tropical ecosystems, which could, in turn, affect origination and extinction rates. If the size of forested areas does indeed control levels of local species diversity, conserving isolated pockets of tropical rainforest may not be sufficient to prevent high rates of extinction in the long run.

The RED (Reducing Emissions from Deforestation) Plan

More scientists have joined the growing chorus to support a plan by developing countries to fight global warming by reducing deforestation rates. Tropical deforestation releases more than 1.5 billion metric tons of carbon into the atmosphere every year, though in some years, like the 1997–1998 el Niño year when fires released some 2 billion tons of carbon from peat swamps alone in Indonesia, emissions are more than twice that.

Writing in the journal *Science*, an international team of scientists argued that the "Reducing Emissions from Deforestation" (RED) initiative, launched in 2005 by the United Nations Framework Convention on Climate Change, is scientifically and technologically sound (Gullison *et al.* 2007). On the scientific and technical side, the authors addressed two main questions. First, can preserving tropical forests make a significant dent in climate-threatening carbon emissions? And second, will these preserved forests be able to survive in an environment altered by the climate change that cannot be avoided?

On the first question, Gullison *et al.* (2007) found that reducing deforestation rates by 50% over the next century will save an average of about half a billion metric tons of carbon every year. This by itself could account for as much as 12% of the total reductions needed from all carbon sources to meet the Intergovernmental Panel on Climate Change (IPCC) target of 450 parts per million of carbon dioxide in the atmosphere by the year 2100. As for the second issue, computer models that link climate effects to changes in the carbon cycle have predicted that tropical forests will survive and continue to act as a "sink" by absorbing carbon, provided that emissions can be kept under control. The efficiency of the tropical forest as a carbon sink might in fact diminish over time, but the authors expect that it will not disappear completely.

The political challenges to reducing deforestation in the tropical developing world are varied and complex. Traditionally, many countries have viewed their forests as an economic resource that they have the right to harvest, much as oil- and ore-rich nations exercise the right to harvest those resources. As such, many proposed solutions are centered on direct economic incentives to reduce rates of tree clearing. However, Gullison *et al.* (2007) described low-cost measures that can enhance the success of carbon-trade systems and subsidized low-carbon development programs.

In conclusion, Gullison *et al.* (2007) stated:

> Providing economic incentives for the maintenance of forest cover can help tropical countries avoid these negative impacts and meet development goals, while also complementing aggressive efforts to reduce fossil fuel emissions. Industrialized and developing countries urgently need to support the RED policy process and develop effective and equitable compensation schemes to help tropical countries protect their forests, reducing the risk of dangerous climate change and protecting the many other goods and services that these forests contribute to sustainable development.

Authors on the paper are Raymond Gullison of the University of British Columbia, Canada; Peter Frumhoff of the Union of Concerned Scientists; Christopher Field of the Carnegie Insitute's Department of Global Ecology at Stanford University; Josep Canadell of the Global Carbon Project, Australia; Christopher Field of the Carnegie Institution; Daniel Nepstad of The Woods Hole Research Center; Katharine Hayhoe of Texas Tech University; Roni Avissar of Duke University; Lisa Curran of Yale University; Pierre Friedlingstein of IPSL/LSCE, France; Chris Jones of the Hadley Centre for Climate Prediction and Research, United Kingdom; and Carlos Nobre of Centro de Previsão de Tempo e Estudos Climaticos, Brazil.

Bioprospecting

The largest potential global market for bioprospecting is the pharmaceutical industry. Global annual sales of pharmaceuticals are approximately US$200 billion, of which an estimated US$40 billion is the result of bioprospecting. The market for seeds is much smaller, as is the contribution made to it by wild biodiversity. In a survey of the source of germplasm for various crop groups for 20 plant breeding and seed companies by the Cambridge University and the World Conservation Monitoring Centre, the companies surveyed obtained 81.5% of all their germplasm from commercial cultivars, with the rest originating from wild species maintained *in situ* and *ex situ*, *in situ* landraces, and landraces maintained in gene banks.

Figures such as those related to the seed industry suggest that companies' interest in investing in bioprospecting is limited, but where they are interested, they look for specific features. According to the (then) Business Council for Sustainable Development, companies look for "macroeconomic and political stability, reliable legal and property systems, an educated work force, and an adequate physical infrastructure (roads, power, communications, and so on)" (Kate 1995). As much as ecological considerations, these factors influence companies' choice of countries in which to prospect.

Between 2000 and 2005, deforestation rates of primary forest rose 25.6% (from 0.67% per year to 0.84% annually) for these 17 countries in the table below.

	Set 1 (17 countries)	Set 2 (62 countries)
Annual deforestation rate 1990–2000	0.67%	0.66%
Annual deforestation rate 2000–2005	0.84%	0.81%
Change in deforestation rate	25.6%	23.7%
Area of annual deforestation 1990–2000	4.97 million hectares	5.41 million hectares
Area of annual deforestation 2000–2005	5.82 million hectares	6.26 million hectares
Data integrity	Acceptable	Poor

(Courtesy of Mongabay)

Top 10 deforesting countries in terms of total forest loss (1995) (area loss in hectares)

Country	Ranking	Annual loss	Country	Ranking	Annual loss
Brazil	1	–2,550,000	Venezuela	6	–500,000
Indonesia	2	–1,080,000	Malaysia	7	–400,000
P.R. Congo	3	–740,000	Myanmar	8	–390,000
Bolivia	4	–580,000	Sudan	9	–350,000
Mexico	5	–510,000	Thailand	10	–330,000

Source: FAO (1997).

Important deforesting countries and regions in terms of annual rate of loss (1995)

Country	Ranking	% Annual loss	Country	Ranking	% Annual loss
Philippines	1	–3.5	Central America	6	–2.1
Sierra Leone	2	–3.0	Caribbean Islands	7	–1.7
Pakistan	3	–2.9	Cambodia	8	–1.6
Thailand	4	–2.6	Ecuador	9	–1.6
Paraguay	5	–2.6	Myanmar	10	–1.6

Source: adapted from FAO (1997).

Developing countries with no remaining large tracts of undisturbed, biologically intact forests

Africa			Latin America and Caribbean	Asia
Angola	Guinea	Senegal	El Salvador	Pakistan
Benin	Guinea-Bissau	Sierra Leone	Haiti	Philippines
Botswana	Kenya	South Africa	Paraguay	
Burundi	Liberia	Tanzania		
Equatorial Guinea	Madagascar	Togo		
Eritrea	Mozambique	Uganda		
Ethiopia	Namibia	Zambia		
The Gambia	Rwanda	Zimbabwe		
Ghana	Sao Tome & Principe			

Source: adapted from Bryant, D. *et al.* (1997), *The Last Frontier Forests – Ecosystems and Economies on the Edge*. Washington D.C.: World Resources Institute, p. 42.

Collectors and the Users

Collectors usually include those from private groups, local communities or individuals from botanical gardens, universities, and research institutes. They should comply with host countries' laws and procedures governing access to biodiversity. As many developing countries do not yet have established policies to protect their biodiversity, collectors are expected to follow certain ethical and moral guidelines voluntarily.

Such companies as the pharmaceutical, seed, agrochemical, and biotechnology industries, whose main interest is commercialization and profiteering, normally send scouts for bioprospecting and/or use collectors to obtain samples. I have discussed the ethical issues involved in a previous book, *Biological and Social Issues in Biotechnology Sharing* (Dronamraju 1998). The biggest users are the pharmaceutical companies and the research institutes employed by various governments in North America and Europe. One example is the National Cancer Institute of the U.S. National Institutes of Health. Although developed countries have been reaping huge profits for several decades, there is no mechanism in place that would ensure the recognition and compensation of indigenous knowledge by local communities. This criminal conduct includes the theft of both physical property and intellectual property.

The wording of Articles 15, 16, and 19 of the Convention on Biological Diversity — which deal with access to biodiversity, the sharing of benefits, and technology transfer — leaves national governments with considerable discretion as to the extent and manner in which they regulate access to biodiversity and involvement in product development and marketing. For developing countries, bioprospecting, if properly managed and exploited, could provide a steady source of income which would pay for the conservation measures that are needed for protecting and perpetuating that biodiversity, and provide employment opportunities for local communities.

Biodiversity Conservation

- *Assessment* of the opportunities, needs, resources and capacities of the country for the sustainable use of its genetic resources, including bioprospecting;
- *National policies and legislation* to require, as a condition for access, a fair share of any benefits arising from the use of genetic resources and knowledge concerning them to be returned to those providing them, and some benefits to be dedicated to conservation;
- Legislation, policies and incentives to increase *national capacities* to add value to genetic resources, thus generating greater income and other social and development benefits from environmentally sustainable activities, and encouraging the protection of biodiversity;
- Identification of institutional links and channels through which the benefits sought by the country could be shared with stakeholders, and mechanisms for negotiating fair and *mutually satisfactory bioprospecting partnerships* with in-country partners, and foreign companies or other provider or user countries;

(*Continued*)

(Continued)

- Creating mechanisms for stakeholders to *participate* in national strategies for the conservation and sustainable use of biodiversity, in the development of international law and policy to support national interests, and in individual bioprospecting agreements;
- Identifying *sources of funding* to initiate sustainable bioprospecting activities and other uses of genetic resources.

(Kate 1995)

Benefit Sharing

I have previously discussed that developing countries could benefit from partnerships with foreign governments, companies, or other groups to increase their revenue-earning capacity and protect their biodiversity. What is important is to start earning benefits right at the beginning, before a foreign entity gains access to the local bioresources, and gradually build a partnership as the development enters the commercial phase. Shaman Pharmaceuticals of San Francisco has set an example in this regard in its collaboration in Africa. In Costa Rica, INBio scientists have entered into an agreement with Bristol-Myers Squibb to develop anti-infective, anticarcinogenic, dermatological, and other medicines.

However, there are many instances where local plant species have been exploited for their medicinal value, but no compensation was paid to the indigenous communities who conserved their germplasm for centuries. For instance, a plant found in Kenya, *Maytenus buchanii*, has been used for centuries by the indigenous Digo community for treating cancer; the National Cancer Institute (NCI) of the U.S. National Institutes of Health seized the whole stock for research purposes, but the rights of the indigenous community were not recognized. Another plant, the rosy periwinkle, a native of Madagascar and Jamaica, is now cultivated extensively in the U.S; it is the source of two anticancerous drugs, vinvristine and vinblastine, but the indigenous communities of Madagascar and Jamaica who conserved this plant for centuries have received no compensation.

A similar situation prevails in another part of the world. In Indonesia and other countries of southeast Asia, a plant called *gaharu* has been used traditionally for treating complications of pregnancy and childbirth. Gaharu is a fragrant, resin-impregnated wood valued mainly for its aromatic, fumigatory, and medicinal properties found in Aquilaria trees. In Indonesia, the estimated stock of the tree is 1.87 trees per ha in Sumatra, 3.37 trees per ha in Kalimantan, and 4.33 trees per ha in Papua. The occurrence of the tree itself does not guarantee the presence of the resin. Scientists estimate that only 10% of the Aquilaria trees in the forest contain gaharu. Indonesia is one of the world's major exporters of gaharu products. With high market demands, many unskilled collectors are attracted to gaharu exploitation; as a result, the gaharu population has suffered destruction across a large part of its range, regardless of the fact that it is listed in the Convention on International Trade in Endangered Species of Wild Fauna and Flora (CITES) Appendix II. Lately, prices for the best-quality gaharu have been quoted at about $400/kg, and much of the material is smuggled and illegally traded out of the country.

Various countries have enacted legislation to protect their biodiversity.

The Philippines

The Republic of the Philippines, in its Executive Order No. 247 ("the PEO"), "Prescribing Guidelines and Establishing a Regulatory Framework for the Prospecting of Biological and Genetic Resources, Their By-products and Derivatives, for Scientific and Commercial Purposes and for Other Purposes", has established a procedure for assessing, obtaining, and verifying the prior informed consent (PIC) of local and indigenous communities in relation to biodiversity prospecting. The policy underlying the PEO is strong and clear: prospecting for biological and genetic resources shall be allowed "within the ancestral lands and domains of indigenous cultural communities only with prior informed consent of such communities;

obtained in accordance with the customary laws of the concerned community."

Latin America

By virtue of Decision 391, "Common Regime of Access to Genetic Resources", of the Commission of the Cartagena Agreement, access to intangible assets (such as traditional ecological knowledge or TEK) owned by a local or indigenous community is prohibited in Andean Pact countries unless their prior consent is obtained. In case of infringement of this requirement for the PIC, the contract will be deemed null (Article 39, Decision 391) and could amount to a criminal offence, as all declarations made by the accessor are affidavits (Article 22, Decision 391).

Africa

The Draft Legislation on Community Rights and Access to Biological Resources ("the draft legislation"), prepared by the Organization of African Unity's (OAU) Scientific, Technical and Research Commission in Addis Ababa, Ethiopa, offers perhaps the most powerful and comprehensive protection for TEK and the natural resource rights of indigenous peoples formulated to date. The principal features of the draft legislation are as follows:

- the requirement for PIC shall not apply to the traditional use and exchange of biological and genetic resources and related knowledge by and between local communities based on their customary practices (Article 3.2);
- provision is made for a two-tiered regime governing the PIC of the State and the local community with regard to access to genetic resources and associated knowledge. The regime subjects the giving of such consent to a detailed set of conditions and requirements that must be met by the applicant (Articles 4.2 and 5.3);
- no access to genetic resources shall be allowed to the accessing country unless the competent national authority in that country

confirms that the PIC has been obtained from the country of origin (Article 4.3). The State in which the collector operates must also guarantee that the collector complies with the mutually agreed terms and conditions provided by the agreement between the country of origin and the local community, and give an indication as to how it intends to enforce it (Article 4.5);

- the competent national authority of the provider State may unilaterally withdraw its consent and terminate an agreement if the mutually agreed terms of the agreement have been violated, or if the overriding public interest so demands (Article 4.4);
- the State shall recognize and protect the rights of local communities to benefit collectively from their knowledge, innovations and practices and to receive compensation for the conservation of genetic resources (Article 5.1). The local communities "shall at all times and in perpetuity be the lawful and sole custodians of the relevant knowledge, innovations and practices" (Article 5.2);
- the State shall take measures to establish a system regulating the collective/communal intellectual rights of the local community members through a process of consultation (Article 5.5);
- the State shall ensure that local communities have the right not to allow the collection of genetic resources and access to traditional knowledge and technologies in their custody (Article 5.7);
- a national "inter-sectoral body at the highest level", including representatives from local communities, should be created as a regulatory body to ensure the proper implementation and enforcement of the legislation (Article 6.1). A technical secretariat (advisory body), also having local community representation, should be created to support the work of the national authority (Article 6.2);
- a National Information System should be established (Article 7) to institute measures for the repatriation of information on the country's traditional knowledge and technologies as well as biodiversity (Article 7(b)); and
- other institutional arrangements should include the establishment of a national (biodiversity) trust fund (Article 8), provision for appeals arising in relation to biodiversity agreements that

allow recourse to the courts if all administrative remedies are exhausted (Article 9), and a scale of sanctions and penalties (Article 10) that should include provision for publicizing any violations through national and international media and their reporting to the secretariats of relevant international agreements/treaties and regional bodies (Article 10.2).

The OAU's draft legislation follows the requirements of Articles 8(j), 10(c), 15.5, 17.2 and 18.4 to the letter. As such, it provides a clear, concise, and plain language model that can be used by indigenous people in Australia in their negotiations with Federal, State and Territory governments to secure appropriate levels of protection for their TEK and traditional natural resources. Besides embedding the right to say no in national legislation, a right critical to the practice of self-determination, the draft legislation also entrenches the expectation that a national government will not allow access to local community genetic resources and associated TEK unless the country from which the collector operates has complementary laws of enforcement.

Australia

The Environment Protection and Biodiversity Conservation Bill 1998 (Cth) addresses the control of access to biological/genetic resources under Chapter 5 Part 13 Division 6. Clause 301 provides the following:

1. The regulations may provide for the control of access to biological resources in Commonwealth areas.
2. Without limiting subsection (1), the regulations may contain provisions about all or any of the following:

   (a) the equitable sharing of the benefits arising from the use of biological resources in Commonwealth areas;
   (b) the facilitation of access to such resources;
   (c) the right to deny access to such resources;
   (d) the granting of access to such resources and the terms and conditions of such access.

While there is scope within the regulations for implementing PIC-type requirements, these will only apply to Commonwealth areas.

Conservation and use of natural resources are, of course, primarily a State and Territory responsibility. However, the Federal Government has constitutional responsibilities in such areas as trade and commerce, corporations, external affairs, export control, and, of course, the power to make laws for the "people of any race". Given Australia's mega-biodiverse status and the interest of foreign-based institutions and transnational pharmaceutical and agricultural corporations in its biological wealth, some Federal monitoring of access to and use of that wealth is necessary. State and Territory governments retain primary responsibility for the management of non-Commonwealth lands and waters. This means that access regimes will have to be negotiated with the States and Territories. The kind of access regimes governing the PIC of the concerned indigenous communities that have been implemented in other countries might therefore be appropriate for indigenous communities/ traditional owner groups on this level. Nevertheless, the monitoring requirements of the Commonwealth might include the setting of minimum standards and the establishment of appropriate "trigger mechanisms", giving rise to the requirement for the PIC of indigenous communities/traditional owner groups. The latter should be in line with the emerging high standards of best practice evidenced in the laws of the countries mentioned above.

State and Territory legislation concerning nature conservation and natural resource use generally establish permit regimes to govern a range of activities, including the collection of genetic resources and research. It is in the best interests of the indigenous communities to negotiate with the State and Territory governments an access regime which will ensure that appropriate PIC procedures regarding access to traditionally used biological and genetic resources and associated TEK are also embedded within this legislation and its regulations.

The outcomes of indigenous peoples' negotiations with Australian State and Territory governments would constitute suitable material for case studies which indigenous organizations have been invited to contribute by the Contracting Parties under Decision IV/9 of the COP. Paragraph 10 of that decision invites indigenous representatives to submit case studies on the following:

(b) The influence of international instruments, intellectual property rights, and current laws and policies on the knowledge, innovations, and practices of indigenous and local communities embodying traditional lifestyles relevant for the conservation and sustainable use of biological diversity; and
(e) Matters of prior informed consent, fair and equitable sharing of benefits and *in situ* conservation in lands and territories used by indigenous and local communities.

Indigenous representatives are encouraged to submit case studies for dissemination through means such as the Convention's clearing-house mechanism "relating to Article 8(j) and intellectual property rights ... for transmittal to the World Intellectual Property Organisation and for use in initiatives on legislating on the implementation of Article 8(j) and related provisions." This invitation to submit such case studies is further evidence of the way in which indigenous peoples are being empowered to contribute to the implementation of the Convention.

The Kuna of Panama

In 1988 the Proyecto de Estudio para el Manejo de Areas Silvestres de Kuna Yala (PEMASKY) and the Asociación de Empleados Kunas (AEK) of Panama produced a manual to regulate scientific research in their area.

The manual requires researchers to:

- Develop a proposal outlining the timing, extent and potential environmental and cultural impact of a research programme, for approval by the Scientific Committee of PEMASKY;

(*Continued*)

(Continued)

- Secure approval for the collection of species from the Scientific Committee of PEMASKY. Collections may not include any endangered species, may not be used for commercial purposes and must be done in a non-destructive manner;
- Undergo an orientation into the culture of the Kuna Yala, and respect the norms of the communities in which they work;
- Include Kuna collaborators, assistants, guides and informants in their research programme and provide them with training in relevant scientific techniques;
- Leave samples of all specimens (for the collections at the University of Panama) and copies of photographs or slides taken during the research programme with PEMASKY;
- Provide written reports of the research, and two copies of any publications, in Spanish and descriptions of all species new to science to PEMASKY;
- Not to introduce exotic plant or animal species or to manipulate genes.

Additionally, research is restricted to certain areas of the reserve, is prohibited in some sites, such as ceremonial or sacred sites, and is controlled in other specific sites, such as some forest areas under community management.

(Chapin 1998)

The New York Botanical Garden

In April 1993, New York Botanical Garden (NYBG) and the Awa Federation, a legal institution which administers the land held under communal title by the Awa of Cachi in Ecuador, signed a two-year agreement for academic scientific research. The agreement includes the following terms:

- All scientists must ask for written permission to carry out studies, setting out a description of objectives, size and composition of

(*Continued*)

(Continued)

the research party, length of research programme, species or object of study, and the manner in which this research will benefit the Awa community;

- The request for permission must be given with a minimum of two months' notice, since dispersed communities only meet four times a year for four days.
- Research groups are limited to five people;
- Local guides and informants must accompany all scientists;
- The removal of any object from Awa territory not approved by the federation is prohibited;
- Payments to the Awa Federation members for their services should be in accordance with a set of prices established by the Federation;
- The Awa Federation must receive acknowledgement in all publications.

(Kate 1995)

Habitat Loss

Human activities and climatic or biological factors, among other causes, have led to the universal phenomenon of ecosystem fragmentation and, in some cases, total destruction. The present rate of human impact on biodiversity is unprecedented and is increasing dramatically. Humans have utilized and transformed over half of the land's surface, and it is difficult to find any area that can be described as untouched by the human hand today.

Habitat loss and fragmentation have greatly increased the threat to a large number of wild species that are facing total extinction. Other species are threatened to varying degrees in different localities. It is precisely the local extinctions that are of concern to farm households. Local populations are adapted to the particular environment, while substitutes are not as well adapted to local conditions.

Fragmentation

Different ecosystems vary widely in their stability or relative invisibility. Undisturbed tropical rainforests and other ecosystems in harsh or extreme environments such as deserts and mangroves have generally remained stable in the face of invasions by "foreign" species. However, some other ecosystems, such as fragmented rainforests and oceanic islands, are readily vulnerable to invasive species which promote further fragmentation. Island ecosystems such as those found in Hawaii are highly vulnerable to devastating invasions by invasive species. Hence, they are carefully protected from any inadvertent invasions by alien species or accidental transport by visitors.

Depletion of Wild Habitat

A characteristic pattern of resource exploitation by many traditional farm communities is the radial depletion of woody plants surrounding villages or settlements, especially plants used for fuelwood or for medicinal purposes. This pattern of depletion can have serious social effects, especially on the poorest members of the community, and on others involved in construction and craft work or medicinal plant utilization.

Increasing interest in alternative medicine, especially herbal medicine and Ayurveda, has greatly magnified the demand for traditional plant-based remedies. This has, in turn, increased pressure on wild plant species, leading to the extinction of several medicinal plants. Many medicinal plants are at risk because of unethical and exploitative practices by pharmaceutical companies which have not put in place any conservation plans.

When commercial exploitation leads to sustained harvesting of wild species, it may lead to genetic erosion. One example is the case of oregano in Turkey, where loss of genetic variability has been recorded for several species, including *Origanum onites*. Similarly, overharvesting of agarwood (*Aquilaria* spp.), one of the most valuable nontimber forest products of Asia, has been reported from certain parts of its range such as India and Vietnam, and is valued for the production of incense and certain traditional medicines. At least two species are particularly at risk: *A. malaccensis* and *A. crassna*.

Invasive Species

Another significant threat comes from invasive species. Several possibilities may occur when exotic species are introduced:

1. The introduction will die out.
2. The introduction will survive with little ecological or economic impact.
3. The introduction will survive with beneficial ecological or economic impact.
4. The introduction will survive with harmful, sometimes devastating, ecological or economic impact.

One of the greatest threats to natural and seminatural vegetation that is often overlooked is the deliberate human introduction of species — trees and fodder crops, for example — which have largely replaced the native ecosystems. Introduced species may also be a threat to productive systems. Examples are the introduced paleotropical grasses which have, since the 1840s, become major agents in facilitating deforestation in Central and South America; and the trees and shrubs introduced into the very diverse Fynbos and Karoo formations of South Africa that have had a devastating effect, putting at risk over 50% of the component species. Many grasslands in both temperate and tropical regions — in Australia, California in the United States, Africa, and Central and South America — are seriously affected by alien invasive species.

The major problems are caused by the following:

- Lack of awareness, until recently, of the threats and consequences of biological invasions;
- Lack of knowledge of the actual and potential effects of invasive species;
- Lack of monitoring systems;
- Lack of early warning systems;

- Lack of concern for the maintenance of native species;
- Lack of appreciation of the value of native species, leading to their deliberate replacement by exotic germplasm;
- Lack of appropriate technologies to combat and control alien and invasive species;
- Lack of legal instruments and other measures (e.g. incentives, fees, and bonding) for dealing with invasive species.

On the other hand, many weedy species are tolerated or even encouraged in traditional farm systems such as home gardens, where they may be an important resource.

Major pest insects (exotic species) introduced into North America, 1940–1990

Year	Pest	Introduction (origin)
1940	Red imported fire ant, *Solenopsis invicta*	Alabama (Brazil)
1940	Mimosa webworm, *Homodaula anisocentra*	Washington, DC (Assam)
1946	Khapra beetle, *Trogoderma granarium*	California (India)
1952	Face fly, *Musca autumnalis*	Nova Scotia (Europe)
1954	Spotted alfalfa weevil, *Therioaphis maculata*	New Mexico (N. Africa)
1962	Cereal leaf beetle, *Oulema melanopus*	Michigan (Europe)
1965	Formosan subterranean termite, *Coptotermes formosanus*	Texas (China)
1975	Blue alfalfa aphid, *Acrythrosipon kondoi*	California (Central Africa)
1983	Asian cockroach, *Blattella asahinai*	Florida
1985	Asian tiger mosquito, *Aedes albopictus*	Texas (Japan)
1986	Russian wheat aphid, *Diuraphis noxia*	Texas (Asia)
1990	Africanized honey bee, *Aphis mellifera scutella*	Texas (Brazil)

From *Am Entomol* (1996) **42**(4): 220.

Impacts of Invasive Species in Australia

The economic, environmental and social impacts of invasive species are reported as well as, where possible, the monetary values of impacts at the national level. State level impacts have been addressed

(*Continued*)

(Continued)

in less detail but a range of examples of impacts is provided. Regional level impacts have not been addressed.

Economic Impacts

1. Invasive species are costing Australia many billions of dollars annually mainly in costs of control and value of production foregone. Estimates of the different costs are incomplete and those that have been made need refinement and further justification if they are to be used for policy purposes in order to prioritise and stimulate further action on invasive species. The estimates made largely exclude the values of environmental or social costs of invasive species. The economic impacts reported in this review for the individual groups of invasive species are summarised as follows.

Summary of Economic Impact Estimates for Invasive Species

Invasive Species Group	Economic Impact Reported	Costs and Benefits Not Included
Weeds	$4.1 billion	Cost and benefits to the environment such as biodiversity; social impacts such as human health costs. Benefits from weeds not included e.g. benefits to the bee industry.
Pest animals — vertebrates	$0.72 billion including some environmental costs.	Social impacts such as human health costs and most environmental impacts not included. Benefits from pests not included eg. sale of goats.

(Continued)

(Continued)

Invasive Species Group	Economic Impact Reported	Costs and Benefits Not Included
Plant diseases and invertebrate pests of plants	At least $0.7 billion per annum. Total is likely to be a multiple of this figure, perhaps at least $2 billion per annum.	$0.7 billion based on studies for wheat, sugar, two horticultural crops, sunflowers, and cotton. Excluded from the $0.7 billion are diseases and invertebrate pests of pastures, nearly all of horticulture, all other grains, cotton, forestry etc. Also excluded are environmental and social impacts. $2 billion estimate made from loss estimate of 7.5% of the $17 m average gross value of plant industries, plus fungicides and insecticides sales of about $0.5 m, plus application costs.
Animal diseases and invertebrate pests of animals	At least $1.2 billion per annum.	Based on production loss of 5% of a $16 billion per annum animal industry plus value of animal health product sales. Excluded are diseases and invertebrate pest impacts on the environment.
Other invertebrate pests	No estimates identified or made.	

Note: The figures in this table for the economic impact of plant and animal diseases and pests are broad estimates only and are not based on published material.

1. The quantitative monetary estimates of economic impact made here are only partial in that they do not include:

   - most values for most environmental or social impacts;

(Continued)

(Continued)

- values of indirect costs of control measures; or
- impacts and potential impacts on industries other than primary industry (eg. the tourism industry).

2. Improved estimates are required in order to provide an authoritative total estimate of what invasives are costing Australia. Trends in these costs over time would also be useful to monitor.
3. The positive economic impacts of invasives are usually neglected in quantitative impact analyses; even in the vertebrate pest economic impact study they were not comprehensively analysed.
4. It is likely that the aggregate monetary costs of these invasives have increased over the past ten years (period 1994 to 2004) through both spatial spread, increased emphasis on managing the problem and increasing costs of control. However estimates of change at this aggregate level are difficult to make and few estimates of aggregate costs by species group have been identified.
5. These increases in monetary impacts would be in spite of reductions in costs being made through successful measures including mechanical, chemical and biological control. As well, large potential future costs have been avoided due to improvements in border protection, surveillance and detection, eradication, and containment and control measures.

Environmental Impacts

1. Assessing and quantifying impacts is difficult as knowledge of the precise contribution of an invasive species to the impact on native species or the wider ecosystem is not always particularly well understood or described. Indirect negative impacts and positive impacts also need inclusion.
2. Valuing environmental impacts and potential benefits from action is desirable due to the potential use of values in ranking and priority setting. An alternative approach is to make assessments qualitatively (eg. impact on biodiversity).
3. There is no commonly accepted method of valuing environmental impacts in dollar terms for purposes of priority setting among

(Continued)

(Continued)

alternative activities and for integration with activities that lessen industry impacts. Willingness to pay methods of valuation have improved recently but are still used only sparingly by planners and policy makers.

4. Tourism deserves more attention with regard to current or potential impact of invasives eg. south west WA, Kakadu. Relationships between tourism experiences and native fauna and flora and invasives are not well understood.

Social Impacts

1. There are few studies that have identified in specific or quantitative terms the health, safety and quality of life/choice impacts of invasive species. A review could be undertaken of the seriousness of these impacts, particularly those involving human health and safety.
2. The most important social impacts are likely to flow from serious economic impacts on regions highly dependent on a narrow range of plant or animal species that are attacked by a new disease or invertebrate pest.

Measuring Australia's Progress. Australian Bureau of Statistics, Canberra, Australia, 2002.

Biofuels

The rapid expansion of biofuels and related industries across the globe is raising new concerns and threats to the environment and society. Valuable ecosystems and biodiversity are being destroyed in the name of biofuel expansion. There is a sinister linkage between agrofuel expansion, large agribusiness industries, and big Wall Street firms.

One of the major causes of global warming is the agroindustrial complex itself and the related industries of global food systems.

It is not widely known that farming is resulting in 14% of greenhouse gas emissions. A large part of this is due to the use of chemical fertilizers which release a huge amount of nitrogen into the soil and nitrous oxide into the air. Deforestation and expansion of crop monocultures result in an additional 18% of the emissions. A further 14% of emissions is caused by the large-scale global transport involved in food distribution and movement.

The rationale for investing in large-scale global agrofuel expansion is the urgent necessity to reverse climate change. However, statistical projections by the U.S. Government indicate that global energy consumption will increase 71% by 2030, and much of that increase will be provided by oil, coal, and natural gas. By that time, all renewable energy including agrofuels will make up only 9% of global energy consumption.

Agrofuel expansion is appropriating huge tracts of land all over the world. Former Indian President, a scientist among politicians, A. P. J. Abdul Kalam, recognized biofuel, and specifically the plant jatropha, as worthy of mention. Discussing the national problems of water scarcity and drought, he stated that "India needs to grow jatropha to tackle dry land and generate bio-diesel." What makes Jatropha especially attractive to India is that it is drought-resistant and can grow in saline, marginal and even otherwise infertile soil, requiring little water and maintenance. It is hearty and easy to propagate — a cutting taken from a plant and simply pushed into the ground will take root. It grows 5 to 10 feet high, and is capable of stabilizing sand dunes, acting as a windbreak and combating desertification. It has been most successful in the drier regions of the tropics with annual rainfall of 300–1000 mm. It grows naturally at lower altitudes (0–500 m) in areas with average annual temperatures well above 200°C, but can grow at higher altitudes and tolerate slight frost.

(*Continued*)

(Continued)

A Choice of Crops

Biodiesel crop	Liters of oil per hectare
Oil palm	2,400
Jatropha*	1,300
Rapeseed (canola)	1,100
Sunflower	690
Soya bean	400

(From: United Nations Development Programme (UNDP). *Jatropha data are from the Planning Commission of India)

Jatropha naturally repels both animals and insects — it can be planted along the circumference of farms to protect other crops. Jatropha seed-cakes, produced as a by-product of pressing the oil, make an excellent organic fertilizer or protein-rich livestock feed, and another by-product is glycerine. The plant lives, producing seeds, for over 50 years.

India is particularly well-suited for the honor of heralding in a green alternative fuel because of its:

(1) Estimated 50 to 130 million hectares of wastelands — saline lands (from mining), degraded forests, and other land unavailable for agricultural use due to overfarming;
(2) Resulting shifting sand dunes and continuing process of desertification;
(3) Fastest growing population rate in the world — increasing the need for food, energy, and employment;
(4) Rural/agricultural population of over 70%: biofuel screw presses are simple to make, and can be produced and maintained by a village blacksmith;
(5) Huge national crude oil bill — second only to defense spending;
(6) Constant battle with drought and shortages of water and electricity;

(*Continued*)

(Continued)

(7) Warm climate, agreeable both to growing biofuels and running engines that use them.

Other desirable qualities that make Jatropha an ideal source for biodiesel in India are the following:

(a) Low cost seeds,
(b) High oil content,
(c) Small gestation period,
(d) Growth on good and degraded soil,
(e) Growth in low and high rainfall areas,
(f) Seeds can be harvested in non-rainy season, and
(g) Plant size is making collection of seeds more convenient.

The current rate of Indian development of biofuels, particularly biodiesel, is just a drop in the bucket when compared to its potential. If 10 million hectares (100,000 square kilometers or 38,000 square miles) of India's vast and sometimes destructive wastelands were used for biodiesel production, with a modest estimate of 1.5 tons of seeds per hectare, 4 million tons of biodiesel would be produced — one tenth of the country's annual oil requirement. If one person was employed per hectare, that would mean 10 million new jobs. And, for use or sale, 11 million tons of organic seedcake fertilizer or livestock feed and 0.4 million tons of technical grade glycerol would be produced.

Ethanol is the most widely used biofuel in the world; technological advances have lowered the cost of its production and processing. Brazil has one of the largest green fuel programs in existence: petrol-only engines have been banned and replaced by engines that use pure ethanol or a 78–22 petrol–ethanol blend. The shift has greatly benefited Brazil environmentally and economically, creating employment and reducing the need for foreign oil. Its hot, wet climate is well-suited to the production of sugarcane (from which ethanol is made), and farmers especially have profited.

(*Continued*)

(Continued)

India is also one of the biggest worldwide producers of sugarcane, but its constant struggle with water shortages in many areas makes growing this crop problematic. However, due to overproduction, sugar prices crashed, and there are actually stockpiles of sugar and spoilt food grain which have no use. These can be used to make ethanol.

The Indian Supreme Court has recently banned the use of undiluted petrodiesel for commercial vehicles in Delhi due to its adverse effects on health, and other cities are reported to have followed suit.

As compared to petrodiesel, biodiesel almost completely eliminates lifecycle carbon dioxide emissions. It reduces emission of particulate matter by 40–65%, unburned hydrocarbons by 68%, carbon monoxide by 44–50%, sulphates by 100%, polycyclic aromatic hydrocarbons (PAHs) by 80%, and the carcinogenic nitrated PAHs by 90% on an average. The biodiesel molecules are simple hydrocarbon chains free of the aromatic substances and sulfur associated with fossil fuels.

It is true that, because of the solvent power of biodiesel, especially older engines or machines can get clogged, but this is because the biodiesel is actually cleaning it, dissolving the residues left by petrodiesel. Rubber gaskets and hoses in vehicles made prior to 1992 may also be degraded, and need to be replaced. Engine efficiency is also increased by its superior lubricating properties, and the more complete combustion of hydrocarbons due to its higher oxygen content (up to 10%). Finally, biofuel is safer to store because of its higher flash point.

One noteworthy drawback of especially undiluted biodiesel (BD100) is its cold-clogging point of 0 degrees Celsius. This is one of the reasons it is usually mixed with conventional diesel, especially in cold countries. This is not a problem, however, in most of India, except in winter in the higher altitudes of the Himalayas.

The argument that biofuels are not energy efficient, due to the oil used to irrigate, fertilize and plow the land is irrelevant in the case of jatropha — both irrigation and fertilization are generally unnecessary,

(*Continued*)

(Continued)

or its own seedcakes can be used as fertilizer. The energy efficiency of the current agricultural and industrial production process is reported (in Nicaragua) to be between 1:3.75 and 1:5.

Another common objection to biomass energy production is that it could divert agricultural production away from food crops in a hungry world. Using wastelands, however, instead of farmlands, solves the "food or fuel" dilemma — these lands are unsuitable for growing other crops. Also, if a biofuel like jatropha is grown, drought and water shortages which would ruin food crops can be survived; if grown in addition to food crops, as mentioned above, it can literally protect them from animals, insects and desertification, and its seedcakes can be used as fertilizer.

Kalam calls for national mission on biodiesel. *The Hindu Business Line*. Chennai, India, June 11, 2006.

Biodiversity Loss with Global Warming

Darwin predicted that species would move with a changing climate, seeking better climates for survival and propagation. But the arctic fox was the exception to that rule. It stayed in the same region and died off when the climate deteriorated. This was confirmed by three scientists at Stockholm University in Sweden: Love Dalen, Anders Gotherstrom, and Anders Angerbjorn (2004).

DNA studies from fossils showed that the Scandinavian arctic fox did not descend from the Central European arctic fox, but rather originated from Siberia, indicating that the Central European arctic fox became extinct at the end of the latest ice age.

Climate constitutes a major component of the dynamics of evolution and therefore is of great importance for the composition of the fauna. When an ice age ends and the temperature rises in an area, species that thrive in a warmer climate spread out to occupy the area, while those who prefer a colder climate occupy less space. Genetics

(*Continued*)

(Continued)

is a key instrument for studying how various species respond to such climate changes. Species that expanded their habitat since the latest ice age have previously been studied in detail as regards their DNA. On the other hand, no species has been studied whose range diminished when Europe grew warmer and the ice receded.

The problem with using genetic methods in a study of this kind has been that one needs DNA samples from areas no longer inhabited by the species in question. However, they analyzed DNA from fossils of arctic foxes that lived in Belgium, Germany, and southwestern Russia during the latest ice age. One of the investigators, Love Dalen, stated: "In a period of rising temperatures the habitat of arctic species is shrinking. If the inability of a species to move in times of adversity represents a general pattern, considerable loss of genetic variation may be expected in the arctic species with global warming."

Traditional studies of Darwinian evolution indicate that species should be able to survive both increases and decreases in habitat availability. Otherwise, they would become extinct during periods of unsuitable climate. Research using ancient DNA techniques examined genetic variation in the arctic fox (*Alopex lagopus*), whose geographic distribution expanded during the last glaciation. The southern Late Pleistocene populations became extinct when the habitat shifted to the north. Today, the Scandinavian arctic fox population is facing extinction because of shrinking habitat, as well as the increasing numbers of a competitor, i.e. the red fox. These results provide new insights into how arctic species respond to climate change, and the mechanisms behind the extinctions.

India

Thousands of rural poor in India move to big cities where there are few environmental policies in place. Though they have come in search of a better life, many eventually end up living in slums, with no access to safe water or sanitation facilities. Yet, they add to the increasing demands of the city population for food and energy. According to UN population surveys, India is likely to have

700 million rural poor moving to its cities by 2050 if the current trend is not reversed in the next few years.

With 45 000 plant and nearly 90 000 animal species, India is considered one of the world's most mega-diverse countries. Experts say the continued growth in its urban population could lead to enormous loss of biodiversity. Yet, so far, the country has failed to show serious planning efforts to address the effects of increased urbanization on the environment.

India: Value of forests

Biomass stock in forest, 2005	
Above-ground biomass (mt)	4,093
Below-ground biomass (mt)	1,085
Dead wood (mt)	570
Total (mt)	5,748
Carbon stock in forest, 2005	
Carbon in above-ground biomass (mt)	1,852
Carbon in below-ground biomass (mt)	491
Carbon in dead wood (mt)	258
Carbon in litter (mt)	222
Soil carbon (mt)	7,181
Change in growing stock, 1990–2005	
Annual change rate (1000 m^3/yr)	
1990–2000	29,900
2000–2005	7,200
Growing stock per hectare, 1990–2005	
Annual change rate (m^3/ha per yr)	
1990–2000	0.08
2000–2005	0.08
Wood removal, 2005	
Industrial roundwood (1000 m^3)	1,252
Wood fuel (1000 m^3)	3,472
Total wood removal 2005 (1000 m^3)	4,724
Total wood removal 2005 (% of growing stock)	n.s.

(Continued)

(Continued)

Plant products, 2005	
Food (t)	61,060
Fodder (t)	45,780
Raw material for medicine and aromatic products (t)	62,700
Raw material for colorants and dyes (t)	4,930
Raw material for utensils, handicrafts & construction (t)	1,315,250
Ornamental plants (t)	—
Exudates (t)	3,380
Other plant products (t)	547,470
Animal products, 2005	
Living animals (units)	—
Hides, skins and trophies (units)	—
Wild honey and bees-wax (t)	—
Bush meat (t)	—
Raw material for medicine and aromatic products (t)	—
Raw material for colorants and dyes (t)	—
Other edible animal products (t)	—
Other nonedible animal products (t)	—
Value of wood and nonwood forest product removal, 2005	
Industrial roundwood (US$)	$208,644,000
Wood fuel (US$)	$8,023,000
Nonwood forest products (US$)	$179,132,000
Total value (US$)	$395,799,000
Total value ($USD/ha)	$6
Employment in forestry, 2000	
Total people employed	4,855,000

Source: Government of India, Ministry of Environment and Forests, New Delhi, India.

India: Forest cover, 2005

Total land area (ha)	297,319,000
Total forest area (ha)	67,701,000
Percent forest cover	22.77%
Primary forest cover (ha)	—
Primary forest, % total forest	—
Primary forest, % total land	—
Other wooded land (ha)	4,110,000

Source: Government of India, Ministry of Environment and Forests, New Delhi, India.

India: Forest types

Tropical (% forest area)	95%
Subtropical (% forest area)	5%
Temperate (% forest area)	0%
Boreal/Polar (% forest area)	0%

Source: Government of India, Ministry of Environment and Forests, New Delhi, India.

India: Breakdown of forest types, 2005

Primary forest (ha | %)	—	—
Modified natural (ha | %)	32,943,000	48.7%
Seminatural (ha | %)	31,532,000	46.6%
Production plantation (ha | %)	1,053,000	1.6%
Production plantation (ha | %)	2,173,000	3.2%

Source: Government of India, Ministry of Environment and Forests, New Delhi, India.

India: Biodiversity — Plants

Growing stock composition	
3 most common species % of total growing stock	21.90%
Growing stock composition	
3 most common species % of total growing stock	27.20%
Number of native tree species	
Native tree species	—
Number of tree species in IUCN red list	
Critically endangered	50
Endangered	98
Vulnerable	98
Vascular plant species, 2004	
Total	18,664
Number endemic	5,000
Number of threatened plant species, 2004	
Species threatened	246

Source: Government of India, Ministry of Environment and Forests, New Delhi, India.

India: Biodiversity — Wildlife

Amphibians	
Total species	233
Endemic species	133
Threatened species	66
Birds	
Total species	1,180
Endemic species	70
Threatened species	79
Mammals	
Total species	422
Endemic species	44
Threatened species	85
Reptiles	
Total species	521
Endemic species	186
Threatened species	25
Wildlife diversity	
Total species	2,356
Endemic species	433
Threatened species	255

Source: Government of India, Ministry of Environment and Forests, New Delhi, India.

World Bank Data

The following data are based on *The Little Green Data Book* of the World Bank for 2007.

The agricultural land of the entire world represents 38% of the total land area. About 30.5% of the world's land area is occupied by forests. The largest percentage of forest area is found in Latin America and the Caribbean, while the least is found in the Middle East and North Africa.

World forest and agriculture areas

Forest area (% of total land area)	Agricultural land (% of total land area)	Region
28.4	51.0	East Asia and Pacific
38.3	27.0	Europe and Central Asia
45.5	36.0	Latin America and Caribbean
2.4	23.0	Middle East and North Africa
16.8	54.0	South Asia
26.5	44.0	Sub-Saharan Africa

Comparison between the number of species in India and the world

Group	Number of species in India (SI)	Number of species in the world (SW)	SI/SW (%)
Mammals	350(1)	4,629(7)	7.6
Birds	1224(2)	9,702(8)	12.6
Reptiles	408(3)	6,550(9)	6.2
Amphibians	197(4)	4,522(10)	4.4
Fishes	2546(5)	21,730(11)	11.7
Flowering plants	15,000(6)	250,000(12)	6.0

Source: World Conservation Monitoring Centre (WCMC) Species Unit.

Globally threatened animals occurring in India by status category

Group	1994 IUCN Red List Threat Category				
	Endangered	Vulnerable	Rare	Indeterminate	Total
Mammals	20	2	5	13	53
Birds	20	25	13	5	69
Reptiles	6	4	5	2	23
Amphibians	0	0	3	0	3
Fishes	0	2	0	0	2
Invertebrates	3	12	2	4	22
Total	49	45	28	24	172

Source: WCMC Species Unit.

Summary of plant conservation status information at the World Conservation Monitoring Centre (WCMC)

IUCN threat category	Number of species
Extinct	19
Extinct/Endangered	43
Endangered	149
Endangered/Vulnerable	2
Vulnerable	108
Rare	256
Indeterminate	719
Insufficiently known	9
No information	1,441
Not threatened	374
Total	3,120

Source: WCMC Species Unit.

IUCN protected-area categories and their objectives

Type of reserve	No. of sites	Area covered (km)	% of total area protected
Reserve for research	4731	1,033,888	5.5
Wilderness protection	1302	1,015,512	5.4
Ecosystem conservation	3881	4,413,142	23.6
Nature conservation	19,833	275,432	1.5
Habitat conservation	27,641	3,022,515	16.1
Landscape conservation	6555	1,056,008	5.6
Sustainable resource	4123	4,377,091	23.3
Multiple objectives	34,036	3,569,820	19.0
Total	102,102	18,763,407	100.0

Deforestation figures for Brazil

Year	Deforestation (sq mi)	Deforestation (sq km)
1978–1988	8,158	21,130
1990	5,332	13,810

(Continued)

(Continued)

Year	Deforestation (sq mi)	Deforestation (sq km)
1991	4,297	11,130
1992	5,322	13,786
1993	5,950	15,410
1994	5,751	14,896
1995	11,219	29,059
1996	7,013	18,160
1997	5,034	13,040
1998	6,501	16,840
1999	6,663	17,259
2000	7,658	19,836
2001	7,027	18,130
2002	9,845	25,500
2003	9,343	24,130
2004	10,088	26,129
2005	7,298	18,900

All figures derived from official National Institute of Space Research (INPE) figures.

Calculating deforestation figures for the Amazon

Period	Estimated remaining forest cover in the Brazilian Amazon (sq km)	Annual forest loss (sq km)	Percent of 1970 cover remaining	Total forest loss since 1970 (sq km)
pre-1970	4,100,000			
1970	4,001,600		97.6%	
1977	3,955,870	21,130	96.5%	144,130
1978–1987	3,744,570	21,130	91.3%	355,430
1988	3,723,520	21,050	90.8%	376,480
1989	3,705,750	17,770	90.4%	394,250
1990	3,692,020	13,730	90.0%	407,980
1991	3,680,990	11,030	89.8%	419,010
1992	3,667,204	13,786	89.4%	432,796
1993	3,652,308	14,896	89.1%	447,692
1994	3,637,412	14,896	88.7%	462,588
1995	3,608,353	29,059	88.0%	491,647

(Continued)

(Continued)

Period	Estimated remaining forest cover in the Brazilian Amazon (sq km)	Annual forest loss (sq km)	Percent of 1970 cover remaining	Total forest loss since 1970 (sq km)
1996	3,590,192	18,161	87.6%	509,808
1997	3,576,965	13,227	87.2%	523,035
1998	3,559,582	17,383	86.8%	540,418
1999	3,542,323	17,259	86.4%	557,677
2000	3,524,097	18,226	86.0%	575,903
2001	3,505,932	18,165	85.5%	594,068
2002	3,484,727	21,205	85.0%	615,273
2003	3,459,576	25,151	84.4%	640,424
2004	3,432,147	27,429	83.7%	667,853
2005	3,413,354	18,793	83.3%	686,646
2006	3,400,254	13,100	82.9%	699,746

These figures are calculated from estimates provided by the Brazilian National Institute of Space Research (INPE) and the United Nations Food and Agriculture Organization (FAO).

Tropical Deforestation Statistics

Forest cover statistics for 62 countries that have some type of wet tropical forest

Region/ Country	Total land area (1000 ha)	Total forest cover						Primary forest cover					
		Area, 2005 (1000 ha)	Area, 2005 (%)	Annual change 1990–2005 (ha)	Total change 1990–2005 (%)	Annual change 1990–2000 (%)	Annual change 2000–2005 (%)	Area, 2005 (1000 ha)	Area, 2005 (%)	Annual change 1990–2005 (ha)	Total change 1990–2005 (%)	Annual change 1990–2000 (%)	Annual change 2000–2005 (%)
Angola	124,670	59,104	47.4	−124800	−3.1	−0.20	−0.21	0	0.0	n/a	n/a	−0.12	−0.07
Belize	2,296	1,653	72.5	n/a	n/a	n/a	n/a	612	26.7	n/a	0.0	−2.9	n/a
Benin	11,262	2,351	21.3	−64733	−29.2	−1.95	−2.42	–	n/a	n/a	n/a	n/a	n/a
Bolivia	109,858	58,740	54.2	−270333	−6.5	−0.43	−0.45	29,360	26.7	−135200	−6.5	−0.59	−0.80
Brazil	851,488	477,698	57.2	−2821933	−8.1	−0.52	−0.63	415,890	48.8	−2974867	−9.7	−0.09	−0.11
Brunei	577	278	52.8	−2333	−11.2	−0.80	−0.69	278	48.2	−2333	−11.2	−4.05	−5.88
Burkina Faso	27,400	6,794	29.0	−24000	−5.0	−0.34	−0.35	0	0.0	n/a	n/a	−0.62	−0.61
Burundi	2,783	152	5.9	−9133	−47.4	−3.15	−4.65	0	0.0	n/a	n/a	n/a	n/a
Cambodia	18,104	10,447	59.2	−166600	−19.3	−1.09	−1.90	322	1.8	−29600	−58.0	−2.06	−2.59
Cameroon	47,544	21,245	45.6	−220000	−13.4	−0.90	−0.98	–	n/a	n/a	n/a	n/a	n/a
Central African Republic	62,298	22,755	36.5	−29867	−1.9	−0.13	−0.13	–	n/a	n/a	n/a	n/a	n/a
Chad	128,400	11,921	9.5	−79267	−9.1	−0.60	−0.64	190	0.1	−1267	−9.1	n/a	n/a
Colombia	113,891	60,728	58.5	−47400	−1.2	−0.08	−0.08	53,062	46.6	−52800	−1.5	−0.19	−0.15
Congo	34,200	22,471	65.8	−17000	−1.1	−0.07	−0.08	7,464	21.8	−5600	−1.1	−5.27	−11.14

(Continued)

(Continued)

Region/ Country	Total land area (1000 ha)	Total forest cover						Primary forest cover					
		Area, 2005 (1000 ha)	Area, 2005 (%)	Annual change 1990–2005 (ha)	Total change 1990–2005 (%)	Annual change 1990–2000 (%)	Annual change 2000–2005 (%)	Area, 2005 (1000 ha)	Area, 2005 (%)	Annual change 1990–2005 (ha)	Total change 1990–2005 (%)	Annual change 1990–2000 (%)	Annual change 2000–2005 (%)
Costa Rica	5,110	2,391	46.8	−11533	−6.7	−0.73	0.13	180	3.5	n/a	−29.4	n/a	n/a
Côte d'Ivoire	32,246	10,405	32.7	12200	1.8	0.10	0.15	625	1.9	n/a	0.0	n/a	n/a
D.R. Congo	234,486	133,610	58.9	−461400	−4.9	−0.38	−0.24	–	n/a	n/a	n/a	n/a	n/a
Dominica	75	46	61.3	−267	−8.0	−0.60	−0.43	27	36.0	n/a	−3.6	n/a	n/a
Dominican Republic	4,873	1,376	28.4	n/a	n/a	n/a	n/a	–	n/a	n/a	n/a	n/a	n/a
Ecuador	28,356	10,853	39.2	−197600	−21.5	−1.43	−1.67	4,794	16.9	n/a	0.0	n/a	n/a
El Salvador	2,104	298	14.4	−5133	−20.5	−1.36	−1.60	6	0.3	n/a	0.0	n/a	n/a
Equatorial Guinea	2,805	1,632	58.2	−15200	−12.3	−0.82	−0.89	–	n/a	n/a	n/a	n/a	n/a
Fiji	1,827	1,000	54.7	1400	2.1	0.21	0.00	894	48.9	n/a	−0.1	n/a	n/a
French Guiana	9,000	8,063	91.8	−1867	−0.3	−0.03	0.00	7,701	85.6	−13867	−2.6	−0.11	−0.36
Gabon	26,767	21,775	84.5	−10133	−0.7	−0.05	−0.05	–	n/a	n/a	n/a	n/a	n/a
Gambia	1,130	471	41.7	1933	6.6	0.43	0.43	–	n/a	n/a	n/a	n/a	n/a
Ghana	23,854	5,517	24.2	−128733	−25.9	−1.82	−1.89	353	1.5	n/a	0.0	n/a	n/a
Guatemala	10,889	3,938	36.3	−54000	−17.1	−1.14	−1.28	1,957	18.0	−26800	−17.0	−1.26	−1.33
Guinea	24,586	6,724	27.4	−45600	−9.2	−0.68	−0.52	63	0.3	n/a	0.0	n/a	n/a

(Continued)

(Continued)

Region/ Country	Total land area (1000 ha)	Total forest cover						Primary forest cover					
		Area, 2005 (1000 ha)	Area, 2005 (%)	Annual change 1990–2005 (ha)	Total change 1990–2005 (%)	Annual change 1990–2000 (%)	Annual change 2000–2005 (%)	Area, 2005 (1000 ha)	Area, 2005 (%)	Annual change 1990–2005 (ha)	Total change 1990–2005 (%)	Annual change 1990–2000 (%)	Annual change 2000–2005 (%)
Guinea–Bissau	3,612	2,072	73.7	–9600	–6.5	–0.43	–0.45	940	26.0	n/a	0.0	n/a	n/a
Guyana	21,497	15,104	76.7	n/a	n/a	n/a	n/a	9,314	43.3	n/a	n/a	n/a	n/a
Honduras	11,209	4,648	41.5	–182467	–37.1	–2.65	–2.88	1,512	13.5	n/a	0.0	n/a	n/a
India	328,726	67,701	22.8	250800	5.9	0.57	0.04	–	n/a	n/a	n/a	n/a	n/a
Indonesia	190,457	88,495	48.8	–1871467	–24.1	–1.61	–1.91	48,702	25.6	–1447800	–30.8	–2.33	–3.05
Kenya	58,037	3,522	6.2	–12400	–5.0	–0.34	–0.34	704	1.2	–2533	–5.1	–2.30	–2.98
Laos	23,680	16,142	69.9	–78133	–6.8	–0.45	–0.47	1,490	6.3	n/a	0.0	n/a	n/a
Liberia	11,137	3,154	32.7	–60267	–22.3	–1.49	–1.74	129	1.2	n/a	0.0	n/a	n/a
Madagascar	58,704	12,838	22.1	–56933	–6.2	–0.49	–0.28	10,347	17.6	–10400	–1.5	–0.07	–0.07
Malawi	11,848	3,402	36.2	–32933	–12.7	–0.84	–0.93	1,132	9.6	–39667	–34.5	n/a	n/a
Malaysia	32,975	20,890	63.6	–99067	–6.6	–0.35	–0.65	3,820	11.6	n/a	0.0	n/a	n/a
Mexico	195,820	64,238	33.7	–318533	–6.9	–0.50	–0.40	32,850	16.8	n/a	–15.3	0.0	n/a
Myanmar	67,658	32,222	49	–466467	–17.8	–1.19	–1.35	–	n/a	n/a	n/a	n/a	n/a
Nicaragua	13,000	5,189	42.7	–89933	–20.6	–1.53	–1.26	1,849	14.2	n/a	0.0	–1.0	–1.1
Nigeria	92,377	11,089	12.2	–409667	–35.7	–2.38	–3.12	326	0.4	–82000	–79.0	–0.60	–0.67
Panama	7,552	4,294	57.7	–5467	–1.9	–0.16	–0.06	3,023	40.0	–45533	–18.4	–0.94	–0.95
Papua New Guinea	46,284	29,437	65	–139067	–6.6	–0.44	–0.46	25,211	54.5	–266600	–13.7	–0.43	–0.45

(Continued)

(Continued)

Region/ Country	Total land area (1000 ha)	Total forest cover						Primary forest cover					
		Area, 2005 (1000 ha)	Area, 2005 (%)	Annual change 1990–2005 (ha)	Total change 1990–2005 (%)	Annual change 1990–2000 (%)	Annual change 2000–2005 (%)	Area, 2005 (1000 ha)	Area, 2005 (%)	Annual change 1990–2005 (ha)	Total change 1990–2005 (%)	Annual change 1990–2000 (%)	Annual change 2000–2005 (%)
Peru	128,522	68,742	53.7	−94267	−2.0	−0.13	−0.14	61,065	47.5	−123000	−2.9	−0.35	−0.34
Philippines	30,000	7,162	24	−227467	−32.3	−2.48	−1.98	829	2.8	n/a	0.0	n/a	n/a
Rwanda	2,634	480	19.5	10800	50.9	0.82	7.91	0	0.0	n/a	n/a	n/a	n/a
Senegal	19,672	8,673	45	−45000	−7.2	−0.48	−0.51	1,598	8.1	−10733	−9.2	−0.80	−0.69
Sierra Leone	7,174	2,754	38.5	−19333	−9.5	−0.63	−0.68	–	n/a	n/a	n/a	n/a	n/a
Solomon Islands	2,890	2,172	77.6	−39733	−21.5	−1.43	−1.68	–	n/a	n/a	n/a	n/a	n/a
Sri Lanka	6,561	1,933	29.9	−27800	−17.7	−1.14	−1.43	167	2.5	−6000	−35.0	−5.13	−10.91
Suriname	16,327	14,776	94.7	n/a	n/a	n/a	n/a	14,214	87.1	n/a	0.0	n/a	n/a
Tanzania	94,509	35,257	39.9	−412267	−14.9	−0.99	−1.10	–	n/a	n/a	n/a	n/a	n/a
Thailand	51,312	14,520	28.4	−96333	−9.1	−0.72	−0.40	6,451	12.6	n/a	0.0	n/a	−0.7
Togo	5,679	386	7.1	−19933	−43.6	−2.91	−4.12	0	0.0	n/a	n/a	n/a	n/a
Uganda	24,104	3,627	18.4	−86467	−26.3	−1.76	−2.13	–	n/a	n/a	n/a	n/a	n/a
Venezuela	91,205	47,713	54.1	−287533	−8.3	−0.55	−0.59	–	n/a	n/a	n/a	—	—
Vietnam	33,169	12,931	39.7	237867	38.1	2.52	2.06	85	0.3	−19933	−77.9	−1.14	−1.28
Zambia	75,261	42,452	57.1	−444800	−13.6	−0.91	−1.00	–	n/a	n/a	n/a	n/a	n/a
Zimbabwe	39,075	17,540	45.3	−312933	−21.1	−1.41	−1.64	–	n/a	n/a	n/a	n/a	n/a
Total	3,745,546	1,629,991	43.5	−10240133	−8.3	−0.57	−0.62						

These figures are derived from data provided in "Forest Resources Assessment 2005" by the Food and Agriculture Organization of the United Nations. A negative number for "Annual change" and/or "Total change" reflects deforestation. A positive number in these columns suggests that forest is regrowing or, more likely, plantations are being established. "n/a" means data are not available.

China

The tables below are taken from the folowing source: Butler, R. A., World deforestation rates and forest cover statistics, www. mongabay.com, November 16, 2005.

China: Forest Cover, 2005

Total land area (ha)	932,742,000
Total forest area (ha)	197,290,000
Percent forest cover	21.15%
Primary forest cover (ha)	11,632,000
Primary forest, % total forest	5.90%
Primary forest, % total land	1.25%
Other wooded land (ha)	87,615,000

China: Forest types

Tropical (% forest area)	3%
Subtropical (% forest area)	59%
Temperate (% forest area)	29%
Boreal/Polar (% forest area)	8%

China: Breakdown of forest types, 2005

Primary forest (ha | %)	11,632,000	5.9%
Modified natural (ha | %)	114,332,000	58.0%
Seminatural (ha | %)	39,957,000	20.3%
Production plantation (ha | %)	28,530,000	14.5%
Production plantation (ha | %)	2,839,000	1.4%

China: Biodiversity — Wildlife

Amphibians	
Total species	340
Endemic species	206
Threatened species	86

(Continued)

(Continued)

Birds	
Total species	1221
Endemic species	92
Threatened species	82
Mammals	
Total species	502
Endemic species	78
Threatened species	80
Reptiles	
Total species	424
Endemic species	113
Threatened species	31
Wildlife diversity	
Total species	2487
Endemic species	489
Threatened species	279

China: Biodiversity — Plants

Growing stock composition	
3 most common species % of total growing stock	29.00%
Growing stock composition	
3 most common species % of total growing stock	64.00%
Number of native tree species	
Native tree species	2,500
Number of tree species in IUCN red list	
Critically endangered	34
Endangered	45
Vulnerable	96
Vascular plant species, 2004	
Total	32,200
Number endemic	18,000
Number of threatened plant species, 2004	
Species threatened	443

China and India

There are some remarkable similarities between India and China with respect to their biodiversity situation. The percentage of land area covered by agriculture and forests is about the same in both countries. China's population is about 20% greater than India's, but the numbers of mammalian species and bird species threatened with extinction is about the same in both countries. The rich biodiversity of the small country of Costa Rica, and the relatively fewer species of mammals and birds at risk, is highly impressive. On the other hand, the high proportion of mammals at risk, about 22%, in Indonesia is alarming!

Biodiversity of India, China, Brazil, Indonesia, and Costa Rica

	India	China	Brazil	Indonesia	Costa Rica
Population (in millions)	1,094.6	1,304.5	186.4	220.6	4.3
Percentage of land area					
Agriculture	61.0	59.0	31.0	26	56.0
Forests	22.8	21.2	56.5	48.8	46.8
Mammalian species, total known	422	580	578	667	232
Mammalian species, threatened	85	80	74	146	13
Bird species, total known	1180	1221	1,712	1,604	838
Bird species, threatened	79	82	120	121	18

Source: Butler, R. A., World deforestation rates and forest cover statistics, www.mongabay.com, November 16, 2005.

Threat to European Mammals: IUCN Report

Recent estimates have shown that 15% of the 250 mammalian species that live in Europe and western Russia are classed as "vulnerable" or worse, according to the IUCN's "Red List" criteria, facing a "high risk of extinction in the wild" if action is not taken. Unless the trend is reversed, the European Union will not be able to meet its self-imposed target of halting biodiversity loss by 2010.

According to Julia Marton-Lefèvre, IUCN director-general, the new assessment proves that many European mammals are declining at an alarming rate. Worst affected is the handful of European mammals classed as "critically endangered", the most serious category. The Iberian lynx is the world's most endangered big cat — only about 150 are surviving today. The Arctic fox and European mink face similar risk. The situation is even more perilous for Europe's marine mammals, of which 22% are classed as vulnerable or worse. Almost half of European marine mammal species do not have enough data to determine their conservation status.

According to the IUCN report, in tropical regions, which have greater numbers of species overall, an average of one in four is officially threatened, mainly due to extensive deforestation.

According to Craig Hilton-Taylor of the IUCN's Red List, the number of threatened species may be set to increase. Currently, some 27% of the mammal species surveyed are thought to be declining, whereas only 8% are growing in number. The main causes of this change are habitat degradation, deforestation, pollution, and excessive hunting. However, the Alpine ibex and the European bison have successfully rebounded in numbers, which is encouraging.

Marine Conservation

The simplest methods, when properly practiced, can be quite effective for biodiversity conservation. In Micronesia, the old taboo system is being exploited to help modern conservation measures by controlling fishing practices. The fact that people still respect their traditional chiefs has been very helpful in this situation.

President Tommy Remengesau Jr., is probably the world's most conservation-minded head of state. He made news recently with his so-called Micronesian Challenge: a call to the rest of the region to set aside for conservation 30 percent of coastal waters and 20 percent of the land area by 2020. Palau is a tiny island state 600 miles east of the Philippines that is internationally known as a site for recreational diving. It is pioneering a worldwide movement to ban fishing in key reefs to allow the return of prized species. It protects an area of reefs and lagoon waters totalling 460 square miles.

(*Continued*)

(Continued)

Palau's challenge was timely. Reef-fishing communities around the world are discovering that setting aside no-fishing areas yields dividends in a few years because the resurgent fish populations spill over into areas where fishing is allowed. In Fiji, local authorities have raised the number of no-take zones to 189 from 2 in 10 years.

Mr. Ratu Aisea Katonivere, a traditional chief, who rules over 7,000 people in the Great Sea Reef, the world's third-largest barrier reef, imposed a no-take zone. Soon the fish were found to be closer and bigger. The protected areas have increased from 2 to 30 in five years in the Solomon islands, and in Vanuatu, they exceeded 100.

In 1994, they banned fishing in a small area of reef that was partly accessible on foot. The village women, who traditionally gather shellfish at low tide, noticed how the fish became more plentiful there in a few years. The reef became locally famous, and other villages started to do the same.

In Indonesia, President Susilo Bangbang Yudhoyono pledged to increase marine protected areas to 24.7 million acres from 18 million acres by 2010.

Leaders and visionaries: Tommy Remengesau Jr., *Time*, April 1, 2007.

IUCN in Micronesia

According to Bill Raynor, the Nature Conservancy's director for Micronesia, in the Antilles, the states of Grenada, the Bahamas, Belize and the Grenadines, which have already protected some reef areas, have committed themselves to a Caribbean Challenge and are trying to persuade the other nations to make similar pledges.

The Nature Conservancy helped establish the Palau Conservation Society (PCS), the nation's first non-governmental environmental organization. In a short time, PCS and the Conservancy encouraged Palau's chiefs to reinstate the age-old bul tradition, allowing fish stocks to reproduce and replenish the waters.

(*Continued*)

(Continued)

The Conservancy conducted Palau's first biological inventory of coastal and marine resources. The information has helped target conservation efforts to the most biologically diverse areas.

PCS and the Conservancy continue to work with local Palauan communities, who now monitor and protect their marine sites independently. The partners established a marine mooring buoy program at key dive sites to protect reefs from anchor damage. They are also developing plans for a small-scale catch-and-release sport fishing industry, which will train local fishermen as guides.

PCS and the international conservation organization RARE conducted an intensive campaign to increase public awareness of environmental issues. With the Conservancy's help, PCS and the Palauan government secured US$6 million from the Japanese government for a new Palau International Coral Reef Center to focus on reef preservation and marine awareness.

The Conservancy and PCS are working with U.S. and Palauan governments to apply the highest environmental standards during road construction on Babeldaob. PCS is focusing on terrestrial conservation efforts near the road to avoid erosion and pollution of coastal waters.

Micronesia Challenge, IUCN, Gland, Switzerland, 2007.

Invasive Species

According to a recent IUCN report (July 2007), a world-renowned lake in Palau is currently being invaded by a prolific species of anemone. It is taking over more than half of Jellyfish Lake, a tourist attraction that lures thousands of visitors to Palau each month. "Invasive species, marine or terrestrial, represent one of, if not the most, dangerous threat to our islands," said the President of Palau, Tommy E. Remengesau.

The invasion of Jellyfish Lake, which lies within a conservation area, is a clear example of the risks associated with tourism. Marine

(Continued)

(Continued)

Protected Areas generate a significant attraction for marine tourism, including recreational boating, yachting, diving and snorkelling, and where allowed, fishing. All these activities increase the risks of introducing non-indigenous marine species.

According to IUCN sources: "Efforts to address marine invasive species in most of the world have focused on ports and harbours, but it is also important to pay attention to high value areas, such as marine protected areas, because these are often the first points where invasive species spread after ports."

The intentional and unintentional transfer of species around the world has boomed in recent decades. Many seas and regions have been invaded by a high number of non-native marine species. Some of these species thrive in their new habitats, displacing native species and changing ecosystems, sometimes dramatically.

There are already several introduced species that have been recorded in Palau during regular dives by members of the Coral Reef Research Foundation. The known introduced hydrozoan, *Eudendrium carneum*, is found in the channel linking Babeldaob to Koror and may have been introduced by the floating bridge that was transported from China in 1995. The species is currently found in two other channels south of Koror.

A prerequisite for efforts to manage the introductions and spread of invasive species in the marine environment of Palau is to know the occurrence, current distribution and abundance of these species. A baseline survey provides this basic information and is critical to enable development of management measures.

Joel Miles, Palau's National Invasive Species Coordinator, commented: "The invasion of Jellyfish Lake is unfortunate, but it is hopefully a wake up call for all of us, and reinforces the importance of prevention, monitoring and early response actions that we need to put in place in order to maintain Palau's most precious marine resources and our livelihoods."

Palau, Global Marine Program, IUCN, Gland, Switzerland, 2007.

Economic Inequality and Biodiversity Loss

Gregory Mikkelson *et al.* from McGill University have demonstrated a striking correlation between economic inequality and biodiversity loss. While their findings cohere with previous work showing links between inequality and human health, they contrast with previous research suggesting that the overall size of an economy (i.e., population times per capita GDP or income) is the primary driver of environmental impacts. According to one cross-country analysis of per capita GDP and threatened species, the numbers of threatened species in most taxa follow a U-shaped pattern: first falling, but then rising, with increasing per capita GDP. This is the opposite of the hump-shaped "environmental Kuznets" relationship expected by many economists between affluence and its environmental impacts. They used very similar data on threatened and total species; and they also allowed for detection of monotonic, U-shaped, and hump-shaped relationships; by adding a quadratic term for GDP PPP per capita. Nevertheless, they did not find any such patterns. This may be partly due to sample size (45 countries in our analysis, as opposed to more than 100). But the previous study also did not include inequality, or allow for a time lag between socio-economic causes and biological effects, as they have.

Future research could test the generality of the link between economic equality and biological diversity, e.g., by examining states or provinces in countries other than the US. Further studies are also needed to establish the degree to which this link arises from common influences on both variables vs. direct effects of equality on biodiversity. In this analysis, they took two steps toward proving a direct causal relationship. First, they controlled for several likely common causes, and second, they incorporated time lags that are more realistic than any instantaneous effect of equality on biodiversity would be. Controlling for other potentially confounding variables — e.g., the degree to which different societies are governed democratically — could further test the extent to which this relationship is causal. But perhaps most importantly, future studies should explore possible mechanisms.

(*Continued*)

(Continued)

If such research confirms a causal relationship, it may help to predict future impacts of the rising inequality that most countries, as well as US states, have suffered over recent decades. For example, given that the Gini ratio in the US rose by 5% from 1989 to 1997, the country-level power model described in Table 1 suggests that we should expect a roughly 9% increase in the number of threatened plant and vertebrate species there by 2012. And we might expect the 3% rise in British inequality from 1989 to 1996 to result in a 5% increase in threatened species there by 2011. In general, unless current trends toward greater inequality are reversed, it may become increasingly hard to conserve the rich variety of the living world. Conversely, if we can learn to share economic resources more fairly with fellow members of our own species, it may help us to share ecological resources more fairly with our fellow species.

They found that among countries, and among US states, the number of species that are threatened or declining increases substantially with the Gini ratio of income inequality. At both levels of analysis, the connection between income inequality and biodiversity loss persists after controlling for biophysical conditions, human population size, and per capita GDP or income. Future research should explore potential mechanisms behind this equality–biodiversity relationship. Their results suggest that economic reforms would go hand in hand with, if not serving as a prerequisite for, effective conservation.

Mikkelson and his colleagues related indicators of income inequality and biodiversity loss on two different scales: among 45 countries worldwide and among 45 states within the United States. They controlled for differences such as area and climate, human population size, and per capita consumption. The same general trend is evident in both scales: societies with more unequal distributions of income experience greater losses of biodiversity.

"While there is often a trade-off between economic growth and environmental quality," says Mikkelson, "this study suggests that there is a synergy between a different kind of economic development — namely, toward a more equitable distribution of wealth and the

(Continued)

(Continued)

conservation of biological diversity." For example, if the US were to achieve levels of income equality comparable to those of Sweden, the pattern reported in their paper implies that 44% fewer plant and vertebrate species in the US would be in danger of extinction.

"In the past, people thought that human population size was the main driver of biodiversity loss, then people showed that the size of the economy provided a better explanation," said co-author Dr. Garry Peterson, an assistant professor in the School of Environment and Department of Geography, and a Canada Research Chair (CRC) in social-ecological modeling. "This study shows that the structure of the economy is also important."

Fig. 1. Texas pond. Courtesy of Michele Wambaugh.

Table 1 Biodiversity and income inequality

Country	Plant and vertebrate species (2004)	Human population (1989)	GDP PPP per capita (1989) (current int. $)	Gini ratio of household income inequality (1989)	Threatened plant and vertebrate species (2004)
Argentina	11,387	32,124,000	$7,666	0.480	176
Australia	19,463	16,638,000	$16,215	0.373	338
Austria	3,698	7,682,000	$17,501	0.260	23
Bangladesh	5,952	101,710,000	$912	0.338	85
Brazil	60,322	146,858,000	$5,446	0.615	663
Bulgaria	4,196	8,789,000	$5,608	0.226	35
Chile	6,222	12,948,000	$4,589	0.524	123
China	35,082	1,139,265,000	$1,229	0.360	769
Costa Rica	13,864	2,999,000	$5,152	0.460	222
Denmark	2,051	5,130,000	$18,616	0.332	24
Dominican Republic	6,250	6,967,000	$3,535	0.515	102
Estonia	2,018	1,586,000	$7,533	0.276	8
Finland	1,653	4,968,000	$17,642	0.302	15
Georgia	4,841	5,469,000	$4,142	0.284	33
Ghana	5,000	15,054,000	$1,238	0.417	160
Guatemala	10,134	8,693,000	$2,627	0.590	200
Honduras	7,119	4,723,000	$1,977	0.565	204

(Continued)

Table 1 (Continued)

Country	Plant and vertebrate species (2004)	Human population (1989)	GDP PPP per capita (1989) (current int. $)	Gini ratio of household income inequality (1989)	Threatened plant and vertebrate species (2004)
Hungary	2,751	10,396,000	$9,463	0.269	26
India	26,769	832,409,000	$1,287	0.354	529
Italy	6,477	56,659,000	$16,755	0.358	56
Jamaica	3,914	2,354,000	$2,790	0.485	262
Japan	7,491	123,076,000	$17,663	0.376	160
Kyrgyzstan	4,803	4,333,000	$1,851	0.260	13
Latvia	1,604	2,710,000	$8,072	0.259	15
Malaysia	17,539	17,384,000	$4,156	0.484	873
Mexico	29,510	82,644,000	$5,838	0.520	707
Moldova	2,080	4,347,000	$3,009	0.254	22
Netherlands	1,857	14,852,000	$17,305	0.321	27
New Zealand	3,094	3,371,000	$13,005	0.366	135
Norway	2,321	4,221,000	$19,778	0.367	24
Panama	11,759	2,362,000	$3,421	0.567	308
Russian Federation	12,667	147,695,000	$8,237	0.265	121

(Continued)

Table 1 (Continued)

Country	Plant and vertebrate species (2004)	Human population (1989)	GDP PPP per capita (1989) (current int. $)	Gini ratio of household income inequality (1989)	Threatened plant and vertebrate species (2004)
Sierra Leone	3,127	4,009,000	$706	0.679	82
Singapore	3,003	2,945,000	$10,944	0.390	84
Slovakia	3,620	5,234,000	$8,878	0.221	29
Spain	5,983	39,154,000	$13,214	0.309	90
Sweden	2,390	8,506,000	$17,319	0.304	23
Tajikistan	5,491	5,171,000	$1,870	0.276	22
Thailand	13,648	53,854,000	$3,264	0.493	220
Uganda	6,541	17,134,000	$671	0.422	115
Ukraine	5,678	51,766,000	$7,191	0.265	41
United Kingdom	2,738	56,595,000	$17,030	0.386	45
United States	22,575	252,934,000	$22,069	0.421	582
Uruguay	3,003	3,085,000	$5,704	0.413	49
Venezuela	23,675	19,250,000	$4,500	0.440	218

Science Daily, May 16, 2007.

Protected Areas in the Work of CBD

Protected areas form a central element of the work in the thematic areas and cross-cutting issues addressed by the Conference of the Parties to the Convention on Biological Diversity.

(a) Marine Protected Areas:
Because oceans and seas cover 71% of the Earth, the under-representation of marine and coastal ecosystems in the current global protected areas system is particularly alarming. At the same time, global and regional assessments indicate that marine biodiversity globally continues to decline rapidly. For example, coral reefs are highly degraded worldwide, approximately 35% of mangroves have been lost in the last two decades, and historical over-fishing has greatly reduced the abundance of large consumer species, including predatory fish. In addition, there are increasing and urgent concerns about the effects of over-fishing and destructive fishing practices on biodiversity. Halting, and perhaps ultimately reversing, this trend presents the global community with a formidable challenge.

The seventh meeting of the Conference of the Parties to the Convention on Biological Diversity agreed in 2004 that marine and coastal protected areas are one of the essential tools and approaches in the conservation and sustainable use of marine and coastal biodiversity (decision VII/5 on marine and coastal biological diversity). The Conference of the Parties also agreed that a national framework of marine and coastal protected areas should include a range of levels of protection, encompassing both areas that allow sustainable uses and those that prohibit extractive uses (i.e., so-called "no-take" areas). The Conference further recognized that protected areas alone could not accomplish everything, and that sustainable management practices are needed over the wider marine and coastal environment.

(b) In the programme of work on the biological diversity of inland water ecosystems (decision VII/4), goal 1.2 calls for the establishment and maintenance of comprehensive, adequate and

(*Continued*)

(Continued)

representative systems of protected inland water ecosystems within the framework of integrated catchment/watershed/river basin management;

(c) The use and establishment of additional protected areas and the strengthening of measures in existing protected areas are identified as some of the necessary target actions for the implementation of the work programme on dry and sub-humid lands (Decision V/23, annex 1, part B, activity 7(a));

(d) The expanded programme of work on forest biodiversity, which was adopted in decision VI/22, contains a number of activities related to protected areas. The programme of work also calls for work on the role and effectiveness of protected areas. Controlling deforestation, including through establishment of forest protected areas, is being considered as a means to avoid greenhouse gas emissions and contribute to the objectives of the United Nations Framework Convention on Climate Change;

(e) Goals 1.1 and 2.3 of the programme of work on mountain biodiversity (decision VII/27) contain provisions on how to plan, establish and manage protected areas in mountain ecosystems, including the buffer zones of protected areas, using, as appropriate, planning or management mechanisms, such as ecological/economic/ecoregional planning/bioregional/hazardous-areas zoning, so as to ensure the maintenance of biodiversity, in particular ecosystem integrity. Actions 1.2.5 and 2.3.1, in particular, call for the establishment and strengthening of adequate, effective national, regional and international networks of mountain protected areas, and the promotion of integrated transboundary cooperation, strategies for sustainable activities on mountain ranges and protected areas;

(f) The programme of work on Article 8(j) on traditional knowledge includes a component on protected areas relating to the management of protected areas by indigenous and local communities (decision VI/10). Specific emphasis is put on the respect of their rights when establishing new protected areas (decisions VII/16);

(*Continued*)

(Continued)

(g) The Guidelines on Biodiversity and Tourism Development, adopted by the world community in decision VII/14 of the Conference of the Parties, include guidelines on how to incorporate sustainable use and equity strategies within and around protected areas;

(h) The value of taxonomic data in assisting protected area site selection is recognized in the programme of work for the Global Taxonomy Initiative, contained in decision VI/8. Protected areas are also mentioned in connection with identification, monitoring, indicators and assessments (decision VI/7) and the Addis Ababa principles and guidelines for sustainable use of biodiversity (decision VII/12);

(i) In the Global Strategy for Plant Conservation (annex to decision VI/9), the Conference of the Parties adopted targets 4 and 5, which specify respectively that by 2010 (i) at least 10% of each of the world's ecological regions should be effectively conserved, implying increasing the representation of different ecological regions in protected areas, and increasing the effectiveness of protected areas; and (ii) protection of 50% of the most important areas for plant diversity should be assured through effective conservation measures, including protected areas.

(j) In decision VIII/28 the COP endorsed voluntary guidelines on biodiversity-inclusive impact assessment and urged Parties, other Governments and relevant organizations to apply the voluntary guidelines on biodiversity-inclusive environmental impact assessment as appropriate in the context of their implementation of paragraph 1(a) of Article 14 of the Convention and of target 5.1 of the provisional framework of goals and targets for assessing progress towards 2010. The COP invited multilateral environmental agreements to take note of and, if appropriate, apply the voluntary guidelines on biodiversity-inclusive environmental impact assessment. These guidelines make environmental impact assessment mandatory for activities in protected areas and activities in threatened ecosystems outside protected areas.

(*Continued*)

(Continued)

(k) In decision VIII/30 on biodiversity and climate change, the COP encouraged Parties and other Governments to integrate biodiversity considerations into all relevant national policies, programmes and plans in response to climate change, taking into account the maintenance and restoration of the resilience of ecosystems, which are essential for sustaining the delivery of their goods and services. The COP further encouraged Parties, other Governments, relevant organizations and research institutions to develop rapid assessment tools for the design and implementation of biodiversity conservation and sustainable use activities that contribute to adaptation to climate change, particularly in vulnerable countries and regions, including small island developing States. The COP requested the Subsidiary Body on Scientific, Technical and Technological Advice to develop draft guidance on how to integrate relevant climate change impacts and response activities into the programmes of work of the Convention, taking into account, *inter alia*, the contributions that protected areas can make in this context.

It is important to note that the planning, establishment, management and monitoring of protected areas need to take into account the ecosystem approach (decision VII/11), biodiversity-inclusive guidelines on environmental impact assessment (decision VIII/28), guidelines for tourism (decision VII/14), provisions of Article 8(j) of the Convention and incentive measures (decision VIII/26).

Convention on Biological Diversity, Montreal, Canada.

The Arctic Experience: Arctic Biodiversity Beyond Polar Bears

Climate Change as a Threat to Biodiversity

Arctic regions are alarming signals of changes in the Arctic ecosystems, which should alert us. We still do not know the trend for many

(*Continued*)

(Continued)

populations of Arctic biodiversity and we also know too little to fully understand the root causes of these trends. Climate change has already started and has shown its impact on Arctic biodiversity.

Climate change is threatening biodiversity on a global scale. The Arctic region is warming 2–2.5 times faster than the global average, due to a thinner atmosphere. These changes in the Arctic are already having a major impact on all other global regions through many changes in the weather cycle and atmospheric changes, but also in its unique biodiversity. Many Arctic species are migratory, connecting the entire globe by its annual migration routes of billions of migratory birds, marine mammals and fish. The changes in the Arctic region will determine the future ecosystem goods and services, mostly its natural resources, marine and freshwater fish and terrestrial reindeer to name a few. They all occupy a vital link in the biodiversity of the entire planet. Because of its unique location and wilderness where human impact has been minimal so far, Arctic region is ideally suited for monitoring the impact of climate change and industrial development on biodiversity, providing us insights about what might happen to the planet as a whole in the future.

Arctic biodiversity and ecosystems are an ideal test case for measuring progress towards the Convention on Biological Diversity's (CBD) 2010 target to significantly reduce the rate of biodiversity loss by 2010, a barometer for the state of biodiversity.

Beyond Polar Bears

Polar Bears represent Arctic biodiversity to most of us. The charismatic Arctic predator has been the focus of the media recently. It will suffer severely if the sea ice continues to disappear in many areas of the Arctic. But little has been mentioned about the fate of other Arctic species. There are four other species which are very much a part of the Arctic scene and are under threat. They include solely Arctic Ivory Gull (*Pagophila eburnea*), the Reindeer and Caribou (*Rangifer tarandus*) which provide the major source of income for the local people, the Arctic Char (*Salvelinus alpinus*), and the Red Knot (*Calidris canutus*).

UNEP Key Polar Center, United Nations Environment Programme (UNEP), Nairobi, Kenya.

3

Bioprospecting or Biopiracy?

Bioprospecting is the perpetuation of the colonial habit of plundering other countries' biological resources without fair and equitable compensation, resulting in environmental, economic, and social detriment.

In her book, *Protect or Plunder?* (2001), noted activist of India, Vandana Shiva, wrote that bioprospecting is mistakenly promoted as the model for relationships between corporations who commercialize indigenous knowledge and indigenous communities which have collectively innovated and evolved the knowledge. However, "bioprospecting is merely a sophisticated form of biopiracy." According to Shiva, bioprospecting, in effect, leads to the enclosure of the biological and intellectual commons. It converts the intellectual heritage of indigenous communities into commodities protected by intellectual property rights (IPRs).

The World Resources Institute (WRI) has defined "biodiversity prospecting" as exploring commercially valuable genetic and biochemical resources. The term "prospecting" usually implies that prior to prospecting, the treasure lies buried, undetected or unknown. However, such is not the case with biodiversity or indigenous knowledge. Hence, the term "bioprospecting" is misleading. It implies that stealing indigenous property is a legitimate activity.

Biopiracy

The term "biopiracy" was coined by the Canadian activist Pat Mooney, and was intended as a counterweight to the claims of the developed countries that their ideas were being pirated by the developing countries. However, for the purposes of this book, biopiracy is meant to focus attention upon the theft of intellectual and traditional knowledge and physical property from the developing countries by large corporations which occupy a central place in the economies of the rich countries. The boards of directors of these corporations consist of the most respected leaders of the developed countries who seem to be using all of their intellect for pirating the intellectual and physical property of the poorer countries of the world.

Biopiracy is defined as biological theft, or the unauthorized and uncompensated collection of indigenous plants, animals, microorganisms, genes, or traditional communities' knowledge on biological resources by corporations that patent them for their own use.

South America

> According to the president of Peru's National Institute for the Defence of Competition and the Protection of Intellectual Property, Santiago Roca, much of the biodiversity of Peru was appropriated by the U.S., Europe and Japan, under the guise of patenting. He was addressing the first meeting of intellectual property officials from the eight Amazon basin countries; Bolivia, Brazil, Colombia, Ecuador, Guyana, Peru, Surinam and Venezuela. (IPS News — Inter Press Service, International Association, Rome, September 18, 2007)

Approximately 500 products based on plants native to Peru are registered in patent offices in the United States, Europe, and Japan. Although the patenting itself may have been done legally according to U.S. patent laws, Peruvian laws were broken in gaining access to their biodiversity and traditional knowledge. The evidence came from the findings of a commission set up by the Government of Peru to examine the patent registries in Europe, Japan, and the U.S.A.

The report laid the foundation for evaluating the legality of the patents. Investigation is continuing to examine if patent applications violated Peruvian legislation for gaining access to these bioresources. Prior consent and compensation for the indigenous communities that possess the traditional knowledge are required. As the investigation proceeds, two or three test cases of proven legal infractions will be selected to demand the revocation of the patents. Based on that experience, a strategy will be devised to challenge all of the patents in due course.

Countries with great biological diversity like those of the Amazon jungle must protect that wealth and the knowledge about it held by traditional indigenous peoples, just as industrialized nations apply pressure around the world to fight the piracy of their products like software, films, and albums. The meeting of intellectual property officials, which was sponsored by the Amazon Cooperation Treaty Organization (ACTO), was a first step towards the sharing and exchange of information on biopiracy, cooperation, and international negotiations on patents.

Achieving effective recognition of collective rights requires an effort by all countries, because national laws generally only protect the copyright and intellectual property rights of individuals or companies, not of the communities that developed the knowledge in the first place. Future discussions will involve all countries of the region, especially those with "mega-biodiversity," like Bolivia, Brazil, Colombia, Ecuador, Peru, and Venezuela. In today's world, new rights are constantly emerging and being defined, such as in the field of aeronautics and space activity. Biodiversity deserves a similar consideration.

A pressing question now is obtaining reparations for indigenous communities for the knowledge that they have collected over centuries and that is being used to develop food products, pharmaceuticals, and cosmetics. This is expected to be a long process, and is a particularly sensitive issue in the Amazon jungle region because of its great biodiversity. Each hectare of this part of South America contains more biological diversity than all of Europe combined.

An emblematic case of the illegal appropriation of biodiversity occurred a few years ago in Brazil, when the Japanese company Asahi Foods patented the name of an Amazon fruit, cupuazú, which is related to cacao. Legal action that was initially led by non-governmental organizations succeeded in getting the patent annulled in Japan and Europe in 2007.

Brazil's National Institute on Industrial Property has begun to inform its counterparts around the world of the names of Amazon jungle plants in order to prevent further cases of inappropriate registration of patents, said Roberto Jaguaribe, secretary of industrial technology in Brazil's Development Ministry. Peru's experience could make a significant contribution to Brazil's struggle in this area, since many species are shared by both countries, like the "quiebra piedra" (literally "stone breaker") plant used to make herbal tea for people suffering from kidney stones or gallstones, said Jaguaribe, who called for joint international legal action against biopiracy.

There is no uniformity in the intellectual property laws among nations. This causes problems in promoting cooperation between them. For that reason, future meetings of intellectual property officials from the Amazon nations will focus on the harmonization of laws.

Globalization

Among the many undesirable consequences of globalization is the appropriation of the genetic, intellectual, and cultural property of the indigenous communities of developing countries by the industrialized countries. Earlier agreement reached at the Uruguay Round of international trade negotiations that came into effect in 1995, i.e. the World Trade Organization (WTO) agreement on the Trade-Related Aspects of Intellectual Property Rights (TRIPS), has defined the nature and rules of engagement regarding intellectual property (IP) in the developed countries. It defines the rules under which the developing countries may access that knowledge and the cost for being allowed to do so by the developed countries.

However, no provision is made to protect the biodiversity and other natural resources as well as the traditional knowledge of the developing countries from illegal or predatory appropriation by various groups, multinational companies, or the governments of developed countries.

UN General Assembly Backs Indigenous Peoples' Rights

United Nations (AFP), September 13, 2007 — The UN General Assembly on Thursday adopted a non-binding declaration upholding the human, land and resources rights of the world's 370 million indigenous people, brushing off opposition from Australia, Canada, New Zealand and the United States.

The vote in the assembly was 143 in favor and four against. Eleven countries, including Russia and Colombia, abstained.

The declaration, capping more than 20 years of debate at the United Nations, also recognizes the right of indigenous peoples to self-determination and sets global human rights standards for them.

Victoria Tauli-Corpuz, the Philippine chair of the UN Permanent Forum on Indigenous Issues, joined UN chief Ban Ki-moon in hailing the vote. "It marks a major victory for Indigenous peoples," said Tauli-Corpuz, adding that the document "sets the minimum international standards for the protection and promotion of the rights" of native peoples.

But Canada, Australia, New Zealand and the United States, countries with sizable indigenous populations, expressed disappointment with the text. But Australia's top rights group, which welcomed the declaration, said it was "a matter of great regret" that it was opposed by Canberra.

The declaration, which recognises the right to self-determination, was "a milestone for the world's indigenous peoples," Tom Calma, of Australia's Human Rights and Equal Opportunities Commission, said. "It also acknowledges that without recognising the collective rights of

(*Continued*)

(Continued)

indigenous peoples and ensuring protection of our cultures, indigenous people can never truly be free and equal," he said.

The New Zealand government said Friday it voted against the UN declaration on indigenous rights because it disadvantaged non-indigenous people and conflicts with the country's laws.

Parekura Horomia, the New Zealand minister responsible for policy on the native Maori people, said his government was committed to protecting the rights of indigenous people.

But Horomia, himself a Maori, said the UN declaration on the human, land and resource rights of indigenous people was incompatible with New Zealand law.

"These articles imply different classes of citizenship where indigenous people have a right of veto that other groups or individuals do not have," Horomia told Radio New Zealand.

New Zealand was far ahead of other countries in promoting the rights of indigenous people, he said.

"Unfortunately, the provisions in the Declaration on lands, territories and resources are overly broad, unclear, and capable of a wide variety of interpretations, discounting the need to recognize a range of rights over land and possibly putting into question matters that have been settled by treaty," Canada's UN Ambassador John McNee told the assembly.

Another bone of contention was an article upholding native peoples' right to "redress by means that can include restitution or when not possible just, fair and equitable compensation, for their lands and resources which have been confiscated, taken, occupied, used or damaged without their free, prior and informed consent."

Opponents also objected to one provision requiring states "to consult and cooperate in good faith with the indigenous peoples . . . to obtain their free and informed consent prior to the approval of any project affecting their lands or territories and other resources, particularly in connection with the development, utilization or exploitation of mineral, water or other resources."

(*Continued*)

(Continued)

Indigenous advocates note that most of the world's remaining natural resources — minerals, freshwater, potential energy sources — are found within indigenous peoples' territories.

A leader of Canada's native community, Phil Fontaine, slammed his government's stance.

"We're very disappointed with Canada's opposition to the declaration on indigenous peoples," said Fontaine, leader of Assembly of First Nations, who came to New York to lobby for adoption of the text.

Canada's indigenous population is about 1.3 million people, out of a total population of 32.7 million.

The declaration was endorsed by the Geneva-based UN Human Rights Council last year.

Ancient Knowledge

The knowledge of ancient and indigenous cultures has been misinterpreted and downgraded by Western countries as primitive, undeveloped, or the product of a racially inferior society. It must be added that the term "traditional", which has also been used in this context, does not represent antiquated, stagnant, or useless knowledge. Nothing is further from the truth. Several patents filed by large multinational companies in the U.S. and Europe recently are directly derived from traditional knowledge regarding medicinal plants or other useful information of practical value. Furthermore, it is precisely traditional knowledge and practices which have created the large variety of food crops and domestic animals that we enjoy today. It must be noted further that traditional knowledge is not necessarily indigenous, whereas indigenous knowledge is traditional knowledge.

Traditional knowledge embraces a wide variety of people, communities, cultures, and niches, originating in traditional lifestyles with diverse geographical and ethnic backgrounds. It differs from the current knowledge in one main respect: its resilience and practical value has already been proved by centuries of human experience.

North–South Debate

The center of debate today between South and North (or East and West) is concerned with the difference between the intellectual property protection of Western inventions on the one hand and the lack of any such protection for the intellectual creations of the indigenous people of developing countries on the other. As Hastings Law Professor Naomi Roht-Arriaza (1996) noted: "(T)he appropriation of the scientific and technical knowledge of indigenous and local peoples, of the products of that knowledge . . . has become both notorious and contested." The theft of biodiversity and related ancient knowledge has led to the biopiracy concept and terminology. Mgbeoji (2006) defined biopiracy as the "unauthorized commercial use of biological resources and/or associated traditional knowledge, or the patenting of spurious inventions based on such knowledge, without compensation" (pp. 11–12). It includes any unauthorized transport or theft of plants and animals or their parts from the developing countries to developed countries through individuals, private institutions, local governments, or any other public or private bodies.

The modern method of appropriation of intellectual and physical property is markedly different from the blatant robberies of the colonial period. It presents itself as a respectable business, utilizing the legal terminology of patents, which were specially created for a subtle and smooth robbery under the cloak of apparent legality. To understand the patent system clearly, one must look at the origins of the system, the terminology employed, the countries involved at its inception, the major beneficiaries of the system, and those who are being victimized by the system.

Biopiracy is being aided by the historical background of the relationship between the "have" and "have-not" nations, the globalization of the patent system, and the racial boundaries of the knowledge as well as of the economic success of the fast-rising biotechnology industry. All of these factors facilitate the phenomenon that has come to be called "biopiracy". It is driven by a hunger, an economic hunger which propels large multinational corporations to become larger and richer, utilizing the biodiversity of poor nations and the tools of the Western capitalistic economies of which they are a part. What began as a simple need to reward individual inventors, many working quietly in their garages back home, rapidly evolved into an international instrument — a public menace — that has been appropriating large segments of traditional knowledge and biological resources from poor countries, which lack the knowledge and resources to protect their biological wealth (Dronamraju 2002).

Of special interest are the relevant sections of the U.S. patent law (and of Europe and Japan) that have facilitated the claims of multinational seed and pharmaceutical corporations. This point of view has been emphasized with reference to the U.S. patent system by Professor Mgbeoji in the following account:

> The reasons for this emphasis on the U.S. patent system are as follows: (1) It is common knowledge that Article 27 of the Agreement on Trade-Related Aspects of Intellectual Property Rights (TRIPS), constituting the global minimum patentability, is an approximation of U.S. jurisprudence and ideology; (2) it is also known that TRIPS is a product of the immense clout of the American pharmaceutical and biotechnology industries; (3) the U.S. patent system accounts for almost half of all patents issued in the world; (4) most of the controversial patents that raise the question of biopiracy were issued by the U.S. patent office; (5) the U.S. has the most appropriative regime on patents; and (6) the pronouncements and decisions of U.S. courts on matters of patent law have immense international influence. (Mgbeoji 2006, 123)

Some examples of the wrongful claims appropriating the biodiversity of the developing countries are provided in the accompanying table. There are many other examples also.

Bioprospecting in Asia

Country of Origin	Crop	Bioprospector
China	Bitter Melon	U.S.
China	Xi Shu/Happytrees	U.S.
Malaysia	Bintangor tree	Singapore
Pacific	Kava	U.S.
Pacific	Nonu (Indian Mulberry)	Europe, U.S.
Papua New Guinea	Coral reef sponges	U.S.
Philippines	Soil microbes	U.S.
Philippines	Ilang Ilang (*Cananga odorata*)	France
Philippines	Banaba (*Lagerstroemia sp.*)	Japan, U.S.
Philippines	*Nata de coco*	Japan, U.S.
Philippines	Snails (Conus)	U.S.
India	Basmati rice (*Oryza sativa*)	U.S.
India	Turmeric (*Curcuma longa*)	U.S.
India	Neem (*Azadirachta indica*)	U.S.
India	Guggul (*Commiphora mukul*)	U.S.
Thailand	Jasmine rice	U.S.
Thailand	Plao-noi (*Croton sublyratus*)	Japan
Samoa	Mamala tree (*Homalanthus*)	U.S.
Sri Lanka	Kothala himbutu (*Salacia*)	Japan, U.S.

(Modified from *Grain*, October 2002)

Biopiracy in Zimbabwe, Patenting by Swiss University Denounced

C. Raghavan, reporting for the Third World Network, explained how the University of Lausanne, in Switzerland, gained access to Zimbabwe's genetic resources via a benefit sharing agreement negotiated involving two Zimbabwe institutions, the University of Lausanne, and a US pharmaceutical company, as well as the grant, to a Lausanne University Professor, of a patent relying on traditional knowledge and the root of a tree, *Swartzia madagascariensis*, found throughout tropical Africa, according to a report in the SEATINI bulletin.

The denunciation has come from the Berne Declaration and two Zimbabwe NGOs, the Community Technology and Development

(*Continued*)

(Continued)

Association (CTDA) and the Zimbabwe National Traditional Healers Association (ZINATHA). SEATINI is the Harare-based NGO, the Southern and Eastern Africa trade and information network, providing technical assistance and training for trade officials of the region.

The three NGOs have cited the case as another example of how current bio-prospecting in countries of the South contradicted the rules defined by the UN Convention on Biodiversity. "While the Swiss government supports the guidelines for Access and Benefit Sharing at the CBD, a Swiss University is engaged in illegal bio-prospecting activities in Zimbabwe," says Francois Meinberg of the Swiss Berne Declaration.

The statement of the NGOs and details of the case are published in the SEATINI Special Bulletin on Intellectual Property Rights, and the issue comes to light on the eve of a WIPO-organized meeting next week where some of these issues are on the agenda of an intergovernmental committee.

The NGOs have condemned the way in which the University of Lausanne gained access to Zimbabwean genetic resources and the way the benefit sharing has been negotiated. They also reject the patent on antimicrobial diterpenes which Prof. Hostettmann of Lausanne University took out on these resources and which is based on traditional knowledge.

The US patent 5,929,124 on anti-microbial diterpenes (according to the report in SEATINI) was granted to Kurt Hostettmann, professor at the University of Lausanne. The patented invention relies on traditional knowledge from Zimbabwe and on the root of the tree *Swartzia madagascariensis*, found throughout tropical Africa. Two years before, in April 1997, an addendum to a material transfer and confidentiality agreement between the American pharmaceutical company Phytera and the University of Lausanne was signed, under which Phytera received an option for an exclusive world-wide license and in return agreed to pay royalties of 1.5% on the net sales of any product marketed under this license. Professor Hostettmann, on the other hand, is obliged to give 50% of any royalties received to the

(*Continued*)

(Continued)

National Herbarium and the Botanical Garden of Zimbabwe and to the Department of Pharmacy at the University of Zimbabwe.

The following consequences resulted from these actions.

CTDT and Berne Declaration demand that the current research agreement between the universities concerning the open access to medicinal and poisonous plants of Zimbabwe be suspended with immediate effect for the following reasons:

(a) The government is not party to this agreement as required by the Convention on Biological Diversity (CBD). The representative of the Ministry of Environment and Tourism confirmed at the meeting, that they are the only legal authority to grant access to any Zimbabwean biological resources.
(b) The benefit sharing mechanisms and frameworks of the agreement are not consistent with common practice. For example, there are no provisions for a future benefit-sharing agreement if a product is commercialized.
(c) Under the current agreement only the University of Zimbabwe is a beneficiary, thus marginalizing other stakeholders such as traditional healers and the government.
(d) The current agreement has no mechanism to acknowledge and compensate the use of traditional knowledge systems.
(e) It seems that access to medicinal plants was granted to the University of Lausanne at less than a fair value. CTDT and Berne Declaration call upon the Ministry of Environment and Tourism of Zimbabwe to take the initiative and a leading role in defining a model agreement for access and benefit sharing. Such an agreement involving all relevant stakeholders should contain provisions for prior informed consent, mutually agreed terms and benefit sharing mechanisms. To avoid shortcomings and loopholes, the new agreement should be accessible for comments by civil society.

Raghavan, C. (2001) Biopiracy in Zimbabwe, patenting by Swiss University denounced. South–North Monitor, reproduced in Third World Network, Penang, Malaysia, April 26, 2001.

Other Examples

In September 1997, the Texan company RiceTec was granted a patent by the U.S. Patent Office on a variety of rice produced by crossing a strain of Indian basmati rice with an American 'semi-dwarf' variety. Under the patent, the company claims ownership of basmati-like rice grown anywhere in the Western Hemisphere and future rights on any new varieties produced by crossing this new variety with traditional Asian strains. The approval of this patent claim caused outrage across the Indian subcontinent. In India, farmers took to the streets in protest against a foreign company laying claim to what they saw as a national resource. Basmati has a worldwide reputation as a high-quality rice grown only in India, Pakistan, and Nepal; they argued that a US rice producer should not be allowed to use the name. (In fact, says the WTO, their fears were based on a misunderstanding, as the patent did not concern the use of the name.) But, there was more than national pride at stake. Basmati rice is a valuable export crop earning India alone around US$800 million a year. Although the USA takes only less than 10% of the annual exports of 480,000 tonnes, the Indian government may fight the patent claim, fearing it might destroy the North American market and the more important European export trade.

In 1995, two researchers at the University of Mississippi Medical Center were granted a U.S. patent on some pharmaceutical applications of turmeric. Turmeric has been used in Indian medicine to treat wounds for thousands of years. The U.S. authorities later withdrew the patent, accepting the objection by the Indian Council for Scientific and Industrial Research that the proposed "new" use of turmeric lacked novelty, an essential requirement for a patent.

In January 1998, two Australian research institutes dropped an attempt to claim monopoly rights on varieties of chickpeas acquired from the International Crops Research Institute for the Semi-Arid Tropics, in Hyderabad, Pakistan. The UN Food and Agriculture Organization and the Consultative Group on International Agricultural Research acted to end any further attempts to claim ownership of varieties of plant held in trust by international institutions.

Other patents which have caused protest are on Neem, a tree with agricultural and medicinal uses cultivated widely in Asia and particularly India.

Vandana Shiva (2001) wrote:

> Biopiracy occurs because of the inadequacy of western patent systems and the inherent western bias against other cultures. Western patent systems were designed for import monopolies, not for screening all knowledge systems to exclude existing innovations and establish prior art in other cultures. Western culture has also suffered from the 'Columban blunder' of the right to plunder by treating other people, their rights, and their knowledge as non-existent. *Terra nullius* has its contemporary equivalent in Bio-Nullius — treating biodiversity knowledge as empty of prior creativity and prior rights, and hence available for ownership through the claim to invention. (p. 49)

On the recognition and protection of indigenous knowledge, Shiva (2001) wrote:

> This diversity of knowledge needs to be recognized and respected, and a pluralistic IPR regime needs to be evolved which makes it possible to recognize and respect indigenous knowledge, and protect the indigenous knowledge systems and practices and livelihoods based on it. We, therefore, need diverse legal regimes to protect the diverse knowledge systems and the diverse communities. The legal regime being universalized through TRIPs and WTO is restricted to western IPR systems reflecting the interest of the dominant economic systems of the west — the MNCs. (p. 51)

With respect to the patenting of the drugs derived from the medicinal plant *Phyllanthus niruri* in curing hepatitis, Shiva (2001) wrote:

> By isolating the application of *Phyllanthus niruri* for the treatment of one form of infective hepatitis only, i.e. hepatitis B, and treating this as a novel application, even though medicines derived from *Phyllanthus riruri* have been used for treating all forms of hepatitis in traditional systems of medicine (of India), scientists of the Fox Chase Cancer Centre have falsely presented an act of piracy as an act of invention. (p. 55)

Patents on the anti-diabetic properties of karela, jamun, and brinjal once again highlight the problem of biopiracy:

> The use of karela, jamun, and brinjal for control of diabetes is everyday knowledge and practiced in India. Their use in the treatment of diabetes is documented in authoritative treatises like the Wealth of India, the Compendium of Indian Medicinal Plants and the Treatise on Indian Medicinal Plants. The claim to the use of karela or jamun for anti-diabetic treatment as an invention is false since such use has been known and documented widely in India. The U.S. patent granted recently to Cromak Research Inc. based in New Jersey for the use of these plants in the cure of diabetes is a clear case of intellectual piracy coupled with the arrogance in assuming that these resources become 'value added' when processed in western laboratories. Such a patent also has a serious negative impact on the possible export market for formulations by Indian drug companies that meet the requirement of Indian systems of medicine. (Shiva 2001, 49)

Colonial Criminals

States have always sought to keep economically useful bioresources out of the reach of other states, for example, the colonial era powers such as France, the Netherlands, and the United Kingdom. During their colonial rule, the British government controlled the production of black peppercorns from India so tightly that it was impossible to obtain them. The French were so determined to control their monopoly of the indigo dye trade that the export of indigo seeds from their colony, French Antigua, was made a capital offence!

Sometimes, draconian measures were enforced by states to maintain their control. The Dutch, in their great desire to maintain control of the supply of nutmeg, destroyed all nutmeg and clove trees in the Moluccas except those on the three islands which were under their control. Other examples are cocoa germplasm from Ecuador, coffee germplasm from Ethiopia, and quinine (from the bark of the Cinchona tree) from Peru and Bolivia.

It is, of course, to be expected that tight controls produced in turn numerous clever and devious or even audacious attempts to circumvent the embargo. One such example is the control of rubber. In 1876, the British government wanted to break the Brazilian monopoly of the global supply of rubber (controlling 95% of global trade) and helped Henry Wickham to smuggle out 70,000 seeds of the rubber plant in a boat. They eventually reached British colonies in Asia and the rest is history!

Biopiracy and the Role of International Agricultural Research Centers

Some people have claimed that international agricultural research centers (IARCs) have played a massive role in transferring plant germplasm from the third world to the industrialized world. Under the guise of conservation and research, such institutes have become the collection centers for storing the germplasm of numerous plant species including many domestic crops and enormous quantities of germplasm.

Rockefeller Foundation

According to the Canadian author, Ikechi Mgbeoji (2006), a historic agreement was reached in 1941 between Vice President Henry Wallace of the United States and Raymond Fosdick, the President of the Rockefeller Foundation. The central part of the agreement was to initiate "a program of agricultural development aimed at Latin America in general and Mexico in particular (which) would have both political and economic benefits" for the United States (cited in Mgbeoji 2006). By 1943, the Rockefeller Foundation had started its Mexican Agricultural Program, with the ostensible goal of improving the quality of wheat and corn.

It was very clear from the beginning, however, that the collection of indigenous germplasm, although an important component of the Rockefeller Foundation's Program in Mexico and Latin America, was but a timely and opportunistic approach to establish a mechanism for funneling plant germplasm from the poor countries of the South to the industrialized north. It was a time when the colonial rule of many third world countries would be coming to an end. It was a historic change in the nature of the relationship between the South and the North. A new mechanism was needed to replace the old colonial exploitation of the biodiversity-rich third world. What was more suitable than a benign-looking research foundation which would, in fact, be the "Trojan horse" of the new post-colonial era?

The two foundations, Rockefeller joined by Ford, served that purpose admirably. The true goal could be achieved with the willing cooperation of the Southern countries in a smooth transfer of not only the germplasm, but all other natural resources and bioresources. With the support of the U.S. Government, several private American foundations led by Rockefeller established several IARCs in the South. The location of the IARCs in southern countries was not a random or fortuitous event, but was closely tied to each region known for its unique germplasm of particular landraces, corresponding to the geographic variation of genetic races which are analogous to Vavilov's "centers of origin". The centers include (but are not limited to) the International Rice Research Institute (IRRI) in the Philippines; the International Center for Agricultural Research in Dry Areas, located in Syria; the West African Rice Development Association in Liberia; the International Potato Center in Peru; and the Guatemala Tropical Research Center, which aims to "search for genes or characters that will improve our corns." Through these centers, the U.S. amassed a huge collection of plant and crop germplasms of all kinds. Their seeds and other means of propagation have enormously benefited developed nations, but the countries of the South — where the biodiversity has originated — received no recognition or compensation while their intellectual and physical property was freely being transferred to the rich countries.

Simultaneously, the patent system of the North was expanded and strengthened to ensure that the plant-derived products and other resources of biodiversity are protected from challenges. Not surprisingly, the largest storage facilities for seed collections and other forms of biodiversity are located in the North, while their source countries (i.e. countries of the South) lack even the minimum facilities required for food storage. Indeed, they are mired in ever-growing debt, and are hardly able to keep up with the interest payments that are required to service their debt.

Storage of Germplasm

Governments	83.00%
IARCs	11.00%
Private sector	1.27%

The IARC collections are by far the most significant, containing about 35% of the samples. Naomi Roht-Arriaza (1996) noted that 68% of all crop seed, 85% of all livestock, and 86% of microbial culture collections are held at the IARCs or in Northern countries. She wrote: "Indeed in a number of crops (wheat, barley, food legumes, potato) the advanced capitalist nations possess more stored germ plasm accessions than do those nations that are the regions of natural diversity for the crop."

According to Kloppenburg (1988),

> The IARCs perform a dual role in the processing of plant germ plasm. They necessarily collect and evaluate indigenous land races and primitive cultivars that are the raw materials from which HYVs (high-yield varieties) are bred. And because their "imported" agricultures are based on the very species that the IARCs are mandated to improve (i.e. corn, wheat, potato), such collection and evaluation are of direct value to the developed nations. The IARCs are . . . vehicles for the efficient extraction of plant genetic resources from the Third World and their transfer to the gene banks of Europe, North America, and Japan. . . . The CGIAR system is, in one sense, the modern successor to the eighteenth- and nineteenth-century botanical gardens that served as conduits for the transmission of plant genetic information from the colonies to the imperial powers. (p. 189)

With respect to the role of seed banks, Roht-Arriaza (1996) wrote:

> Seed banks and gene banks collect Southern germ plasm and distribute it to gene-poor Northern countries; thus a large proportion

of commercially used genetic material moves to the Northern countries via the IARCs. Studies estimate, for example, that 21 percent of the US wheat crop was derived from material stored at the International Maize and Wheat Improvement Center, the IARC for wheat. (p. 944)

Patentability

The law regarding the patentability of purified biological/biochemical substances derived from nature is inconsistent. While patents on purified biological substances which do not normally occur in nature are recognized, patents on naturally occurring but purified metallic substances have not been allowed. This bias is in favor of the pharmaceutical industry, and has facilitated the appropriation of traditional knowledge and biodiversity of the developing countries. One such example is U.S. Patent No. 4,673,575, issued on 16 June 1987, which was granted to Fox Chase Cancer Center on a derivative of *Phyllanthus amarus*, a medicinal plant used in India for treating various liver ailments including jaundice. The medicinal value of the plant has been well known to practioners of Ayurvedic medicine for several centuries. However, scientists at Fox Chase Cancer Center in Philadelphia have claimed to have "discovered" that the plant extract was an effective treatment for viral hepatitis B and E.

Another well-known example is the medicinal plant of India, Neem, or *Azadirachta indica*, which has been used in various forms to cure several ailments for millennia in India. It has been mentioned in ancient Indian texts, written 2000 years ago, as an air purifier and as an insect and pest repellent which can influence over 200 species of insects. In the last few years, growing opposition to chemical sprays in Western countries has increasingly turned its attention towards India and its medicinal biodiversity including the Neem tree. In 1971, a U.S. timber importer Robert Larson, noting the pesticide properties of Neem, began importing its seeds to his company headquarters in Wisconsin. After several tests, he developed a pesticidal Neem extract called Margosan-O which was cleared by the Environmental Protection Agency (EPA) in 1985. Three years later, he sold the patent to the multinational chemical giant, W.R. Grace & Co. In the following years, more than a dozen patents have been taken out by several U.S. and Japanese companies on Neem-based solutions and other derivatives including a Neem toothpaste.[a]

The molecular formula for azadirachtin is $C_{35}H_{44}O_{16}$. The demand for large quantities of Neem seeds by W.R. Grace & Co. had three primary effects (Shiva 2001):

1. The price of Neem seed has gone up beyond the reach of the ordinary people who need it most; for instance, Neem oil for lighting lamps is no longer available.
2. Almost all the seed, which was until then freely available to the farmers and the Ayurvedic practitioners, is bought by the company.
3. Poor people have lost their access to a resource vital to their survival.

It is a matter of some concern that the exploitation of Neem had the blessing of the National Academy of Sciences (NAS) itself. An entire report dedicated to the potential commercialization of Neem was published in 1992 by the National Research Council of the NAS:

> Beyond all the possible pesticides and pharmaceuticals, neem provides many useful and valuable commonplace materials. For instance, oil extracted from the seeds goes into soaps, waxes, and lubricants, as well as into fuels for lighting and heating. The solid residue left after the oil is removed from the kernels is employed as a fertilizer and soil amendment. In addition, wood from the trees is valued for construction, cabinetry, and fuel. The bark is tapped for gum and extracted for tannins and dental-care

---

[a] As a child growing up in India, I was encouraged to use Neem twigs to brush my teeth every morning. Several members of my family, including my grandmother, regularly brushed their teeth every morning with a Neem twig which was simply broken off the tree in our backyard.

> products. The leaves are sometimes used for emergency livestock feed. And the profuse flowers are a prized source of honey. (U.S. National Research Council 1992)

The commercial appeal of this document has not escaped attention. Others have noted this point, and opposed it on ethical and moral grounds.

Monsanto has taken out patents on neem products, claiming broad fungicidal and insecticidal uses. The Indian government challenged patents granted to the U.S. company W.R. Grace for extraction and storage processes for neem, but withdrew its complaint upon realising that in these cases the processes were in fact new inventions rather than traditional knowledge.

Contrast

There is a huge gulf between the great concern and interest that is evidently displayed in reaction to biodiversity theft in India and other developing countries on the one hand and the absolute deafening silence in the United States and Europe on the other. How do we account for this difference in public reaction?

First, the public can only react if they are fully aware of a particular situation. The popular press in Western countries has by and large ignored the alarming problem of the theft of biodiversity from the developing countries by the developed countries. This situation is comparable to the era of colonial empires, when many of the atrocities committed by the developed countries in their colonies were either ignored or suppressed in the popular press of the developed countries. Second, the obvious fact is that the public in the developed countries has been accustomed to enjoying the bounty from the third world countries, and they do not seem to be interested in delving deeply into the sources and methods used to obtain that bounty. Third, there is a total lack of interest in the U.S. about learning the problems of other countries, any country. This apathy extends across all aspects of life in other countries.

Broad Patents on Plants

The granting of broad patents for all genetically engineered varieties of a species, regardless of the genes involved or how they are transferred or modified, empowers an inventor to control our food supply and our health. The utility patents are very broad-based, allowing monopoly control over large sections of the genome. A company could file several patents covering large external parts such as flowers, fruits, and seeds; smaller internal organs or parts such as cells and genes; and the novel processes employed to understand or manipulate these parts.

It is obvious that patent protection excludes farmers' rights. A landmark case was the 1985 judgement in the U.S. which granted patents on the tissue culture, seed, and whole plant of a corn line derived from that tissue culture.

Hibberd Patent

A landmark judgement in the U.S. for the patenting of plants was the so-called *ex parte* Hibberd of 1985. Geneticist Kenneth Hibberd and his colleagues were granted broad patents on the tissue culture, seed, and whole plant of a corn line that was selected from tissue culture. It covered 260 separate claims, excluding the right of others in all their aspects. It had important implications for the agricorporate business as well as the individual farmers and seed growers.

Amazingly, Hibberd continued as the primary precedent without court challenge from 1985 until J.E.M. Ag Supply v. Pioneer Hi-Bred in 2001. There is, of course, a reason for this: in patent battles between seed companies, no one wanted to "kill the goose that laid the golden egg." Put another way, while it was always lurking in the background that perhaps the Patent and Trademark Office (PTO) decision exceeded its authority or was at least on questionable ground, no one challenged the entire concept of issuing the patents in the first instance. Rather, in the many litigations seed companies used more conventional affirmative defenses such as noninfringement, invalidity for inadequate written description or claim definiteness under 35

U.S.C. § 112, lack of novelty under 35 U.S.C. § 102, or the ubiquitous defense of obviousness to one of ordinary skill in the art under 35 U.S.C. § 103.

The highly visible decision in the Hibberd case led to the issuance of some 1,800 utility patents for plants between Hibberd (1985) and J.E.M. Ag Supply (2001).

Impact of Patenting on Agriculture

No one has stated the consequences and implications of patenting in agriculture more eloquently than Shiva (2001):

> It will encourage and facilitate corporate monopolies from the west to control crop materials in the developing countries. Of primary importance is the control of seed supply, with far reaching consequences for the world's food supply which is being increasingly controlled by the large agri-business corporations. It will have other undesirable effects as well. Guided by profit motive alone means a narrowing of the variety of crops planted or a definite reduction in agricultural biodiversity and rising prices of seeds and food supply. Smaller farmers will be driven out of business which has been happening already due to various reasons. Dependence on a few major crops will increase the likelihood of the occurrence of famines.
>
> Changes in the agricultural practices will have other adverse consequences in the economics, employment and other aspects of the society. Smaller and poorer developing countries which depend on agriculture for their economic and nutrition needs will be more severely affected. Their national debt to developed countries will increase manifold, imposing a crushing burden. (p. 76)

Industrialized Agriculture

Industrialization of agriculture has greatly expanded arable land as well as food production. However, it has also resulted in the destruction of biodiversity, in particular agricultural biodiversity. It tends to promote monocultures, which facilitate greater efficiency and profitability. For instance, excessive commercialization of the potato in the U.S. has reduced the range of cultivars to twelve out of a total of about 2000 known to farmers. Indeed, even among the potato varieties, over 40% are of one kind only, namely the Russet Burbank, which is suitable for food processing because of its size and quality. There are specific requirements of size when potatoes are sliced into fries. Forty per cent of McDonald's fries must be 2–3 inches long, another 40% must be over 3 inches, and the remaining 20% is less than 2 inches. Such considerations tend to drive out other potato breeds from the mainstream markets. Indeed, with the onset of strictly enforced intellectual property rights (IPRs) and plant breeder rights (PBRs), many varieties of food grains, fruits, and vegetables have been driven out of the commercial markets. Commercialization, which is being enforced and driven by these new market forces, has been the deathknell of agricultural biodiversity.

It has been estimated that raising US$1 million worth of cattle for market requires the destruction of 100 km^2 of Amazon forest. On the other hand, extracting US$1 million worth of rubber destroys only 6.8 km^2 of Amazonian forest.

It has been estimated that 97% of the vegetable varieties, 87% of pear varieties and 86% of apple varieties have become extinct since the beginning of the 20th century. In recent years, the pace of bioprospecting and biopiracy has picked up speed, mainly because of the pressure from the pharmaceutical industry to discover and market new drugs. These explorations are often wasteful, destroying large quantities of biodiversity in return for meager or no benefits of any kind. However, when a big discovery is made, which only happens very rarely, the benefits to the pharmaceutical industry can be astronomical. However, the cost incurred for one such discovery can be enormous. One such case is the production of the anticancer drug taxol, which is made from large quantities of the bark of the Pacific Yew tree. The production of 1 kg of taxol requires 20,000 pounds of bark or 2500 to 4000 Pacific Yew trees.

With the rise of India, China, Brazil, and other countries which are rapidly developing their own biotechnology and pharmaceutical industries, there is not enough biodiversity on this planet to sustain such wasteful bioprospecting programs. Other, more efficient, methods should be employed.

Rice patents

As part of a growing worldwide campaign against "biopiracy", a coalition of US and Indian groups are taking legal aim at US companies who are selling US rice they have falsely labelled under the unique names of Indian and Pakistani basmati rice and Thai's jasmine rice.

The Washington-based International Center for Technology Assessment (ICTA) and the Research Foundation for Science, Technology and Ecology (RFSTE), based in New Delhi, want to stop US rice millers, producers and trade associations from marketing low quality US aromatic rice under the terms "basmati" and "jasmine" in order to receive a premium price.

"The current US policy of allowing virtually any aromatic rice to be labelled basmati or jasmine is nothing short of criminal," says Andrew Kimbrell, executive director of the Center for Food Safety, a project of ICTA.

The Texas company, RiceTec, Inc. for example, sells US grown rice as "Texmati" as "American basmati" and "jasmati" as "American Jasmine".

Both groups filed legal petitions this month with two US government agencies to revise their laws to protect the jasmine and basmati rice types grown in Asia. The petitions say current US regulations allow US companies to deceive consumers and threaten the livelihoods of millions of Indian and Pakistani farmers who grow basmati rice and Thai farmers who grow jasmine rice.

Current US rice standards allow companies to use the terms "basmati" and "jasmine" as generic terms that can apply to rice grown anywhere.

One petition, filed with the US Department of Agriculture, demands that it amend its rice standards on "aromatic" rice to clarify that the

(*Continued*)

(Continued)

term "basmati" can only be used for rice grown in India and Pakistan, and the word "jasmine" grown in Thailand.

"Since American consumers and farmers correctly believe that "Basmati" rice can only be produced in India and Pakistan and "Jasmine" rice in Thailand, the use of the descriptors "basmati" and "jasmine" in current US rice standards is misleading," says the petition.

The groups' proposal would make those of the US consistent with those of Saudi Arabia and the United Kingdom's Code of Practice for Rice, says Kimbrell.

The other petition, filed with the Federal Trade Commission (FTC), demanded the agency initiate a trade regulation to prevent US grown rice from being advertised or otherwise represented as "basmati" or "jasmine."

The groups are making their case under the FTC Act which prohibits "unfair or deceptive acts of practices in or affecting commerce."

Basmati rice is critical to the economies of India and Pakistan, says the petition. Each year India sells approximately $300 million' worth of Basmati rice. And it is counted among the nation's fastest growing exports. In Thailand — dependent on its rice exports to alleviate its economic downturn — jasmine fetches the steepest price among all Thai rice.

"When American companies steal the very names and strains of our indigenous rice, they threaten the lives of millions of farmers and their families in India, Pakistan, and Thailand," says Vandana Shiva, executive director of RFSTE.

The petitions do not address the disputes over rice breeding patents held by RiceTec. The Indian government is currently contesting a patent held by the company for a US hybrid rice that was crossed with traditional Indian Basmati, which has been cultivated by farmers for centuries on the foothills of the Himalayas — in Punjab, Haryana, and Uttar Pradesh.

Knight, D. (2000) Groups take legal action to end U.S. biopiracy. Penang, Malaysia: Third World Network, April 25, 2000.

India: Registration of Plant Varieties

The Indian government has initiated the registration of plant varieties with the Protection of Plant Variety and Farmers' Rights Authority (PPV&FRA) to provide them internationally-recognised protection against piracy.

Starting the third week of February, 12 varieties of crop will be registered under the sui generis system (specifically evolved own system) of plant variety protection mooted in the Plant Variety and Farmers' Rights Act, 2001.

The PPV&FRA will, to begin with, undertake the documentation and registration of varieties of rice, wheat (bread wheat types), maize, sorghum (jowar), pearl millet (bajra), chickpea (chana), pigeon pea (arhar), green gram (moong), black gram (urad), lentil (masur), field pea (matar) and kidney bean (rajmah).

Under the TRIPS (Trade-Related Aspects of Intellectual Property Rights) Agreement of the World Trade Organisation (WTO), countries must provide for patenting of plant varieties or enact their own law. India opted for the latter and passed the Protection of Plant Variety and Farmers' Rights Act in 2001. The Plant Variety and Farmers' Rights Authority was set up in November 2005 under the Act, for the registration of plant varieties. It has evolved detailed rules and regulations and crop-specific guidelines for seeking patent protection.

Besides allowing for variety protection, Indian law also ensures the rights of farmers and farm communities to evolve, preserve and refine crop varieties. Germplasm from seeds cultivated indigenously by farmers for generations is often used by researchers to develop newer varieties. Now that the varieties are registered, farmers will get the benefit when a trait is used from this germplasm to develop new varieties.

A new programme of giving recognition to these communities is also being launched simultaneously. Called Plant Genome Saviour Community Recognition, this programme will be financed from a proposed gene fund to be set up under this Act.

(*Continued*)

(Continued)

Farmers and communities will have to provide documentary evidence to prove that they conserved, improved and made available or shared material with active plant-breeding programmes for the development of a new plant variety. Details of the new variety and its contribution to the advancement of agriculture will also have to be provided to claim recognition and rewards. The law also upholds the traditional rights of farmers to use or exchange seeds they have grown, between themselves.

www.centad.org/, February 22, 2007.

Do India's Seed and Biodiversity Policies Promote Biopiracy?

National laws and policies relating to biodiversity have immense implications for the livelihoods, food security and health of nation. S. Bala Ravi of the M.S. Swaminathan Research Foundation pointed out that two laws enacted by the Indian Parliament in the past five years provide a loophole which encourages the unfair misappropriation of Indian genetic resources. India is one of the world's eight major centres of crop diversity and the original home of 163 fruit tree and crop species.

The Protection of Plant Varieties and Farmers' Rights Act of 2001 (PPVFR Act) complies with the World Trade Organization's agreement on Trade-Related Aspects of Intellectual Property Rights. The act protects the rights of plant breeders, farmers and researchers over plant varieties. The Biological Diversity Act of 2002 (BD Act) seeks to establish India's sovereignty over its biological resources and associated traditional knowledge. Following the CBD, the act aims to establish a system for managing access to biodiversity and how benefits from its use are shared.

Under the BD Act access to biological resources by non-Indian people or companies and by non-resident Indians requires prior

(*Continued*)

(Continued)

approval of the National Biodiversity Authority. For resident Indian citizens and companies, the State Biodiversity Board must grant permission for access, while for local communities none of these restrictions applies. Intellectual property rights over innovations based on Indian biological resources or traditional knowledge can be established only with the prior approval of the National Biodiversity Authority, which will notify the public of approvals. During such granting of permission, a mutually agreed decision on benefit sharing is made.

According to Ravi, a major problem arises from a provision in the BD Act that allows the government to exempt certain items "including biological resources normally traded as commodities" from the remit of the act. In the case of seeds, which are tradable commodity, such an exemption in the absence of other laws to regulate seed exports opens a legitimised door for biopiracy. Ravi also claimed that the BD Act has other deficiencies that undermine its provisions on access and benefit sharing. The terms 'commercial utilisation', 'use' and 'utilisation' are critical to the way the act restricts access to biological resources. But the act defines neither use nor utilisation. And although it defines 'commercial utilisation' as any activity that generates economic gain, this definition excludes "conventional breeding or traditional practices in use in any agriculture".

Therefore access to Indian genetic resources for use in conventional breeding or other traditional practices followed in agriculture, even by the non-Indian entities does not require prior approval under BD Act.

Seed Piracy

If seeds were exempted, this would mean that the only law controlling access to them would be the PPVFR Act, which allows anyone conducting research free access without prior informed consent to any genetic resource, including varieties protected by plant breeders' rights.

(*Continued*)

(Continued)

The PPVFR Act does not differentiate the nationalities of people or organisations accessing Indian genetic resources, including varieties protected by plant breeders' rights, for breeding new varieties. The only exception is the need for prior informed consent for repeated use of such a protected variety as a parental line for the commercial production of a new variety.

This means that non-Indian entities can freely access plant genetic resources and associated knowledge for use in breeding or for bio-surveys within India. Secondly, having freely accessed the genetic resources of choice to develop breeding lines or new varieties or nothing, seeds of this material can be taken out in different pretexts as 'exports'.

Mr. Ravi further wrote: "The lack of a legal system regulating seed exports and of an informed customs system with the capacity to verify what is exported leaves a wide open door for the unchecked outflow of the planting material of virtually any genetic resource — including farmers' varieties, land races and pre-bred material. Once these resources are taken out through the trade route and used in conventional or non-conventional breeding, there is virtually no way to ensure that benefits are shared equitably to the communities that generated and conserved these resources."

He concluded: "The irony is that laws established to protect these resources and promote their conservation are in fact legitimising their piracy and misappropriation from the holder community."

From: Science and Development Network (online), September 14, 2007.

4

Global Appeal Against Patents on Conventional Seeds and Crops

Major corporations have been using the patent system quite effectively to gain control of the world's food supply.

World's Top 10 Seed Companies (US$ million)

1.	Monsanto (US)	$4,028
2.	Dupont (US)	$2,781
3.	Syngenta (Switzerland)	$1,743
4.	Groupe Limagrain (France)	$1,035
5.	Land O'Lakes (US)	$756
6.	KWS AG (Germany)	$615
7.	Bayer Crop Science (Germany)	$430
8.	Delta & Pine Land (US) (acquisition by Monsanto pending)	$418
9.	Sakata (Japan)	$401
10.	DLF-Trifolium (Denmark)	$352

Source: ETC Group (based on 2006 seed revenues).

According to Context Network, the global proprietary seed market was worth $19,600 million in 2006. Based on the table above, Monsanto — the world's largest seed company — accounts for one fifth of the global proprietary seed market. The top three companies — Monsanto, Dupont, and Syngenta — account for $8,552 million, or 44% of the total proprietary seed market. The top four companies account for $9,587 million, or almost half (49%) of the total proprietary seed market. The top ten companies account for $12,559 million, or 64% of the total proprietary seed market.

Challenging Industrial Patents

Monsanto's soybean seeds and traits accounted for about 90% of the genetically modified (GM) soybeans planted worldwide in 2005. What's more, genetically engineered soybeans reportedly account for almost 60% of the global soybean area — an increasingly dominant share of one of the world's most important food and commodity crops.

ETC Group, an international civil society organization, and the environmental group Greenpeace (supported by 19 other civil society organizations worldwide) successfully challenged Monsanto's control over the soybean. Among other projects, ETC documented the growing dominance over the global seed market by a handful of firms: Monsanto, Syngenta, and Dupont.

With respect to the control of the world's seed supply, Hope Shand, Research Director of ETC, wrote:

> Patents, exclusive legal monopolies granted by governments, are a favorite tool of big business to exercise power over the little guy. Giant pharmaceutical firms have most notoriously used patents to price anti-HIV drugs out of the reach of poor people in the global South. Less familiar are biotech battles in the agricultural sector, where multinational seed companies are using patents to deny farmers — or entire nations — the right to use and sell seeds from patent-protected crops. Patents, we are told, are designed to promote innovation. Instead, they are allowing giant seed companies to secure exclusive monopolies that undermine the economic security of farming communities and jeopardize access to seeds — the first link in the food chain. And lest we forget: Whoever controls the seeds controls the food supply.

> Instead of fostering agricultural research, breathtakingly broad patents are shutting down competition and stifling research. Perhaps no patent symbolizes the brokenness of the patent system more than Monsanto's European patent on all genetically engineered soybean varieties and seeds — European Patent No. 301,749. Critics call it a "species-wide" patent because its claims extend to all biotech soybean seeds — irrespective of the genes used or the genetic engineering technique employed — unprecedented in its broad scope.
>
> The livelihoods of Argentina's soybean farmers are directly affected by this patent because Monsanto, the world's largest seed corporation, is using its exclusive monopoly to deny Argentine soybeans from entering European markets. Why? Because Monsanto alleges that Argentine farmers aren't paying royalties to Monsanto for using the company's patented soybean seeds. Over the course of a single decade, Monsanto devoured dozens of seed companies (and their patents) to become the largest seed corporation in the world and the only soybean seed superpower. [Quoted from a press release by ETC (2007)]

According to ETC, Monsanto's legal defense of its patent is hugely hypocritical. Before Monsanto acquired it in 1996, the company vigorously opposed the patent, which was then owned by the US-based biotech company Agracetus. In 1994, Monsanto submitted an exhaustive, 292-page opposition statement to the European Patent Office (EPO) that shredded the technical merits of Agracetus's soybean patent. Monsanto's lawyers wrote that the soybean patent should be "revoked in its entirety," is "not ... novel," "lacks an inventive step," and "sufficient disclosure [of scientific method] is woefully lacking." However, after Monsanto acquired Agracetus in April 1996, Monsanto withdrew its challenge, reversed its position, and announced that it would defend its newly acquired patent!

In 2003 — more than nine years after the patent was first awarded and legally challenged — an EPO patent tribunal heard legal arguments against the notorious patent. Opponents were shocked when the EPO upheld Monsanto's monopoly in

2003. Today, nearly two thirds of the patent's 20-year term has expired.

Monsanto Patent Revoked in Munich

Munich — On 3rd May 2007, the European Patent Office put the brakes on Monsanto's over-the-top corporate greed by revoking its species-wide patent on all genetically modified soybeans (EP0301749) — a patent unprecedented in its broad scope. ETC Group, an international civil society organization based in Canada, won its 13-year legal challenge against Monsanto's species-wide soybean patent when an EPO appeal board ruled that the patent was not new or sufficient (i.e., the invention claimed was not sufficiently described for a skilled person to repeat it). The patent challenge was supported by Greenpeace and "No Patents on Life!" Dr. Ricarda Steinbrecher of UK-based EcoNexus also joined the opposition team in Munich as a scientific expert.

The patent was vigorously and formally opposed by Monsanto itself until the company purchased the original patent assignee (Agracetus) in 1996. The technology related to the now-revoked patent has been used, along with other patents in the company's portfolio, to corner 90% of the world's GM soybean market. (For more information, see ETC Group News Release, "Monsanto's Soybean Monopoly Challenged in Munich," April 30, 2007, http://www.etcgroup.org/en/materials/publications.html?pub_id=616)

"It's shameful that it took the European Patent Office 13 years to kill Monsanto's immoral patent, which was ultimately revoked on technical grounds. Though we're relieved that the species-wide patent on all genetically modified soybeans — both seeds and plants — was not allowed to stand, the delay of more than a decade demonstrates just how broken the patent system is. The patent had barely a year to go before expiring!" said Hope Shand, who represented ETC Group in Munich today.

"It was particularly satisfying," said Shand, "that Monsanto's own blistering 1994 arguments against the patent were ultimately key in

(*Continued*)

(Continued)

defeating it." One of Monsanto's top scientists testified in 1994 that the genetic engineering process described in the patent was insufficient to allow a skilled scientist to replicate the procedure — a necessary criterion for patentability.

ETC Group, which first challenged the patent in 1994 (as RAFI), was represented in Munich by UK barrister Daniel Alexander and patent attorney Tim Roberts of Brookes Batchellor, LLP.

According to patent attorney Tim Roberts, "It is very satisfying that the European Appeal Board has completely revoked this patent. This decision sends a message to greedy patentees — don't claim more than you are entitled to."

Patent expert Dr. Christoph Then of Greenpeace commented on the outcome of today's hearing, "The EPO's decision to throw out the patent will have implications for Monsanto and the EPO. It is now shown that the Patent Office is granting patents covering broad sectors of agricultural diversity with no real invention to back them up," said Then.

Ruth Tippe from the European-wide initiative, "No Patents on Life!" asserts, "This is an important step against patents on seeds because it shows that civil society will keep on fighting and can ultimately succeed against powerful multinationals."

According to Dr. Ricarda Steinbrecher of EcoNexus, "Monsanto's patent couldn't even survive on its scientific merits. It was a thoroughly bad patent — from both a technical and moral perspective."

Multinational firm Syngenta also made oral arguments today opposing the patent. While their technical expertise may have contributed to the patent's ultimate downfall, their opposition is viewed by civil society as cynical. In January 2005, ETC Group reported on three Syngenta patent applications that also make breathtakingly broad claims — multi-genome patents with claims on gene sequences that extend to 40 plant species. Despite assurances from Syngenta that the company would let the patents lapse, all three applications appear to be active still at the World Intellectual Property

(*Continued*)

(Continued)

Organization (WIPO). [See ETC Group Communiqué, "Syngenta — The Genome Giant?" January/February 2005, http://www.etcgroup.org/en/materials/publications.html?id=73]

This isn't ETC Group's first successful battle against species-wide patents. Most notably, another Agracetus patent — this one granted by the US Patent and Trademark Office in 1992 and claiming all genetically engineered cotton varieties — was eventually revoked in India and the US in 1994.

Other overly broad, unjust patents have yet to be revoked, however. The formal challenge to the notorious "Enola Bean" patent, US Patent No. 5,894,079, granted on a yellow bean genetically identical to a pre-existing Mexican bean variety, has entered its seventh year. [See ETC Group Genotype, "Whatever Happened to the Enola Bean Patent Challenge?" December 21, 2005, http://www.etcgroup.org/upload/publication/41/01/genotypeenola05.pdf]

ETC, press release, May 3, 2007.

Global Prohibition of Patents

A new international coalition of farmers' unions as well as developmental and environmental NGOs has made an appeal for a global prohibition of patents on seeds and farm animals. Farmers will become increasingly dependent on multinational corporations, which own patents on seeds and animals. In reaction to the recent stepped-up activity by the EPO on patenting in agriculture, farmers' unions from India joined similar groups from Italy and Argentina and, with the support of Greenpeace, launched a global appeal against the decision by the EPO to grant hundreds of patents on genetically modified and conventional normal plants. The EPO is also preparing a dramatic increase in the number of general patents to be approved on conventional breeding methods as well as normal animals and plants.

The data on patents which were filed and approved by the EPO since the year 2000 are given in the following table.

Year	GE Plants		Non-GE Plants	
	Applications	Granted	Applications	Granted
2000	530	27	20	1
2001	536	41	20	2
2002	471	40	18	6
2003	384	60	21	10
2004	315	68	19	5
2005	295	110	24	4
2006	285	151	44	5

Source: European Commission. (2006) European Regional and Urban Statistics. Luxembourg: European Commission.

The following global appeal was launched by the farmers' unions.

Keep Out Patents on Conventional Seeds and Animals

For several years, patents on genetically modified seeds and animals have been granted worldwide. The damaging impacts on farmers, who are deprived of their rights to save their seeds, and on breeders who can no longer use the patented seeds freely for further breeding, are well known.

In Canada and the US, for example, the multinational seed company Monsanto has sued many farmers for alleged patent infringements. The same company has also filed court cases against importers of Argentinean soy to Europe. Furthermore, the possibility of patenting seeds has fostered a highly concentrated market structure with only 10 multinational companies controlling about half of the international seed market. Many farmers organisations and NGOs around the world are fighting against these patents. Because genetically modified organisms (GMOs) are still not grown in most countries, or only used in a small number of crops, the negative impacts of these patents are not being felt everywhere.

(*Continued*)

(Continued)

However, there is an alarming new trend for patents not only to be claimed on GMOs (such as Round-up Ready soybeans), but also on conventional plants. For example, patent claims have been made for soy beans with a better oil quality covering parts of the plant genome when used in conventional breeding and technologies to improve conventional breeding (such as marker assisted breeding).

Some of the most threatening examples in this context are patent applications from Syngenta which claim huge parts of the rice genome and its use in breeding of any food crops that have similar genomic information to rice (such as maize and wheat). The European Patent Office has also granted a patent on aphid resistant composite plants which are based on marker assisted breeding. Other recent patent applications by Monsanto on pigs are also related to normal breeding methods, indicating the increasing danger of agricultural genetic resources becoming monopolised by a few multinationals on a global scale.

Soon the Enlarged Board of Appeal of the European Patent Office will decide on another patent of this kind — for a method of increasing a specific compound in Brassica species. This decision will determine the patentability of conventional seeds in Europe.

Whereas patents on conventional plant varieties are normal practice in the US, many other countries, especially developing countries, do not grant patents on plants or animals. But as the recent history shows, the standards defined and used at the European, Japanese and US patent offices influence international regulations (the WTO Agreement on Trade Related Aspects of Intellectual Property Rights, TRIPS, and the World Intellectual Property Organisation, WIPO). Patent offices all over the world are pushed to adapt their regulations and practices either through the international regulations or by bilateral agreements. India, for example, has just passed a third patent amendment in order to adapt its law to the TRIPS regulations.

(*Continued*)

(Continued)

This frightening new trend in patent policy will affect many more farmers and breeders, than has been the case with GMO patents. Any remaining farmers' rights and breeders' access to plant varieties and animal breeds for breeding purposes, will disappear everywhere. These patents will destroy a system of farmers' rights and breeders' privileges that has been shown to be crucial for the survival of farmers and breeders, for food sovereignty, and for the preservation of biodiversity in agriculture. The vast majority of farmers in developing countries are small-scale farmers, completely reliant on saving and exchanging their seeds.

In order to secure the continued existence of independent farming, breeding and livestock keeping and hence the food security of future generations, we, the undersigned farmers, researchers, breeders and civil society organisations from all over the world, restate our rejection of any patents on life, and urge policy makers and patent offices to act swiftly to stop any patents being granted on conventionally bred plants and animals and on gene sequences for use with conventional breeding technique, as well as on methods for the conventional breeding of plants and animals. We also urge companies not to apply for any patents of this kind.

European Commission. (2006) European Regional and Urban Statistics. Luxembourg: European Commission.

Biopiracy, Crops, and Seeds

What is biopiracy? When a government or a multinational corporation patents a plant or a part of it, it is also entitled to (and is likely to) claim all substances and products containing the plant's genome. The farmers and indigenous people who have conserved the germplasm and protected these plants and seeds for centuries are not among the beneficiaries, although the ultimate profits could not have been generated without their pioneering and continuing hard work. The biodiversity of developing countries, both agricultural and domesticated plants as well as the "wild" species, has become the target for multinational raiders and foreign governments who are determined to possess and control all biological wealth of the planet.

Some Important Disputed Patents

Corn

The European Patent Office granted a patent in August 2000 to DuPont to cover all varieties of corn (maize) containing higher amounts of oil and oleic acid. The patent covers any corn plant with these qualities, both genetically and nongenetically engineered varieties, and their derivatives. If the patent remains unchallenged, it would give DuPont a complete monopoly over several maize varieties. Opposition to DuPont's claim has come from the Mexican Government and Greenpeace.

Contrary to DuPont's claim, some corn breeds containing high proportions of oil and oleic acid occur in nature and have been known to exist naturally for centuries to farmers in several countries. They have also been produced by conventional methods of breeding. For farmers cultivating such breeds of corn, the patent will force them to buy new seeds and pay royalties each year. Finally, it must be noted that corn is much more than a mere food crop in many countries: it is a vital part of their culture and religion, and has been connected with social and spiritual rituals for centuries. Furthermore, a broad patent of that nature would discourage other investigators from pursuing further research.

DuPont and Pioneer filed many more patent applications based on false inventions, such as isolating genes and claiming gene sequences as inventions even though they are not created by human ingenuity.

EPO Reconsiders DuPont Patent on Maize

Faced with a broad alliance of protestors from a broad alliance of groups, the EPO was obliged to reconsider if the DuPont patent (EP 744888) issued in August 2000 is valid in view of the objections submitted by the Mexican government, church groups, and Greenpeace. The EPO patent (EP 744888) currently covers naturally and traditionally cultivated maize with higher oil content. DuPont claims patent rights on all products derived from this maize.

Greenpeace's expert on patents, Ulrike Brendel, commented:

> This is daylight robbery and the EPO is helping in the theft. Obviously this maize is not a DuPont invention, but that is what they have a nerve to claim in order to rob the people in Latin America who have cultivated and planted maize seeds for thousands of years. We ask the EPO to halt this patent right away and recognize that the current practice of granting them clearly goes too far. (Greenpeace International, press release, Amsterdam, February 12, 2003)

So far, DuPont has applied for more than 250 patents on seeds in Europe alone, and some 40 have been granted. Altogether, the EPO has granted more than 300 patents on plants and seeds. However, this EPO practice is not supported by the new European Union (EU) patent directive, which contradicts the legal foundation of the EPO. The majority of EU member states are still undecided about the course they should follow.

Syngenta's Rice Monopolies

Syngenta, the giant global corporation of agribusiness, is famous for inventing the so-called "Golden Rice", which was specially designed to overcome malnutrition, especially vitamin A deficiency. Recently, Syngenta filed a far-reaching patent application for genetically engineered "Golden Rice", including far-reaching global patent applications on major parts of the rice genome and other crop plants.

Recently, a new generation of "Golden Rice" was developed by Syngenta that emphasized the humanitarian nature of the project in the following statement: "Syngenta has no commercial interest in Golden Rice. . . . Golden Rice2 transgenic events will be donated for further research and development through license under certain conditions." However, Syngenta's worldwide patent (WO 04/085656) was filed in March 2004 under the international Patent Cooperation Treaty (PCT), assuring worldwide windfall profits for Syngenta. Indeed, the entire project was driven by commercial interests.

Syngenta is also the largest producer of pesticides in the world and the third-biggest seed company, with sales of US$7.1 billion in 2004. Syngenta holds more than 65 patents on seeds and plant breeding in Europe.

Genome Monopoly

Syngenta's project to analyze the rice genome was completed in 2001, but the full data were only published in 2004. Simultaneously, a team from China published similar data. Syngenta applied for extremely broad patent monopoly claims on the genome of rice plants that would not only restrict further research by others, but would also pose a serious threat to farmers and breeders. In addition, several world patent applications were filed for more than 100 countries, including North and South America, Europe, and many African and Asian countries. In a whole range of patent applications, three huge applications cover over 1,000 gene sequences, most of them from the rice and corn genomes. The data collection itself was based on previous research by many others, and the technical methods were already published and well-established. The sequences themselves cannot be said to have been invented by Syngenta because they occur naturally. So, neither the methods nor the genetic information obtained meet the criteria required for patenting.

The applications for the three major gene sequences that were filed by Syngenta in 2002 are (1) WO 03/008540, for gene sequences for plants under stress; (2) WO 03/000905, for gene sequences concerned with the nutritional value of grains; and (3) WO 03/000906, for gene sequences responsible for defense mechanisms against plant diseases. Related claims involving several other plant species were also filed because they can be genetically engineered by inserting one or more of the gene sequences mentioned in the previous paragraphs. The most important crops mentioned are rice, corn, wheat, and other cereals. Ultimately, Syngenta's patents involve most of the major crops. The major points of their claims appear to be (1) sweeping claims of functions of gene sequences without adequate evidence; (2) a monopoly on all similar gene sequences in several crop species; and (3) sweeping claims of genes in not only genetically engineered crop plants, but also those obtained by conventional breeding methods.

Syngenta extended its claims worldwide by world patent application WO 03/048319. It includes the database for the rice genome that was obtained in collaboration with the American company, Myriad.

Major multinationals such as DuPont, Monsanto, and Syngenta all have something in common: they were in the business of chemicals, but later switched to agribusiness. Greenpeace, Swissaid, and the Berne Declaration have demanded for no patents on life, for a worldwide ban on patents on seeds, and for Syngenta to drop its worldwide patent applications on parts of the genome of food crops.

> The first strains developed had only 1.6 micrograms of beta-carotene per gram of rice, which would mean that a person would have to eat 1.5–2 kg of the rice per day to get the recommended daily allowance of provitamin A. With this apparently solved by the development of lines with increased beta carotene the other objections are still standing. Greenpeace, for instance, opposes all genetically modified organisms, and is concerned that golden rice is a Trojan horse that will open the door to more widespread use of GMOs. (Shiva 2001, 21)

Vandana Shiva (2001), an Indian anti-GMO activist, argued that the problem was not particular deficiencies in the crops themselves, but rather problems of poverty and loss of biodiversity in food crops. These problems are aggravated by the corporate control of agriculture based on genetically modified foods. By focusing on a narrow problem (vitamin A deficiency), Shiva argued, the "golden rice" proponents were obscuring the larger issue of a lack of broad availability of diverse and nutritionally adequate sources of food.

Similarly, other groups have argued that a varied diet containing vitamin–A–rich foods like sweet potato, leafy green vegetables, and fruit would provide children with sufficient vitamin A. While this is true, others also contend that a varied diet is beyond the means of many of the poor, which they say is why they subsist on a diet of rice. To suggest a more varied diet for the poorest to combat micronutrient deficiencies is the present-day equivalent of saying "Let them eat cake!"

The aleurone layer that surrounds the rice endosperm is removed by a process called milling or polishing in most countries to improve the shelf life of the rice. Brown rice with the aleurone intact contains more vitamin B, iron, manganese, selenium, zinc, and phosphorus than milled rice. The Institute of Science in Society claims that if rice was not milled, then supplementation would not be necessary. However, United States Department of Agriculture (USDA) data shows that brown rice does not contain any more beta-carotene than milled rice. Scientists at the International Rice Research Institute are screening rice germplasm, and are trying to use conventional breeding approaches for breeding varieties with increased beta-carotene in the aleurone.

Wheat

The baking properties of Indian Nap Hal wheat make it particularly suited to producing crisp bakery products such as biscuits. About 20% of Nap Hal plants are homozygotic for the responsible genes. These plants were developed through the application of natural breeding methods over the centuries. They are the result of the labors of breeders and farmers in India, who originally grew these plants for their own regional requirements.

Now, Monsanto holds a monopoly on the farming, breeding, and processing of this type of wheat. A patent on traditional wheat varieties from India with a special baking quality was granted for Monsanto in 13 European countries as well as for Japan, Australia, Canada, and the US.

As it is natural for farmers to freely swap seeds, it comes as no surprise that this wheat seed has been stored in various international gene banks outside India for many years. Thus, samples of the seed can be found in the collections held by the US agricultural administration as well as in Japan and Europe.

Monsanto's patent was filed as a world patent application in 1991 from Unilever (WO 91/11905). Monsanto bought the wheat research department from Unilever in 1998. After that, the patent was granted in the US in 1999 and in Europe in 2003 as EP 445929.

In 2004, a Greenpeace press release crowed, "Monsanto's wheat patent withdrawn in Europe following Greenpeace opposition." It continued, "In a clear victory for Greenpeace and Indian farmers, European Patent office (EPO) on 23rd September revoked the patent on Indian 'Nap Hal' wheat variety following a legal opposition filed by Greenpeace at the EPO in February" (Ronald Bailey, Greenpeace declares victory against wheat patent, *Reason Online*, December 8, 2004).

However, new crop varieties are treated differently. According to Jeff Rodwell, an intellectual property rights lawyer in London with the firm Reed Smith, new plant varieties are not patented, but given "plant variety rights" (PVRs). Patents are awarded for novel inventions, whereas PVRs are awarded for the creation of a new strain of plant exhibiting attributes not hitherto seen in a created thing. The genes in the new wheat variety were certainly not invented, but they were put together in novel combinations never before seen in nature. (This can be done by traditional cross-breeding as well as by isolating and inserting specific genes directly to create a new variety.) However, for brevity's sake, let's just call them patents.

Biopiracy

The justification for the patent as an invention is questionable. Companies like Monsanto and Unilever have unrestricted access to many seed banks. They were aided by the research results of various scientists who identified the corresponding gene regions for quite some time. These natural combinations of genes have been patented by Monsanto as an "invention", when in fact the patents are a monopoly on Nap Hal plants themselves and on all wheat plants and derivatives which have been crossed with the Indian varieties. It involves the theft of the results of the breeding efforts of Indian farmers, those small farmers who make a decisive contribution to agricultural diversity and ensure sufficient food supplies by freely swapping seeds and breeding various forms of crops.

Patent legislation ignores the need for cultural and ethnic diversity in the creation of innovations, the sharing of benefits derived from such activities, and the needs and concerns of the farmers of developing countries. It can be correctly described as the silent theft by developed countries of knowledge which has been amassed over the centuries in developing countries (United Nations Development Program (UNDP) 1999).

Farmers' Organizations

Greenpeace joined hands with one of the biggest farmers' organizations in India. The Supreme Court of India ruled against the patent. The European farmers' organization COPA also filed an opposition. After the patent was challenged in 2004, Monsanto sold it to a French seed company, R.A.G.T, together with other parts of the European wheat business.

History of Wheat Production in India

The production and productivity of Wheat crop were quite low, when India became independent in 1947. The production of Wheat was only 6.46 million tonnes and productivity was merely 663 kg per hectare during 1950–51, which was not sufficient to feed the Indian population. The Country used to import Wheat in large quantities for fulfilling the needs of our people from many countries like USA under PL-480. The reasons of low production and productivity of Wheat at that time was (a) the tall growing plant habit resulting in lodging, when grown under fertile soils, (b) the poor tillering and low sink capacity of the varieties used, (c) higher susceptibility to diseases, (d) the higher sensitivity to thermo & photo variations, etc., resulting in poor adaptability, and (e) longer crop duration resulting in a long exposure of plants to the climatic variations and insect pest/disease attacks.

The Government of India appointed a commission in 1961 to assess the feasibility of increasing the crop productivity under prevailing Indian ecological conditions. The Commission consisted of Dr. M.S. Swaminathan, Dr. N.E. Borlaug and many others and concluded

(*Continued*)

(Continued)

that production level of Wheat could be increased, if suitable and superior germ-plasm/varieties were available in the country. The discovery of genes responsible for dwarfing and non-lodging habits in 'Norin' Wheat varieties of Japan opened the doors to evolve high yielding varieties of Wheat. The dwarf Wheats, besides having stiffer and shorter straw, were relatively photo-insensitive and were capable of giving high yields at high doses of fertilisers, irrigation and other inputs. The Harvest Index (i.e. grain:straw ratio) was also more favourable in terms of grain production.

After assessing the possibility of increasing the Wheat production in India, Wheat scientists introduced five dwarf Wheat varieties, *viz*, Lerma Rojo 64-A, Sonora 63, Sonora 64, Mayo 64 and S 227 alongwith around 200 other breeding lines of dwarf Wheat through the courtesy of Rockefeller Foundation and Mexican Ministry of Agriculture in 1963. These varieties were extensively tested in all the Wheat growing states of the country and it was concluded that the varieties such as 'Lerma Rojo 64-A', 'Sonora 64' and 'PV 18' developed in Mexico and carrying the genes for dwarfism having [*sic*] a high potential for yield in our country too. These findings lead [*sic*] the Government of India to undertake massive import of 18,000 tonnes seeds of 'Lerma Rojo 64-A' and to some extent of 'Sonora 64' in 1966 for planting nearly 4 lakh hectares area in the country. As a result of this, a major breakthrough in Wheat productivity and production started to be visualized. These dwarf varieties were high yielding and disease and lodging resistant. But some how the farmers and consumers were not very much satisfied with the deep red colour of grains of these varieties and were reluctant to accept them. Then this lead [*sic*] to selection of many promising lines such as S 227, S 308, S 307, etc. from the advance breeding material received from Mexico and these selected varieties/lines were found to have amber grain colour with a very high yield potential and good degree of lodging and disease resistance. With the identification/development of amber or white seeded genotypes like Kalyan Sona, Sonalika, Safed Lerma and Chhoti Lerma in 1967, the 'Wheat Revolution' in India got

(*Continued*)

(Continued)

a real momentum. Out of these, Kalyan Sona and Sonalika varieties became very much popular among the farmers because of their high yield, rust resistance, amber grains and adaptability to different soil and climatic conditions of the country and occupied nearly 10 million hectare area in the country and made the 'Wheat Revolution' happen.

Subsequently, a number of dwarf Wheat varieties like Sarbati Sonara, Pusa Lerma, Arjun, Pratap, Janak, Mukta, Shera, etc. were evolved and released for general cultivation in the country. For popularizing these varieties, the Government of India had started High Yielding Varieties Programme (HYVP) in Wheat during 1966–67 with a humble coverage of 0.54 million hectares (4.2%) of the total area of 12.8 million hectares. It has slowly and steadily gained the strength over next 10 years and the area under HYVs reached to 12.5 million hectare (62%) of area under Wheat. Thereafter, large number of high yielding, input responsive and disease resistance varieties of Wheat were released for various ecological and growing conditions of the country. The important Wheat varieties, which have become very popular include C-306, UP-262, HD-2009, WL-711, HP-1102, HUW-206, HUW-234, HD- 2189, HD-2204, HD-2285, VL-616, VL-421, HS-42, WH-147, WH-157, WH-542, HD-2329, UP-2003, UP-2338, LOK-1, RAJ-1555, RAJ-3765, RAJ-3077, UP-2425, PBW-154, PBW-343, PBW-443, PBW-373, HI-8381, HI-8498, HD-2687, KRL-19, HUW-468, GW-273, etc.

Several policy decisions and actions were taken by Government of India from time to time to increase production and productivity in the country. The Ministry of Agriculture, Govt. of India launched and implemented various Centrally Sponsored/Central Sector Schemes, namely, IADP (1960–61), IAAP (1964–65), High Yielding Varieties Programme (1966–67), Wheat Minikit Demonstration Programme (1974–75), SFPP-Wheat (1988–89) and ICDP-Wheat (1994–95). Prior declaration of Minimum Support Price, construction of large grain handling facilities to procure the surplus grain and several other market promotional steps, establishment of seed production chains, fertilizer factories and farm machinery units, increased public

(*Continued*)

(Continued)

investments in irrigation and agricultural research by Govt. of India created an environment for an all round production increase of Wheat crop in the country.

Currently, India is second largest producer of Wheat in the world after China with about 12% share in total world Wheat production. India has been exporting Wheat in the International Market, earning foreign exchange.

Biopiracy

Wheat the Golden grain, is called "Kanak" in North Western India. It is the staple of a large majority. Wheat diversity has been evolved by Indian farmers over millennia for taste, for nutrition, for ecological adaptation to cold climates and hot climates, dry regions and wet regions.

The present condition of Indian agriculture is the heritage of experience handed down from time immemorial by a people little affected by the many changes in the government of the country. The present agricultural practices of India are worthy of respect, however strange and primitive they may appear to Western ideas. The attempt to improve Indian agriculture on Western lines appears to be a fundamental mistake. What is wanted is rather the application of Western scientific methods to the local conditions so as to improve Indian agriculture on its own lines.

Millennia of breeding by millions of Indian farmers is however now being hijacked by Monsanto which is claiming to have "invented" the unique low-elasticity, low gluten properties of an indigenous Indian wheat, rice lines derived from such wheat and all flours, batters, biscuits and edible products made from such wheat.

On 21st May, 2003, the European Patent Office in Munich granted a patent to Monsanto with the number EP 445929, with the simple title "plants", even though plants are not patentable in European Law. The patent covers wheat exhibiting a special baking quality, derived from native Indian wheat. With the patent, Monsanto holds a monopoly

(Continued)

(Continued)

on the farming, breeding, and processing of a range of wheat varieties with low elasticity. Earlier in a patent (EP 518577) filed in 1998 Unilever and Monsanto have claimed "invention" of an exclusive claim to the use of flour to make traditional kinds of Indian bread such as "chapattis".

And it is not just in Europe that Monsanto has filed and obtained patents based on the biopiracy of Indian wheat. In the U.S on May 3, 1994 patent number 5,308,635 was given for low elasticity wheat flour blends, on June 9, 1998 patent number 5,763,741 was given for wheat which produce dough with low elasticity, and on January 12, 1999, patent number 5,859,315 another patent was granted for wheats which produce dough with low elasticity.

Monsanto's wheat biopiracy has been challenged in the Indian Supreme Court, and in the European Patent Office in Munich by Greenpeace. The following is a summary of the challenge submitted to the EPO on 17th February, 2004.

> The patent is a blatant example of biopiracy as it is tantamount to the theft of the results of endeavours in cultivation made by Indian farmers. In the countries of the southern hemisphere, it is frequently the small farmers who make a decisive contribution to agricultural diversity and secure sufficient food supplies by freely swapping seeds and breeding regionally modified forms of crops.
>
> Monsanto is now unscrupulously exploiting the fruits of their labour. The company is able to restrict not only the farming and processing of crops, but also trade in them, in the countries for which the patent has been granted. At the same time it can block the free exchange of the seed, thus preventing other growers and farmers from working with the patented seeds.
>
> The wheat exhibiting these special baking qualities is the result of the labours of cultivators and farmers in India who originally grew these plants for their own regional requirements, growing them to bake traditional Indian bread (chapatis). As it

(*Continued*)

(Continued)

is natural for these farmers to freely swap seeds, it comes as no surprise that this wheat seed has been stored in various international gene banks outside India for many years.

The traits of low elasticity, low gluten which are being patented are not an invention, but derived from an Indian variety. The crossing with a soft milling variety is an obvious step to any breeder. The patent is based on piracy, not on non-obvious novelty, and hence needs to be challenged to stop legal precedence being created on false claims to invention.

The broad scope of the patent covering products made with Indian wheat robs Indian food processes and biscuit manufacturers of their legitimate export market and could in future affect our domestic food sovereignty. The Government's 2020 vision refers to making India a "global food factory".

With an estimated annual turnover of US$1.5 billion, the baking industry in India is one of the largest manufacturing sectors in India, production of which has been increasing steadily in the country. The two major bakery industries, *viz*. bread and biscuit account for about 82 percent of the total bakery products; with overall annual growth estimated at 6.9%. According to ASSOCHAM India, a business support services firm, there are almost 85,000 bakeries in the country. Approximately 75,000 of these operate in the unorganised sector, which has a 60% market share. The remaining 1,000 bakeries operate in the organised sector, which has a 40% market share.

Packaged Food in India, a recently released report from Euromonitor, recorded year 2000 volume sales of the organised biscuit sector at 500,000 MT, or approximately US$492 million in value terms. The unorganised sector, which supplies 60% of total production, has an annual turnover of nearly US$718 million. If combined, the two sectors would bring overall biscuit sales to more than US$1.2 billion annually, or 1.3 MMT, making India the world's second largest biscuit manufacturer and consumer behind the US.

(*Continued*)

(Continued)

Further, the patent covers not just biscuits but all edible products and flours with low elasticity. Indian *chapattis* are in effect covered by the patent. If such biopiracy based patents are not challenged, crop lines and products based on unique properties evolved through centuries of indigenous breeding become the monopoly of multinational corporations (MNCs).

Directorate of Wheat Development, India.

Soybean

China is home to more than 6000 wild soya varieties, over 90% of the global total. Soybeans (*Glycine max* L.) are a multibillion dollar commodity crop, particularly important for oil and animal feed, but also used in numerous industrial products. The multinational agrochemical corporation Monsanto — which also owns the world's most widely grown genetically engineered (GE) crop, herbicide-resistant Roundup Ready Soya — has been trying to obtain exclusive rights over soya genetic resources through a broad species patent.

Monsanto's Patent Application

Monsanto's worldwide soybean patent application (WO/0018963) was published on 6th April 2000, under the title "Methods for breeding for screening of soybean plants with enhanced yields." Monsanto's soya germplasm came from the US National germplasma collection, although its primary country of origin was China. Some of the molecular markers could also be found in other wild soya found in Russia and China, and domesticated soya varieties found in Japan. Monsanto's objective was to breed soya varieties using the molecular markers in order to obtain elite soya plants with enhanced yield characteristics. Furthermore, the option to introduce these markers into other plants through genetic engineering was also foreseen.

The patent claims included the following:

- the markers associated with enhanced yield;
- all domesticated and wild plants with these markers, as well as their progeny;
- the methods employed to locate these markers;
- the plant breeding methods employed for the production of elite varieties; and
- the breeding method for the production of enhanced-yield *Glycine max* plant and many other plant species where the marker can be introduced.

Biopiracy

- The very broad nature of the patent would restrict the research, breeding, and use of soya, and would have adverse effects on scientific research and food security.
- Soya farmers would be prohibited to exchange seeds with other farmers or conduct any further breeding.
- Importing Chinese soya with these markers could be a patent infringement.
- According to Prof. Carlo Leifert of the Department of Agriculture, Newcastle University, UK: "This patent would restrict molecular marker assisted breeding in soybean for enhanced yield once the Monsanto got the patent. . . . How severely could only be assessed once it was known what physiological trait the marker is linked to" (Christophe Then, Biopiracy and seeds, Greenpeace, December 2007).

After this patent and the issues involved were made public by Greenpeace in China and Europe, the application is no longer followed by Monsanto.

Primates

Of the world's 394 primate species, 114 are classified as threatened with extinction by the World Conservation Union. The report by Conservation International and the International Primatological Society in Hainan, China, focuses on the plight of the 25 most endangered primates, including China's Hainan gibbon, of which only 17 remain.

Habitat loss due to the clearing of tropical forests for agriculture, logging, and fuel wood continues to be the major factor in the declining number of primates, according to the report. In addition, climate change is altering the habitats of many species, leaving those with small habitat ranges even more vulnerable to extinction, it says. Hunting for subsistence and commercial purposes is another major threat to primates, especially in Africa and Asia. Capture of live animals for the pet trade also poses a serious threat, particularly in Asia, the report found.

According to Russell A. Mittermeier, President of Conservation International, "You could fit all the surviving members of the 25 species in a single football stadium; that's how few of them remain on Earth today."

"The situation is worst in Asia, where tropical forest destruction and the hunting and trading of monkeys puts many species at terrible risk," said Mittermeier, who is also chairman of the World Conservation Union's Primate Specialist Group, which prepared the report with the International Primatological Society.

The 25 most endangered primates include 11 from Asia, 11 from Africa, and 3 from South and Central America. The list includes well-known primates like the Sumatran orangutan of Indonesia and the Cross River gorilla of Cameroon and Nigeria, as well as lesser known species such as the greater bamboo lemur from Madagascar.

Six species are in the report for the first time, including a recently discovered Indonesian tarsier that has yet to be formally named and the kipunji from Tanzania, which was discovered in 2003. "Some of the new species we discover are endangered from the get go," Mittermeier said. "If you find a new species and it's living in an area heavily impacted by habitat destruction and hunting, you recognize it's in trouble."

(*Continued*)

(Continued)

Four primates on the list from Vietnam have been decimated by hunting for their meat and bones, according to Barney Long, a conservation biologist based in Vietnam for the WWF Greater Mekong Program. "All four species are close to extinction," Long said of the Delacour's langur, golden-headed langur, grey-shanked douc, and Tonkin snub-nosed monkey. "The key populations have been stabilized. But there needs to be a lot more law enforcement and work to persuade local communities to support conservation for those numbers to increase."

Nine primates from the last report in 2004 were taken off the list, mostly because of bolstered conservation efforts to save their populations. Among them are the eastern gorilla from Africa, the black-faced lion tamarin and the buffy-headed tufted capuchin from Brazil, and the Perrier's sifaka from Madagascar. Mittermeier commented, "If you invest in a species in a proper way and do the conservation measures needed, you can reduce risk of extinction. If we had resources, we would be able to take every one of the species off the list in the next five or 10 years."

Primates in danger of extinction. Yahoo News, October 26, 2007.

Impact of Intellectual Property Rights (IPRs) on Agricultural Biodiversity

Modern agricultural biotechnology depends on the use of several processes and products which are likely to be patentable subjects. For instance, the development of genetically modified (GM) crops involves individual genes affecting such characters as disease resistance and herbicide tolerance. The patents may also involve DNA sequences which control the expression of these genes (e.g. promoters). In addition, patents may include methods for transferring foreign DNA or methods for identifying the recipient cells which have successfully incorporated the foreign genes. It is clear that several patents are involved in the development of even the simplest steps in genetic modification. Another complication is that the patent holder of the GM plant may be required to share the revenues with the developers of the original plant variety.

Golden Rice

One of the most famous failures of a GM crop is perhaps the case of "golden rice", which involved several complicated patenting processes. Three foreign genes were introduced into a rice plant to modify it to produce provitamin A; two of these genes came from the daffodil, and one came from a bacterium. Scientists from Switzerland and Germany collaborated to create the new variety, which was meant for farms in the developing countries. The patenting process was quite complicated, involving 70 patents and 32 owners.

However, "golden rice" turned out to be a famous failure. Critics have listed several reasons for abandoning "golden rice".

"Golden rice' exhibits all the undesirable, hazardous characteristics of existing GM plants, and in added measure on account of the increased complexity of the constructs and the sources of genetic material used. The hazards are highlighted below.

- It is made with a combination of genes and genetic material from viruses and bacteria, associated with diseases in plants, and from other non-food species.
- The gene constructs are new, and have never existed in billions of years of evolution.
- Unpredictable by-products have been generated due to random gene insertion and functional interaction with host genes, which will differ from one plant to another.
- Over-expression of transgenes linked to viral promoters, such as that from CaMV, exacerbates unintended metabolic effects as well as instability (see below). There are at least two CaMV promoters in each transgenic plant of the 'golden rice', one of which is linked to the antibiotic resistance marker gene.
- The transgenic DNA is structurally unstable, leading to instability of the GM plants in subsequent generations, multiplying unintended, random effects.

(*Continued*)

(Continued)

- Structural instability of transgenic DNA increases the likelihood of horizontal gene transfer and recombination.
- Instability of transgenic DNA is enhanced by the CaMV promoter, which has a recombination hotspot, thereby further increasing the potential for horizontal gene transfer.
- The CaMV promoter is promiscuous in function and works efficiently in all plants, in green algae, yeast and *E. coli*. The spread of genes linked to this promoter by ordinary cross-pollination or by horizontal gene transfer will have enormous impacts on health and biodiversity. In particular, the hygromycin resistance gene linked to it may be able to function in bacteria associated with infectious diseases.
- Horizontal transfer of transgenic DNA from GM plants into soil fungi and bacteria has been demonstrated in laboratory experiments. Recent evidence suggests that it has also taken place in a field-trial site for GM sugar-beets, in which transgenic DNA persisted in the soil for at least two years afterwards.
- Prof. Hans-Hinrich Kaatz from the University of Jena, has just presented new evidence of horizontal gene transfer within the gut of bee larvae. Pollen from GM rapeseed tolerant to the herbicide glufosinate were fed to immature bee larvae. When the microorganisms were isolated from the gut of the larvae and examined for the presence of the gene conferring glufosinate resistance, it was found in some of the bacteria as well as yeast cells.
- All cells including those of human beings are now known to take up genetic material. While natural (unmanipulated) genetic material is simply broken down to supply energy, invasive pieces of genetic material may jump into the genome to mutate genes. Some insertions of foreign genetic material may also be associated with cancer.
- Horizontal transfer of genes and constructs from the 'golden rice' will spread transgenes, including antibiotic resistance genes to bacterial pathogens, and also has the potential to create new viruses and bacteria associated with diseases.

Golden Rice Fact Sheet. Golden Rice Humanitarian Board, International Rice Research Institute, The Philippines, November 2005.

5

Patenting Life

Patenting in the life sciences is a relatively recent phenomenon. Some of the greatest inventions of the past that have benefited humanity have not been patented. One outstanding example is the process called "pasteurization", which Louis Pasteur refused to patent. The process was invented by Pasteur, who discovered that spoilage organisms could be inactivated in wine by applying heat at temperatures below its boiling point; the process was later applied to milk, and remains the most important operation in the processing of milk. Numerous other examples in the life sciences can be cited. For instance, several important methods in genetics were not patented, such as gene mapping in Drosophila by A.H. Sturtevant, or the methods to estimate mutation rates of human genes and early steps in human genome mapping by J.B.S. Haldane. The measurement of genetic load in human populations, which played an important part in assessing the damage resulting from ionizing radiation due to the open-air testing of nuclear bombs, was not patented by Haldane or H.J. Muller. Without such early pioneering inventions, medical genetics would not exist today.

Patenting life forms for commercial purposes received a major boost in 1980 when the patent claim filed by an Indian scientist working for General Electric, Ananda Chakrabarty, was granted by the U.S. Supreme Court. The patent involved a genetically modified bacterium, *Pseudomonas aeruginosa*, which was

Fig. 1. Photo of Dr Ananda Chakrabarty, taken about 1979 when he was working at the General Electric Research and Development Center in Schenectady, New York. He invented a genetically engineered bacterium, *Pseudomonoas aeruginosa*, to break down multiple compounds of crude oil. It was the first example of a patent for a man-made microorganism, which was approved in 1980 by the U.S. Supreme Court in a 5-4 decision in the case of Chakrabarty v. Diamond. It provided a great economic stimulus to the patenting of microorganisms and cells, laying the foundation for the growth of a huge and successful biotechnology industry. Reproduced with the kind permission of Ananda Chakrabarty.

specially designed to break up oil spills in oceans and rivers. The Supreme Court stated that "anything under the sun that is made by man," including living organisms, could be patented. That single step opened the door for an avalanche of patent claims for living organisms, tissues, and cells and various related processes, launching the hugely successful field of biotechnology.

Another example is the so-called oncomouse. In 1988, the United States Patent and Trademark Office (USPTO) granted U.S. Patent 4,736,866 to Harvard College, who claimed "a transgenic non-human mammal whose germ cells and somatic cells contain a re-combinant activated oncogene sequence introduced into said mammal" (U.S. Patent Office, Washington, D.C., 1988). The claim explicitly excluded humans, apparently reflecting moral and legal concerns about patents on human beings and about modification of the human genome. Two separate patents were issued to Harvard College covering methods for providing a cell culture from a transgenic nonhuman animal (U.S. 5,087,571) and testing methods using transgenic mice expressing an oncogene (U.S. 5,925,803). The latter patent was to expire in July 2016; however, in 2002, the Supreme Court of Canada rejected the patent in Harvard College vs. Canada (Commissioner of Patents), overturning a Federal Court of Appeal verdict which ruled in favor of the patent. The patent was amended to omit the "composition of matter" claims on the transgenic mice. The Supreme Court had rejected the entire patent application on the basis of these claims, but Canadian patent law allowed the amended claims to be granted under pre-GATT rules; and the patent remains valid until 2020.

In order to derive maximum economic return from the substantial federal investment in basic research, the U.S. Congress enacted the Bayh–Dole Act in 1980, encouraging universities to patent and commercialize inventions arising out of research supported by Government grants. In the following years, federal grants for biomedical research dramatically increased, and the number of patents assigned to universities increased from 264 in 1979 to 3,259 in 2003.

In 1982, the U.S. Congress created a national appeals court with exclusive jurisdiction for patent cases, which brought some uniformity to the patenting process. Federal Circuit decisions have largely strengthened the rights of patent holders. Its impact has been more acutely felt in biotechnology than in any other field. During the last 25 years, numerous patents have been issued on biotechnology inventions.

The coalescence of these events greatly stimulated an explosive growth in the U.S. biotechnology industry. The number of Food and Drug Administration (FDA) approvals for biotech-derived products increased enormously, and the number of biotechnology companies increased from 225 in 1977 to 1,457 in 2001.

European patent application 85304490.7 for oncomouse was filed in June 1985. It was initially refused in 1989 by an examination division of the European Patent Office, among other things, on the grounds that the European Patent Convention (EPC) excludes patentability of animals per se. The decision was appealed; and the Board of Appeal held that *animal varieties* were excluded of patentability by the EPC [and especially its Article 53(b)], but *animals* (as such) were not excluded from patentability. The examination division then granted the limited patent in 1992.

Beginning under such controversial circumstances, commercial biotechnology has come a long way. At its worst, it has encouraged biopiracy, exploitation, and destruction of the natural resources of the developing countries, most of which are too poor and underdeveloped to possess the know-how and resources to develop their biological wealth. Although patenting has clearly encouraged innovation and development by industries and individuals, it is also realized that it has gone too far, beyond ethical boundaries, as in the case of so-called gene patenting which merely involves discovering certain genes in our bodies. Several years ago, biologist and DNA pioneer James Watson expressed his protest by resigning from his position at the National Institutes of Health.

No one has voiced this point of view more eloquently than the author Micheal Crichton (2007):

> Humans share mostly the same genes. The same genes are found in other animals as well. Our genetic makeup represents the common heritage of all life on earth. You can't patent snow, eagles or gravity, and you shouldn't be able to patent genes, either. Yet by now one-fifth of the genes in your body are privately owned. The results have been disastrous. Ordinarily, we imagine patents promote innovation, but that's because most patents are granted for human inventions. Genes aren't human inventions, they are features of the natural world. As a result these patents can be used to block innovation, and hurt patient care.
>
> Why should people or companies own a disease in the first place? They didn't invent it. Yet today, more than 20 human pathogens are privately owned, including haemophilus influenza and Hepatitis C. And we've already mentioned that tests for the BRCA genes for breast cancer cost $3,000.

The following is an unofficial translation of a Greenpeace Germany press release, dated July 9, 2007:

Who Owns Sunflower and Broccoli?

European Patent Office Is About to Prepare a Far Reaching Decision

Hamburg/Munich, July 9, 2007 — Greenpeace Germany filed opposition against a patent on a sunflower variety. The US company Pioneer received a patent (EP 1465 475 B1) on sunflowers derived from conventional breeding (without genetic engineering) in October 2006. The plants are resistant against a certain parasite in the soil which can damage the roots. Parallel to this new opposition procedure the European Patent Office (EPO, based in Munich, Germany) is preparing a far reaching general decision on the issue of patents on normal plants and animals.

"The European Patent Office has eroded systematically many legal borders of patentability in the last years," says Greenpeace expert Christoph Then. "It is an alarming signal that this office which is financed completely by fees from industry, will now even decide on this very fundamental issue."

In June 2007 the EPO asked its highest legal institution, the so called Enlarged Board of Appeal, to make a decision to what extent essentially biological processes for the breeding of plants and animals are excluded from patentability. In the year 2000 the Enlarged Board already decided that genetically engineered plants can be subjected to patents. Since that hundreds of patents were granted in this field. In year 2002 the EPO granted a patent to a company in the Netherlands on "normal" broccoli (a vegetable) without genetic engineering. Against this patent several oppositions were filed by seed companies because patents on essentially biological processes for the breeding of plants and animals are not allowed according to European Patent law.

"If normal plants such as broccoli or sunflowers are declared to be inventions, in future nearly each and every plant or animal can be

(*Continued*)

(Continued)

patented. The multinationals will then try to gain complete control on the whole chain of food production by actions from their patent attorneys," warns Greenpeace. "Even patent application such as from Monsanto on normal pigs then hardly can be prevented from being granted."

In the last months more than 30 farmers organisations worldwide have joined a coalition against patents on seeds (www.no-patents-on-seeds.org). These organisations will now mobilise public and politicians against patents on seeds and clear exclusion from patents on living beings in international patent laws.

TRIPs and WTO

Patenting* and intellectual property rights have serious implications beyond national boundaries. They impact on all aspects of our lives and culture, but especially on medicine and agriculture and on our ethical and religious outlook. Internationally, they have become a major policy issue in economic and trade relations, a major landmark being the controversial negotiations leading to the agreement on "Trade-Related Aspects of Intellectual Property Rights (TRIPs)", an outcome of the Uruguay Round of multilateral trade negotiations in 1994. TRIPs is now a part of the legal obligations of the World Trade Organization (WTO).

In the 1980s and the '90s, patenting became so widespread that there seemed to be no limits to what is patentable, especially in the life sciences. Human gene patenting seemed to have ignored one of the fundamental rules of patenting; "created by human ingenuity". Genes have been patented although they occur naturally inside the cells of individuals. They were neither created by human ingenuity nor their function or applications known at the time of patenting. There were other problems. Claims overlapped, several applicants attempting to claim priority for the same gene, and in some cases the patent claim was so broad that it bordered on ridiculous! The claim

(*Continued*)

(Continued)

for a gene therapy patent included all and any insertion of foreign DNA into a host's cells (Patent No. 5,399,346, filed in 1994, by W.F. Anderson, R. Michael Blaese, and Steven Rosenberg). "This invention relates to the use of primary human cells as vehicles for human gene transfer. More particularly, this invention relates to the use of human cells (such as, for example, but not limited to, human blood cells) as vehicles for the transfer of human genes encoding therapeutic agents and/or genes encoding detectable markers." Such broad patents are challenged, often successfully.

The Battle Over Patent Legislation: TRIPs and the CBD

The question of whether plant and animal (and human) genetic materials can and should be patented, or regulated by some other type of intellectual property regime, will be hotly debated over the next few years, in fields and forests, law courts, boardrooms, academic and intergovernmental fora. Clear and internationally agreed principles and definitions should be established, in the light of current technological possibilities, globalisation, pressures from corporations, and global welfare and conservation goals.

It will not be easy to achieve international agreement on issues of such importance, in which several different international organisations are involved. *Much current debate focuses on the roles of two of these, the Convention on Biological Diversity (CBD) and the World Trade Organisation's International Agreement on Trade-Related Aspects of Intellectual Property Rights (TRIPs). Many analysts believe these two agreements contradict each other.*

The CBD, which came into force in 1993, states some basic principles. Genetic material found within a country is owned by that country (as opposed to the previous position, that natural substances are a global common heritage, owned by nobody). Benefits from genetic materials and biotechnology should be shared — with the communities where the substances are found or who know and use their properties, or with farmers who have developed valuable crop plant

(Continued)

(Continued)

varieties. Intellectual property regimes should be devised that do not conflict with the CBD's over-riding goals, the conservation and sustainable use of biodiversity.

There are many issues to be worked out before these general principles can be turned into effective laws, regulations and monitoring systems. For instance, how to define a 'community' that 'owns' a plant which may grow in several different areas and different countries? Or, what are the criteria for 'equitable sharing' of the benefits of commercial exploitation? Many research and academic bodies, NGOs and inter-governmental expert institutions are working on these problems.

The TRIPs agreement is part of the WTO, established in 1994. TRIPs article 27.3(b) requires all member countries to introduce legislation allowing patenting of microorganisms, biotechnological processes and products, and patents or some other type of protection (*sui generis,* or unique) for plant varieties. Developing countries have until 2000 to enact this legislation, least developed countries until 2005. There is some ambiguity and differing interpretations of these requirements. For instance, what sort of *sui generis* legislation will be acceptable to the WTO? Article 27.3(b) is due to be reviewed in 1999, in advance of a review of the whole of TRIPs in 2000. It is not clear what the scope of the reviews will be, and governments, NGOs and industry worldwide are preparing to promote their — widely differing — views during the negotiations.

Some critics of patenting are calling for Article 27.3(b) to be amended; they say the push for private ownership of biological resources is unethical, inequitable and contrary to the goals of the CBD. Others suggest that TRIPs as it stands already allows countries to exclude biological materials from patentability. The Canadian development organisation RAFI points out that under international patent law, governments can refuse to allow patents on any inventions which they believe are morally unjustifiable, using the *Ordre public* (Public Order) clause. This covers inventions which it is felt would be harmful to human, animal or plant life or health, or could cause serious prejudice to the environment. RAFI argues that patents

(*Continued*)

(Continued)

contrary to national morality include any claims "usurping national or indigenous knowledge", i.e. biopiracy. Meanwhile, patents on GM crop plants could be rejected on the grounds of being damaging to the environment and threatening national food security.

One of the fora struggling to debate the issues and formulate more detailed rules, definitions and mechanisms is the Commission on Genetic Resources for Food and Agriculture of the Food and Agriculture Organisation (FAO). Its current revision of the International Understanding on Plant Genetic Resources has already reached its fourth draft without agreement. On the whole the Understanding emphasises the need for national-level and farmer-based development of varieties (by conventional methods as well as biotechnology, though the Understanding does not articulate this). It proposes global monitoring of the acquisition and development of genetic materials. Areas of disagreement include a proposal to promote *sui generis* plant variety protection systems rather than patent law; and the whole chapter on farmers' rights (their right to benefit from their own breeds, and to save seed from year to year).

Representatives of the Organisation of African Unity (OAU) meeting in Addis Ababa, Ethiopia in March 1998 were determined that national Intellectual Property rules should support the CBD. They drew up draft legislation for OAU member states which seeks to regulate access by outside interests (such as foreign companies) to the biological resources of the community and ensures that any benefits resulting from the commercial exploitation are equitably shared.

On the other side, many northern governments, especially the US, under pressure from the agricultural and pharmaceutical industries, are pushing for a tightening up of the TRIPs agreement. *The US would prefer a uniform patents system and has been criticised for using heavy handed tactics to persuade other nations to enact laws favourable to its own commercial interests.* There are reports of several instances of American diplomats using threats of trade

(*Continued*)

(Continued)

penalties or cancelled cooperation agreements against countries including Thailand, Ethiopia, Panama, Paraguay, India and even one Northern country ally, Denmark. In other situations the persuasion is more gentle, but according to a report by the Gaia Foundation and GRAIN, "Developing countries are being told that patents and other forms of IPR are the key to attracting investment in biotechnology which will uplift their economies and improve food security. These claims are utterly false. The only motivation behind the global IPR campaign is to increase profits for transnational corporations housed in the North."

The European Union moved closer to the US position in 1998 when it adopted a new Directive on the Patenting of Biotechnological Inventions, despite opposition from many organisations and individuals, including the Green Party group within the Parliament and many NGOs, consumer and patient groups within Europe. Much of the debate focused on the ethics of allowing human DNA to be patented by companies working on new medical treatments for inherited diseases, but there were other important concerns, including the effects of the proposed system on farmers. In its final form the Directive, which aims to harmonise national patent law systems across Europe, aims to bring the scope of patents in the EU more or less into line with US practice, beyond what is required by the TRIPs agreement. The debate continues, with the Netherlands government challenging the Directive at the European Court of Justice, partly on the grounds that it conflicts with the existing European Patent Convention.

Underlying the technicalities of law and patent claims is a fundamental question: whether private ownership of materials for genetic engineering is necessary to drive innovation and feed the world, as the industry claims; or whether human welfare would be better served by keeping control over such basic issues as agricultural sustainability and food security in the public domain.

*There is a widespread misconception that patenting is concerned with only major advances in science and technology. Many patents are, in fact, concerned with incremental changes. From an academic point of view, it is surprising how trivial a patentable subject appears to be. In

(*Continued*)

(Continued)

a British survey, which was conducted by Packer and Webster (1996), scientists saw the advances in small gradual steps, not in grandiose terms of intellectual history. Many great discoveries in several sciences are not patentable. Examples are the discovery of a new species, the discovery of gravity by Newton or the theory of relativity by Einstein.

There is also the unaccustomed process for scientists that their work is not being judged by their scientific peers, but by lawyers who will determine the criteria by which a claim is judged. There are two different worlds: scientific research and patenting process, each with its own set of standards, concepts and rules. Furthermore, a research summary for a patent claim is written in legal terminology for an entirely different audience, not a scientific summary for scientific colleagues.

(World Trade Organization 1994)

U.S. Rejects Monsanto's Crop Patents

The U.S. Patent and Trademark Office has recently (August 2007) rejected four Monsanto patents in genetically engineered crops. The decision stopped short of revoking the patents entirely.

Joe Mendelsen, legal director at the Center for Food Safety, commented: "This calls into question Monsanto's behavior; the patent office has found significant flaws in what Monsanto has presented to them. If, indeed, these patents are invalidated, then Monsanto has essentially brought lawsuits against farmers throughout this country based on patents that aren't valid."

The Patent Office action rejected the claims Monsanto made in applying for the patents. Patent Office will re-examine a patent when it determines there is a substantial new question of patentability.

(Hileman 2007)

Biopiracy

Now, I come to international claims which have been rightly described as "biopiracy". Many of these were claims on traditional intellectual properties which have been practiced for millennia by the indigenous people of so-called developing countries. They used to be called "third world countries" in a bygone era, when the world was polarized into two major power blocks and some preferred not to join either block. The developing countries are often called "poor" countries for obvious reasons; however, they were not poor in the precolonial period, until several European countries colonized them and plundered their wealth while mistreating and slaughtering the local populations. As a social activist from India, Vandana Shiva (2001, 18), commented, Europeans invaded several countries of Asia, Africa, and the Americas precisely because they possessed wealth and great natural resources. Who would want to colonize a poor country?

Today, old-fashioned colonization is replaced by a new form of plunder in the name of "intellectual property rights" and what has come to be called "biopiracy". In a book entitled *Global Biopiracy*, Ikechi Mgbeoji (2006) of York University Law School in Canada wrote:

> I indict patents on indigenous knowledge on the uses of plants and plant genetic patents as mechanisms for the appropriation of Third World resources. I demonstrate that the contemporary patent system of powerful states has been significantly manipulated and retrofitted to suit the interests of seed merchants and pharmaceutical companies. These factors operate within a social culture of prejudice and disrespect for non-western forms of epistemology. Consequently, indigenous peoples and local communities have been exploited and impoverished both economically and culturally.

On the role of patent systems designed by powerful countries in appropriating the world's natural resources, Mgbeoji (2006) wrote:

> A handful of multinational corporations with primary interests in the stock markets control the global supply of seeds and related agricultural inputs. More significantly, agricultural inputs are being "tied in" with the global food supply to create an oligopoly. The implications of this trend for global food supply

> and agricultural security are matters of extant debate and concern. . . . [T]here are also concerns about the implications of the emerging dispensation for human health and the safety and integrity of the environment. . . . Since the lowering of the threshold for patentability of TKUP (Traditional Knowledge of the Uses of Plants) and the emergence of genetic patents, there is increasing evidence of patents on subjects that constitute 'novelty without innovation', especially in the pharmaceutical industry.

Furthermore, according to Mgbeoji (2006), patents on plants seem to create a regime that enables the emergent biotechnology industry to integrate the corporate food chain with agribusiness and the products of chemical multinational corporations. There is an increasing undesirable trend which empowers "seed giants" who also own subsidiaries in the chemical, fertilizer, and pesticide corporations. Some examples are such multinational corporations as Hoescht, ICI, Sandoz, and Shell.

The role of the patent system as a catalyst in the commercialization of agriculture has been facilitating the domination of the global food and agricultural market by a handful of seed corporations. For example, in 1995, the Hoechst group held 86,000 patents and patent applications; and in 1997, Novartis held more than 40,000 patents worldwide. Only three dominant food chains primarily control the world's food supply: Cargill/Pharmacia, ConAgra, and Novartis/ADM. Here are some other facts: Monsanto, DuPont, Aventis, Novartis, and AstraZeneca control most of the world's trade in genetically modified crops. The top ten global seed companies control about one third of the US$30 billion annual seed trade. The world's top ten agrochemical corporations account for 91% of the US$31 billion agrochemical market worldwide. The top five vegetable seed companies control 75% of the global vegetable seed market. Only four companies control 69% of the North American maize (corn) seed market. Only one company controls 71% of the U.S. cotton seed market (RAFI Report).

The patent system makes arbitrary allowances for chemical and pharmaceutical inventions of a biological nature, but not for other purified natural products. This process facilitates the appropriation of traditional knowledge of the uses of plants. For example, U.S. Patent No. 4,673,575, issued on June 16, 1987, which was granted to Fox Chase Cancer Center in Philadelphia, is on an extract of a medicinal plant from India, *Phyllanthus amarus*, which has been in use for thousands of years by Ayurvedic healers for treating various illnesses related to the liver, including jaundice. Scientists at Fox Chase Cancer Center were supposed to have "discovered" that the plant extract was effective for treating viral hepatitis B and E. Another U.S. patent, No. 5,900,240, was granted to Cromak Research Inc., based in New Jersey, on the diabetic properties of an Indian plant known as "Jamun", which was known in India for many years. Yet another example is an African tree, *Swartzia madagascariensis*, which has long been known to the people of Zimbabwe for treating fungal infections. In July 1999, two Swiss scientists "announced" medicinal properties for curing drug-resistant fungal infections, including athlete's foot and some types of eye infections; a product patent was granted on derivatives of the same plant to the two Swiss scientists in the United States (U.S. Patent No. 5,929,124). In addition, two scientists at the University of Wisconsin were awarded U.S. patents for a protein isolated from the plant, *Pentadiplandra brazzeana*, which grows in Gabon, Central Africa; the berries of the tree contain a protein that is 2,000 times sweeter than sugar.

Neem Patent Controversy

Neem is a fast-growing native tree of India (*Azadirecta indica*) that is well known for its many medicinal properties and has been used in India for centuries to cure many ailments. All parts of the tree, including the seeds, leaves, flowers, bark, and roots, are used for preparing medicinal compounds. Neem oil is used for preparing cosmetics. Neem twigs have been used in India for brushing teeth, perhaps the earliest practice of dental care practiced by humanity.

Neem tissues contain terpenoids, which have been shown to have significant pesticidal properties. Multinationals in Western companies had their eye on the many medicinal applications of Neem tree products. One company in particular, W.R. Grace & Co, filed a patent claim.

Vandana Shiva (2001) eloquently narrates the Neem patent controversy in the following account:

> In the last 70 years, there has been considerable research upon the properties of neem carried out in institutes (in India) ranging from the Indian Agricultural Research Institute and the Malaria Research Centre to the Tata Energy Research Institute and the Khadi and Village Industries Commission (KVIC). Much of this research was fostered by Gandhian movements, such as the Boycott of Foreign Goods movement, which encouraged the development and manufacture of local Indian products. A number of neem-based commercial products, including pesticides, medicines and cosmetics, have come on the market in recent years, some of them produced in the small-scale sector under the banner of the KVIC, others by medium-sized laboratories. However, there has been no attempt to acquire proprietary ownership of formulae, since, under Indian law, agricultural and medicinal products are not patentable.
>
> **Patent appeal**
>
> For centuries the Western world ignored the neem tree and its properties: the practices of Indian peasants and doctors were not deemed worthy of attention by the majority of British, French and Portuguese colonialists. However, in the last few years, growing opposition to chemical products in the West in particular to pesticides, has led to a sudden enthusiasm for the pharmaceutical properties of neem.
>
> In 1971, US timber importer Robert Larson observed the tree's usefulness in India and began importing neem seed to his company headquarters in Wisconsin. Over the next decade he conducted safety and performance tests upon a pesticidal neem extract called Margosan-O and in 1985 received clearance for the product from the US Environmental Protection Agency (EPA).

Three years later he sold the patent for the product to the multinational chemical corporation, W R Grace and Co. Since 1985, over a dozen US patents have been taken out by US and Japanese firms on formulae for stable neem-based solutions and emulsions and even for a neem-based toothpaste. At least four of these are owned by W R Grace, three by another US company, the Native Plant Institute, and two by the Japanese Terumo Corporation.

In 1992, the US National Research Council published a report designed to 'open up the Western world's corporate eyes to the seemingly endless variety of products the tree might offer'.

According to one of the members of the NRC panel, 'In this day and age, when we're not very happy about synthetic pesticides, [neem] has great appeal.'

This appeal is blatantly commercial. The US pesticides market is worth about $2 billion. At the moment biopesticides, such as pyrethrum, together with their synthetic mimics, constitute about $450 million of this, but that figure is expected to rise to over $800 million by 1998. 'Squeezing bucks out of the neem ought to be relatively easy,' observes *Science* magazine. [p. 57–61]

Plagiarism or Innovation?

Vandana Shiva's eloquent appeal inspired a huge international following. With respect to the questionable intellectual property justification, Shiva (2001) wrote:

> Grace's aggressive interest in Indian neem production has provoked a chorus of objections from Indian scientists, farmers and political activists, who assert that multinational companies have no right to expropriate the fruit of centuries of indigenous experimentation and several decades of Indian scientific research. This has stimulated a bitter transcontinental debate about the ethics of intellectual property and patent rights....
>
> However, neither azadirachtin, a relatively complex chemical, nor any of the other active principles have yet been synthesised in laboratories. The existing patents apply only to methods of extracting the natural chemical in the form of a stable

> emulsion or solution, methods which are simply an extension of the traditional processes used for millennia for making neem-based products. 'The biologically active polar chemicals can be extracted using technology already available to villages in developing countries,' says Eugene Schulz, chair of the NRC panel. 'Villagers smash em (the seeds) up, soak (them) in cold water overnight, scoop the emulsion off the top and throw it on the crops.'
>
> In short, the processes are supposedly novel and an advance on Indian techniques. However, this novelty exists mainly in the context of the ignorance of the West. Over the 2,000 years that neem-based biopesticides and medicines have been used in India, many complex processes were developed to make them available for specific use, though the active ingredients were not given Latinised scientific names. Common knowledge and common use of neem was one of the primary reasons given by the Indian Central Insecticide Board for not registering neem products under the Insecticides Act, 1968. The Board argued that neem materials had been in extensive use in India for various purposes since time immemorial, without any known deleterious effects. The US EPA, on the other hand, does not accept the validity of traditional knowledge and has imposed a full series of safety tests upon Margosan-O.

Dr R.P. Singh of the Indian Agricultural Research Institute asserts:

> Margosan-O is a simple ethanolic extract of neem seed kernel. In the late sixties we discovered the potency of not only ethanolic extract, but also other extracts of neem. . . . Work on the neem as pesticide originated from this division as early as 1962. Extraction techniques were also developed by a couple of years. The azadirachtin-rich dust was developed by me. (Singh *et al.* 1996)

The reluctance of Indian scientists to patent their inventions, thus leaving their work vulnerable to piracy, may in part derive from a recognition that the bulk of the work had already been accomplished by generations of anonymous experimenters. This debt has yet to be acknowledged by the US patentors and their apologists.

Landmark Victory in World's First Case Against Biopiracy

European Patent Office Upholds Decision to Revoke Neem Patent

Munich, March 8, 2005 — In a landmark decision today, the European Patent Office upheld a decision to revoke in its entirety a patent on a fungicidal product derived from seeds of the Neem, a tree indigenous to the Indian subcontinent. The historic action resulted from a legal challenge mounted ten years ago by three Opponents: the renowned Indian environmentalist Vandana Shiva, Magda Aelvoet, then MEP and President of the Greens in the European Parliament, and the International Federation of Organic Agriculture Movements (IFOAM). Their joint Legal Opposition claimed that the fungicidal properties of the Neem tree had been public knowledge in India for many centuries and that this patent exemplified how international law was being misused to transfer biological wealth from the South into the hands of a few corporations, scientists, and countries of the North. Today the EPO's Technical Board of Appeals dismissed an Appeal by the would-be proprietors America and the company Thermo Trilogy its Opposition Division five years ago to revoke the Neem patent in its entirety, thus bringing to a close this ten-year battle in the world's first legal challenge to a biopiracy patent.

European Patent Office upholds decision to revoke Neem patent. *Navadanya*, New Delhi, March 8, 2005.

The following table summarizes the many medicinal applications of Neem tree parts.

Disease/Condition	**Neem Application**
Blood poisoning	Seed oil/capsules
Diabetes	Seed oil/capsules
Kidney problems	Seed oil/capsules
Gastritis	Seed oil/capsules
Heartburn/Indigestion	Seed oil/capsules

(*Continued*)

(Continued)

Disease/Condition	Neem Application
Hemorrhoids	Lotion/Soap/Seed oil/capsules
Peptic/Duodenal ulcer	Seed oil/capsules
Allergies	Lotion/Soap/Spray/Seed oil/capsules
Arthritis	Seed oil/capsules
Bronchitis	Seed oil/capsules
Cancer	Seed oil/capsules
Conjunctivitis	Seed oil/capsules
Immune system disorders	Seed oil/capsules
Inflammation	Seed oil/capsules
Tuberculosis	Seed oil/capsules
Pain	Seed oil/capsules
Rheumatism	Seed oil/capsules
Tuberculosis	Seed oil/capsules

The Basmati Rice Controversy

The Basmati rice patent controversy is an example of biopiracy that is often cited in India. In September 1997, a Texas-based company, RiceTec Inc., filed a successful US patent (No. 5,663,484) on basmati rice. For centuries, Basmati rice has been cultivated in the Punjab region of India. Farmers in this region have selected and conserved the germplasm of Basmati rice varieties. They are characterized by a fragrant aroma, a long and slender grain, and a unique taste that is admired all over the world. The patent is all-inclusive, encompassing 22 basmati varieties from India.

According to a report issued by India's Confederation of Indian Industry (CII), annual exports of Basmati rice reached Rs 7 billion in 2006. The report further stated that the biggest victory notched up by the basmati rice exporters in India is the decision of the US Patent and Trademark Office to prevent the American company Ricetec Inc. from using the term basmati for its rice products. Basmati is as unique to India as champagne is to France. Consumers of basmati rice from around the world are fully aware that the long-grained, aromatic rice strain from India is the only genuine product.

In view of the widespread piracy of Indian genetic resources by Western corporate interests, experts have urged for the need to immediately register basmati under geographical indications. In this context, the attempts being made by the New Delhi-based NGO, Navadanya, to collect indigenous strains of rice are timely and appropriate. Navadanya, in association with farmers from the nine Indian states, has developed a register documenting over 2,000 indigenous rice varieties. As a result of these protests, no new patents have been given to RiceTec, and no new right has been given to market their varieties as equivalent or superior to Basmati. RiceTec has been forced to give up its far-reaching and false claim of having invented a very broad range of Basmati rice lines and plants.

In the words of Vandana Shiva (2001):

> The Research Foundation (Navadanya) along with other citizens groups launched a global campaign against RiceTec's Basmati patents. Organisations and individuals bombarded the USPTO with protest letters, demanding the US Patent Office not to protect biopirates. The fact that USPTO struck down 15 claims out of 20 in spite of Government of India asking for withdrawal of only 3 and the U.S. Government insisting that they would never drop the generic claim to basmati shows that once again people proved more powerful than corporations and governments (p. 57).

TRIPs

The battle over biodiversity, biopiracy, and IPRs is also central to the concerns of developing countries for reform of the Trade-Related Intellectual Property Rights (TRIPs) agreement rather than its implementation. From the viewpoint of developing countries, it is best to eliminate the eligibility of living organisms and their components as well as all biodiversity as patentable subjects from TRIPs. Instead, recognition of traditional knowledge should be made a priority.

Protecting Farmers, Freeing the Breeders

In the following account, Suman Sahai (2003) discusses the controversies of patents and plant variety protection:

One of the most controversial agreements that resulted from the Uruguay Round of trade talks (GATT, preceding WTO) is that relating to the granting of Intellectual Property Rights on biological materials through the Trade Related Intellectual Property Rights (TRIPs) system. Under TRIPs member nations are required to grant patents on microorganisms, non/biological and microbiological processes as well as effective IPR protection for plant varieties. TRIPs provides a choice for protecting plant varieties. Members may choose from patents or a sui generis system (particular to the nation) or a combination of the two.

Most developing countries including India have decided not to have patents for plant varieties and have instead chosen the *sui generis* option. The sui generis system (translating roughly into self generating) means any system a country decides on, provided it grants effective Plant Breeders Rights. TRIPs does not specify what kind of Breeders Rights and it does not say what else a member state can put in its law, apart from Breeders Rights. In short, TRIPs is a flexible system leaving a lot to the discretion of members. As a response to the TRIPs agreement, India has started enacting a series of domestic laws to implement the commitments it has made. The Protection of Plant Variety and Farmers, Rights Act, 2001, is the Indian sui generis legislation.

Hailed as a progressive, pro/developing country legislation, this law provides for well-defined Breeder's Rights as well as strong and proactive Farmers Rights. Its intent is the establishment of an effective system for protection of plant varieties, the rights of farmers and plant breeders and to encourage the development of new varieties of plants. The Act recognises the necessity of protecting the rights of farmers in respect of their contribution made in conserving, improving and making available plant genetic resources for the development of new plant varieties. In addition, there are

(*Continued*)

(Continued)

clauses to protect the rights of researchers as well as the public interest. The Indian legislation is the first in the world to grant formal rights to farmers in a way that their self-reliance is not jeopardized.

On Breeders Rights

On registration of a particular variety, the plant breeder has rights of commercialization for the registered variety either in his/her own person or through a designated person. These rights include the right to produce, sell, market, distribute, import or export a variety, in short, full control over formal marketing. Violation of the breeder's right can be construed at several levels. It applies to the variety itself as also to its packaging. Infringement will be established if the packaging is the same or even similar, such that the package could appear to be that of the Breeder. Legally, a similar looking package will be considered "Passing Off" and so actionable. Any one other than the breeder can not use the registered name or denomination. The use of the same or similar name in any way, by action or even suggestion, will constitute a violation and will be punishable. Penalties are prescribed for applying false denomination and for selling varieties to which false denomination is applied.

The breeders rights have been strengthened to the extent that if there is mere suspicion of violation or infringement, the onus of proving innocence is placed on the alleged violator. This is somewhat excessive and needs to be toned down. The normal course in law is for the accuser to furnish proof for the accusation and so it must remain in this case too. Penalties can range from Rs. 50,000 to Rs. ten lakh as well as a jail term ranging from three months to two years, depending on the severity of the damage caused. If the violator is actually selling, offering for sale or merely in the possession of a registered variety belonging to someone else, the punishment is somewhat worse. Repeat-offenders face more severe sentencing and penalties.

(*Continued*)

(Continued)

Protecting breeder's rights ensures that there is sufficient incentive for the seed industry to invest. At the same time, it is important to recognize that IPR (Intellectual Property Rights) protection alone does not necessarily deliver a successful product. To be bought, a particular variety must decisively provide an advantage. Otherwise, it will fool the farmers for a few seasons and then fail.

An IPR system in a country should not grant such strong rights to breeders that farmers suffer and their livelihoods are threatened. On the other hand, the breeders' innovation should be rewarded so that they continue to breed useful varieties to benefit agricultural and food security.

On Farmers Rights

The Act recognises the farmer not just as a cultivator but also as a conserver of the agricultural gene pool and a breeder who has bred several successful varieties. There are provisions for such farmers' varieties to be registered with the help of NGOs so that they are protected against being scavenged by formal sector breeders. The law allows the farmer to sell seed in the way he has always done, with the restriction that this seed can not be branded with the Breeder's registered name. In this way, both farmers and breeders rights are protected. The breeder is rewarded for his innovation by having control of the commercial market place but without being able to threaten the farmers' ability to independently engage in his livelihood, and supporting the livelihood of other farmers.

The pivotal importance of the farmer having the right to sell (not save, not exchange, but sell) seed has to be seen in the context of seed production in India. In India, the farming community is the largest seed producer, providing about 87% of the country's annual requirement of over 60 lakh tons. If the farmer were to be denied the right to sell, it would not only result in a substantial loss of income for him but far more importantly, such a step

(*Continued*)

(Continued)

would displace the farming community as the country's major seed provider. Legal sanction for farmers rights keeps the farming community alive and well as viable competitors and an effective deterrent to the take over of the seed market by the corporate sector. Control over seed production is central to food security which is in the forefront of national security.

Apart from the right to sell non-branded seed of protected varieties, the rights of farmers and local communities are protected in other ways too. There are provisions for acknowledging the role of rural communities as contributors of landraces and farmer varieties in the breeding of new plant varieties. Breeders wanting to use farmers varieties for creating Essentially Derived Varieties (EDVs) can not do so without the express permission of the farmers involved in the conservation of such varieties. Any one is entitled to register a community's claim and have it duly recorded at a notified center. This intervention enables the registration of farmer varieties even if the farmers themselves cannot do this due to illiteracy or lack of awareness. If the claim on behalf of the community is found to be genuine, a procedure is initiated for benefit sharing so that a share of profits made from the use of a farmer variety in a new variety goes into a National Gene Fund.

Despite its good intentions of protecting the interests of the farming community, the formulation of this particular section is likely to create problems in implementation because the drafting is poor. The Gene Fund should be the recipient of all revenues payable to the farming community under various heads. Farming communities should collectively, rather than individually, access this money, except in clear cases where an identifiable farmer's variety has been used. Farmers should have the right to decide how this money that they have earned will be spent. The use of the money should not be restricted to conservation or for maintaining *ex situ* collections.

The method for fixing and realising benefit sharing should be made simpler and easier to implement. One approach to fixing benefit sharing could be a system of lump-sum payments, based for example on (projected) volume of seed sale.

(*Continued*)

(Continued)

Protection Against Bad Seed

In providing a liability clause in the section on Farmers Rights, the farmer in principle is protected against the supply of spurious and/or poor quality seed leading to crop failures. But at present there is too much left to the discretion of the Plant Variety Authority which will fix the compensation. This could lead to arbitrary decisions and should be amended. (Sahai, 2000, 2001)

Rights of Researchers

All IPR systems must strike a balance between the monopoly granted to the IPR holder, in this case the plant breeder, and the benefits to society, in this case the farmers and consumers. Since nobody concerned with public interest would want plant breeding to shift into just a few hands, it is important to maintain competition and vitality in the plant breeding sector. That is why freedom and rights for other researchers to use all genetic material, including IPR protected material, is important. The Bill has provisions for researchers rights which allows scientists and breeders to have free access to registered varieties for research. The registered variety can also be used for the purpose of creating other, new varieties. The breeder can not stop other breeders from using his/her variety to breed new crop varieties except when the registered variety needs to be used repeatedly as a parental line. In that case authorisation is required.

There is however some difference of opinion. Some view that the Indian law actually grants very restricted rights to researchers because of the acknowledgment of Essentially Derived Varieties, EDV. It is felt that all kinds of research will become subject to the breeders authorization if a protected variety is used for research. In the Indian Act, the Breeders authorization is needed for making EDVs.

(*Continued*)

(Continued)

Protection of Public Interest

The PPV legislation includes public interest clauses, like exclusion of certain varieties from protection and the grant of Compulsory Licensing. To secure public interest, certain varieties may not be registered if it is felt that prevention of commercial exploitation of such variety is necessary to "protect order or public morality or human, animal and plant life and health or to avoid serious prejudice to the environment".

The Act also provides for the granting of compulsory license to a party other than the holder of the Breeders certificate if it is shown that the reasonable requirements for seeds have not been satisfied or that the seed of the variety is not available to the public at a reasonable price. The breeder is entitled to file an opposition but should the charge be valid, the breeder may be ordered by the Authority to grant a compulsory license under certain terms and conditions including the payment of a reasonable license fee. Compulsory License however will not be awarded if the Breeder can demonstrate reasonable grounds for his inability to produce the seed.

(Sahai 2003)

IPRs, TRIPs, and CBD

TWN Statement to the 2nd Meeting of the Panel of Experts on Access and Benefit Sharing

Montreal, 19–22 March 2001

The Third World Network welcomes the re-convening of the Panel of Experts on Access and Benefit Sharing. We hope that, during this second meeting, the Panel of Experts will be able to conclude its work on outstanding issues from its first meeting.

(*Continued*)

(Continued)

In particular, we believe the Panel of Experts must take this opportunity to seriously consider the means of addressing the implications of intellectual property rights and the WTO-TRIPs Agreement for the conservation and sustainable use of biological diversity.

There are inherent tensions between the granting of intellectual property rights under TRIPs with the objectives of the CBD. Article 16(5) of the CBD, in fact, recognises that IPRs can have a negative effect on the implementation of the CBD provisions, and thus, urges Parties to cooperate to ensure that IPRs are supportive and do not run counter to the CBD objectives.

We list below the vital matters for consideration of the Panel, with regard to the manner in which the TRIPs Agreement may, and will, undermine the CBD:

1. Conflict of rationale, origins and overall framework

There are differences in rationale, origins and overall framework of the CBD and the TRIPs Agreement. TRIPs is a commercial treaty with commercial objectives that largely benefit strong private firms. On the other hand, the establishment of the CBD was prompted mainly by the growing concern over the rapid worldwide loss of biodiversity, a recognition of the important role of traditional knowledge and the rights of local communities that develop and hold the knowledge, and the need to regulate access to and the sharing of benefits deriving from the conservation and sustainable use of biodiversity.

2. National sovereignty vs. rights of IPR holders

Based on the principle of national sovereignty enshrined in the CBD, countries have the right to regulate access of foreigners to biological resources and knowledge, and to determine benefit sharing arrangements. TRIPs enables persons or institutions to patent a country's biological resources (or knowledge relating to such resources) in countries outside the country of origin of the resources or knowledge. In this manner, TRIPs facilitates the conditions for misappropriation of

(*Continued*)

(Continued)

ownership or rights over living organisms, knowledge and processes on the use of biodiversity to take place. The sovereignty of developing countries over their resources, and over their right to exploit or use their resources, as well as to determine access and benefit sharing arrangements, is compromised.

3. Community rights vs. private, individual rights

In the preamble of TRIPs, it is recognised that "intellectual property rights are private rights". Patents confer exclusive rights on its owner to prevent third parties from making, using, offering for sale, selling or importing (for these purposes) the patented product, and to prevent third parties from using the patented process (and from using, selling or importing the product obtained from the patented process). In TRIPs, the award of IPRs over products or processes confers private ownership over the rights to make, sell or use the product or to use the process (or sell the products of that process). This makes it an offence for others to do so, except with the owner's permission, which is usually given only on license or payment of royalty.

IPRs, therefore, have the effect of preventing the free exchange of knowledge, of products of the knowledge, and their use or production. This system of exclusive and private rights is at odds with the traditional social and economic system in which local communities make use of, and develop and nurture, biodiversity. For example, seeds and knowledge on crop varieties and medicinal plants are usually freely exchanged within the community. Knowledge is not confined or exclusive to individuals but shared and held collectively, and passed on and added to from generation to generation, and also from locality to locality.

The patent system endorsed by TRIPs favours private individuals and institutions, enabling them to acquire "rights", including rights over the products or knowledge, whose development was mainly carried out by the local communities. TRIPs and the enactment of patent laws relating to biological materials in some countries have

(*Continued*)

(Continued)

facilitated the misappropriation of the knowledge and resources of indigenous and local communities, and the number of "biopiracy" cases has been increasing at a rapid rate. This misappropriation is counter to the principles and provisions of the CBD that oblige countries to recognise local community rights and fair benefit sharing. Indeed, one of the main objectives of establishing the CBD was to counter the possibility of misappropriation or "biopiracy", whilst one of the effects of TRIPs has been to enable the practice of such misappropriation.

4. Prior informed consent of states and communities vs. unilateral patents

Article 15.4 of the CBD states that "access to genetic resources shall be subject to prior informed consent of the Contracting Party providing such resources, unless otherwise determined by that Party." Thus, intending collectors of biological resources or of knowledge relating to these have to provide sufficient information of their work and how it is intended to be used, and obtain consent, before starting the work.

In TRIPs, there is no provision that applicants for patents or other IPRs over biological resources have to obtain prior informed consent. There is thus no recognition in TRIPs of the rights of the country in which the biological resource or knowledge of its use is located. Thus, patent applicants can submit claims on biological resources or knowledge to patent offices in any country (that recognises such patentability) and the patent offices can approve the claims without going through a process even of checking with the authorities of the country or countries of origin.

5. Benefit sharing arrangements

A key aspect of the CBD is that it recognises the sovereign rights of states over their biodiversity and knowledge, and thus gives the

(*Continued*)

(Continued)

state rights to regulate access, and this in turn enables the state to enforce its rights on arrangements for sharing benefits. Access, where granted, shall be on mutually agreed terms (Article 15.4), shall be subject to prior informed consent (Article 15.5), countries providing the resources should fully participate in the scientific research (Article 15.6) and, most importantly, each country shall take legislative, administrative or policy measures with the aim of "sharing in a fair and equitable way the results of research and development, and the benefits arising from the commercial and other utilisation of genetic resources with the contracting party providing such resources. Such sharing shall be upon mutually agreed terms".

Under TRIPs, there is no provision for the patent holder on claims involving biological resources or related knowledge to share benefits with the state or communities in countries of origin. In fact, there is little that a country of origin can do to enforce its benefit-sharing rights (recognised in CBD) if a person or corporation were to obtain a patent in another country based on the biological resource or related knowledge of the country of origin.

6. Patents on life

IPRs over biological resources and patents on living forms will have serious and adverse implications for access to genetic resources and the equitable sharing of benefits.

TRIPs Review

In the review of TRIPs (which is provided for in Article 27.3(b)), amendments should be made in Article 27.3(b) to bring the scope of exclusion of biological materials and processes in line with environmental and ethical considerations as well as the need for preventing biopiracy; and an interpretation can be made that the sui generis

(Continued)

(Continued)

option for plant varieties can include the protection of traditional knowledge and local community rights, in line with the CBD.

Amendments can also be made to TRIPs, in the context of the review under Article 71.1, to strengthen the obligations of developed countries to ensure the transfer of technology to developing countries, or to operationalise the implementation of technology transfer. Consideration can also be given to revise TRIPs to allow for exclusion or relaxation of standards of IPRs relating to environmentally sound technologies, and to technologies that relate to the use of biodiversity. This would bring TRIPs more in line with the spirit of the CBD, and with Article 16 provisions, including those dealing with technology transfer on concessional and preferential terms (para 2) and with the need to ensure that IPRs are supportive of and do not run counter to CBD objectives (para 5).

CBD Review

In a review of the CBD, Article 16 could be amended to remove the tensions in Article 16, so that the important objectives and principles of access to and transfer of technology to developing countries are not so constrained, as with the present CBD, by the references to the need to be consistent with adequate and effective protection of IPRs and international law. The obligations on technology transfer can also be strengthened and the implementation made more operational.

It should also be recognised that the present provisions in the CBD on access to genetic resources now place the onus of implementation on national policies and legislation. However, measures by national authorities are insufficient to enable effective implementation of access and benefit sharing arrangements. For example, in its national legislation, the state of a country of origin may require as part of its access contract that the collector cannot patent the product or knowledge (or that such a patent can be applied for only under certain conditions or benefit-sharing arrangement); but that state would require the cooperation of patent authorities or Biodiversity Authorities of other states to be able to monitor or effectively implement that contract.

(*Continued*)

(Continued)

An international protocol would be required to establish guidelines and standards for access and for fair and equitable sharing of benefits, as well as to establish international cooperation to facilitate implementation of the access and benefit-sharing arrangements.

(Correa 2001)

TWN

The relationship between the TRIPs Agreement and the Convention on Biological Diversity and the issue of enforcement of TRIPs provisions dominated the discussions at a meeting of the TRIPs Council on Tuesday 13 February.

On the TRIPs/CBD relationship, a group of developing countries and Norway reiterated their call for the TRIPs Agreement to be amended so that patent applicants are required to disclose the origin of genetic material and traditional knowledge. However, several other countries were of the view that negotiations to amend the TRIPs Agreement were premature.

The issue of the enforcement of TRIPs provisions was included on the agenda at the request of the US which presented a paper on its experience on border enforcement.

According to trade officials, China, supported by several developing countries, however stressed that this issue could not be a permanent agenda item. They argued that the TRIPs Agreement gives countries the right to choose how to implement and enforce its provisions.

On the other hand, some developed countries supported continuing an exchange of information on this issue.

On the relationship between the TRIPs Agreement and the CBD, a group of countries that include Brazil, Ecuador, China, India, Cuba, Thailand, Kenya, Pakistan, Norway, Venezuela, Colombia and Turkey repeated their call for the TRIPs Agreement to be amended so that patent applicants are required to disclose the origin of genetic material

(*Continued*)

(Continued)

and traditional knowledge and show that they have complied with the host country's requirements on obtaining consent and sharing benefits.

(One proposal for an amendment is in document IP/C/W/474, whose sponsors are Brazil, China, Colombia, Cuba, Ecuador, India, Pakistan, Peru, South Africa, Thailand and Tanzania. Another is from Norway in document IP/W/473.)

However, the US, Korea, Canada, Australia and New Zealand argued that negotiation to amend the TRIPs Agreement is premature and not the appropriate way to deal with bio- piracy.

According to trade officials, the EU said that the issue is best discussed in the World Intellectual Property Organization (WIPO). Some countries such as Canada and New Zealand said that they are still discussing the issue domestically.

Some countries were satisfied to focus on technical aspects in these TRIPs Council discussions, leaving to separate consultations under Deputy Director-General Rufus Yerxa the question of whether or not to negotiate, and whether or not to amend the TRIPs Agreement.

TWN/CBD enforcement dominate talks at TRIPS Council. Penang, Malaysia: Third World Network. February 21, 2007.

Relationship Between CBD and TRIPs

The objectives of the Convention on Biological Diversity (CBD), which was adopted in 1992, are the conservation of biological diversity, the sustainable use of its components, and the fair and equitable sharing of the benefits arising out of the use of genetic resources. Sharing of the benefits from the use of genetic resources is defined to include, *inter alia*, "the appropriate transfer of relevant technologies, taking into account all rights over those resources and technologies" (Convention on Biological Security 1992). Technology transfer is highlighted as a method for achieving one of the CBD's three principal objectives, and intellectual property rights are identified as a significant aspect of technology transfer.

The agreement on Trade Related Aspects of Intellectual Property Rights (the TRIPs Agreement) was concluded in the package of agreements in the World Trade Organisation (WTO) in 1993. The TRIPs Agreement sets minimum standards for patents and other intellectual property rights (IPRs) in the 134 WTO member countries. The complex relationship between intellectual property rights and the conservation of biodiversity and genetic resources is particularly evident in biotechnology.

Genetic resources provide a store of knowledge and the raw material for the biotechnology industry. Balancing private and public interests in intellectual property which before the conclusion of the TRIPs Agreement was the responsibility solely of national authorities has now become an international concern.

2. Relationship between CBD/TRIPs

Two sets of rights are identifiable in the Convention in respect of genetic resources. The first set can be exercised over the genetic resources per se. The second set relates to the technologies that are based on those genetic resources.

The language used by the CBD lays down standards for access to biological resources, signifying a flow from the biodiversity-rich developing countries, while emphasizing the obligations of technology-rich developed countries.

The TRIPs Agreement aims to provide a multilateral framework for promoting effective and adequate protection of intellectual property rights both to reduce distortions and impediments to international trade and to ensure that measures and procedures to enforce intellectual property rights do not themselves become barriers to trade.

Three primary issues are relevant in evaluating the relationship between CBD and the TRIPs Agreement:

a) promotion of environmentally sound technology, access to and transfer of such technology;
b) provision of incentives for conservation and sustainable use of biological resources;
c) handling of technology that may adversely affect the environment.

(*Continued*)

(Continued)

2.1 Access to and transfer of environmentally sound technology

Article 16 of the CBD states that both access to and transfer of technology among contracting parties are essential elements for the attainment of the Convention's objectives. States should provide and/or facilitate access to and transfer to other Convention parties of technologies relevant to the conservation and sustainable use of biological diversity. Article 16.5 provides that the Convention parties, recognizing that patents and other intellectual property rights may have an influence on the implementation of the CBD "shall cooperate in this regard subject to national legislation and international law in order to ensure that such rights are supportive of, and do not run counter to its objectives". The caveat "subject to national legislation and international law" suggests that the cooperative arrangements between Convention parties are subject to the TRIPs Agreement (part of international law). This raises the question of which system is to prevail should a conflict arise? Would the objectives of the CBD be paramount? Would non-compliance with intellectual property rights obligations be justified if they cannot be supportive of the objectives of the CBD?

For instance the US government has pointed to this provision as potentially indicating authority under the terms of the Convention to compromise American patent rights on technology through compulsory licensing.

The consensus appears to be that the Convention itself does not support a rejection of patent rights through some form of compulsory licensing. Such an interpretation can only be sustained through interpretation by individual parties themselves. Indeed, this kind of interpretation would require sustained pressure by developing countries to show that intellectual property rights are in fact working against the conservation of biodiversity. The issue may actually be academic because only patented technologies held by the Convention parties themselves are likely to be transferred in a manner that does not adequately protect the patent

(*Continued*)

(Continued)

rights and this can only be done on agreed terms. It is generally acknowledged that the Convention could not require technology transfer over and above that which is allowed by the TRIPs Agreement.

Article 16.3 of the CBD addresses the issue of access to and transfer of technology, which makes use of genetic resources to those countries, particularly developing countries, which provide the genetic resources. It provides for parties to take measures to provide access to and transfer of such technology on mutually agreed terms. Article 15 supports this by providing that sharing of results of research and development, and the benefits arising from the commercial and other utilisation of genetic resources should take place in a fair and equitable way, and upon mutually agreed terms, with the party providing such resources.

The TRIPs Agreement seeks to balance the objectives of promoting technological innovation and facilitating access to and transfer of technology through the provision of appropriate standards of intellectual property protection. It therefore reinforces the right of governments to adopt measures to prevent abuse of intellectual property rights by rights holders or practices that adversely affect technology transfer. The TRIPs Agreement provides for the minimum standards of protection, meaning that WTO members are free to adopt higher standards of intellectual property rights protection if they deem fit. Furthermore, WTO members are free to determine the appropriate method for implementing the Agreement within their own legal system and practice.

Article 8 appears to support this by providing that while formulating their intellectual property laws, WTO members can adopt "measures necessary to protect public health and nutrition, and to promote the public interest in sectors of vital importance to their socio-economic and technological development". The article appears to give fairly broad discretion to WTO members to evolve national legislation that suits their development (and environment) needs.

TWN/CBD enforcement dominate talks at TRIPS Council. Penang, Malaysia: Third World Network. February 21, 2007.

Below is a report of a TRIPs meeting, which was published in the SUNS on February 15, 2007:

Norway replied to Switzerland's questions on its proposal. The questions were in relation to the implications for international and regional agreements, the meaning of "supplier" country, country of "origin" and "holders" for genetic material and traditional knowledge, and what would happen if applicants did not know the origin or holders.

Norway clarified these issues and said that applicants would not be penalized if they did not know, but if they refused to supply the information, the patents would not be registered.

Later at the meeting, Brazil expressed its regrets that members still cannot agree on additional observers for the TRIPs Council, particularly the Secretariat of the Convention on Biological Diversity.

The US presented a paper on its experience with border enforcement (IP/C/W/488).

The US said that its paper discusses recent data on the growing scale of IPR infringement detected in the United States, as measured by seizures of infringing imported goods by US customs authorities.

The paper said that the value of infringing goods seized by the US Customs and Border Protection (CBP) in Fiscal Year 2006 reached $155.4 million, and the number of seizures reached 14,675. The Fiscal Year 2006 value and number of seizures are the highest in the history of CBP. This represents an increase of two-thirds by value, and 83% in the number of seizures, compared to the previous fiscal year.

It also said that footwear accounted for a significant proportion of goods seized in Fiscal Year 2006, at 41% of total value. This was a major increase over the previous year, when footwear accounted for only 10% of total seizures, by value.

Wearing apparel, handbags, computers, and consumer electronics (including power strips, DVD players, and cell phones) also figured prominently in the seizures made by CBP during the past fiscal year. Consumer electronics seizures represent a particular concern as these fake products may pose health and safety risks, the US maintained.

(*Continued*)

(Continued)

The US paper also outlined methods used by US customs authorities to use "risk modelling methods" (which look at variables to calculate the risk that a shipment contains counterfeit products), and "post-entry verification audits" (which identifies and penalizes importers after counterfeit goods are sold).

According to trade officials, China led a number of countries in expressing their discomfort, even before the agenda was adopted. They stressed that they were not blocking the agenda, but said that this could not be a permanent agenda item.

They stressed that the TRIPs Agreement gives countries the right to choose how to implement and enforce its provisions, and that enforcement cannot be considered separately from other provisions, including those on non-discrimination and on avoiding creating unnecessary barriers to trade.

They also argued against duplicating work in the World Customs Organization and World Intellectual Property Organization.

A number of countries supported China, either before the agenda was adopted or during the discussion. These were India, Argentina, Cuba, South Africa and Brazil.

On the other hand, Canada, El Salvador, New Zealand, Australia, the EU, Japan and Switzerland supported more exchange of information.

As to other issues, the least-developed countries (Bangladesh speaking) asked the Secretariat to help identify the technical assistance needs under their extended transition period.

Also, five countries have now accepted the amendment of the TRIPs Agreement (on compulsory licensing of pharmaceutical patents for export to countries unable to make the products).

The Chair urged other countries to follow in order to meet the year-end deadline.

At the end of the meeting, Chairperson Ambassador Trevor Clarke of Barbados handed over the Chairmanship to Ambassador Yonov Frederick Agah of Nigeria.

TWN/CBD enforcement dominate talks at TRIPS Council. Penang, Malaysia: Third World Network. February 21, 2007.

Globalization under WTO Has Become Global Robbery

The Basmati case is not an isolated one. Biopiracy has included Darjeeling tea, Karela, jamun, tamarind, haldi, neem, ginger, anar, pepper, amla, and Alphonso mangoes, which have all been patented. Patenting of life forms has become synonymous with piracy of traditional knowledge in agriculture and medicine.

For this, the patent paradigm needs change. The WTO made patent laws global, but they need to be brought back to national sovereign space. Patents need limits and boundaries. Life forms and traditional knowledge cannot be treated as inventions; they need to be excluded from patentability, in India and every other country. Alternative *sui generis* systems need to be evolved that suit the protection of biodiversity and indigenous knowledge. Patents on our crops are a new form of biocolonialism. They need to be fought by changing patent and IPR laws, and TRIPs. Cases like the Basmati and Neem victories highlight what is at stake. But the place to stop biopiracy is where it happens, through perverse IPR systems. Stopping biopiracy demands shaping the appropriate laws for seeds, biodiversity, and patents, nationally and internationally for the defense of our biological and intellectual wealth.

Doha Round (2001)

The goal of the Doha Development Round of WTO negotiations was to lower trade barriers around the world, permitting free trade between developed and developing countries. The Doha round began with a ministerial-level meeting in Doha, Qatar, in 2001, with subsequent ministerial meetings in Mexico (2003), and China (2005). Related negotiations have taken place in Geneva and Paris. Among many points of the program which were agreed to be negotiated further the following are particularly relevant to developing countries.

(*Continued*)

(Continued)

On developing countries, "We agree that special and differential treatment for developing countries shall be an integral part of all elements of the negotiations and shall be embodied in the schedules of concessions and commitments and as appropriate in the rules and disciplines to be negotiated, so as to be operationally effective and to enable developing countries to effectively take account of their development needs, including food security and rural development.

On Biodiversity, "We recognize that under WTO rules no country should be prevented from taking measures for the protection of human, animal or plant life or health, or of the environment at the levels it considers appropriate..."

On TRIPs and CBD, "We instruct the Council for TRIPs, in pursuing its work programme including under the review of Article 27.3(b), the review of the implementation of the TRIPs Agreement under Article 71.1 and the work foreseen pursuant to paragraph 12 of this declaration, to examine, *inter alia*, the relationship between the TRIPs Agreement and the Convention on Biological Diversity, the protection of traditional knowledge and folklore..."

On technical assistance to developing countries, "We recognize the importance of technical assistance and capacity building in the field of trade and environment to developing countries, in particular the least-developed among them.... We agree to an examination, in a Working Group under the auspices of the General Council, of the relationship between trade and transfer of technology, and of any possible recommendations on steps that might be taken within the mandate of the WTO to increase flows of technology to developing countries."

However, after some five years of negotiations, the World Trade Organization (WTO) free-trade talks were halted on 23 July 2006 as ministers of key countries left the negotiating table. But key intellectual property issues will be kept on the table for whenever they return.

A successful outcome of the Doha round has become increasingly unlikely, because the broad trade authority, which was granted

(*Continued*)

(Continued)

by the U.S. Congress, under the Trade Act of 2002 to U.S. president George W. Bush expires in 2007.

Ministers of six key members — Australia, Brazil, the European Union, India, Japan and the United States — discussed the trade round until 23 July 2006 in Geneva, but when agreement on how to proceed could not be reached, it was decided to discontinue the round.

Developing countries stand to lose the most, as this was intended to be a "development round". The round is "between intensive care and the crematorium," Indian Commerce and Industry Minister Kamal Nath said in a 24 July WTO press conference. "It is a loss for everybody. One hundred and fifty countries have missed the bus."

The end of the talks means that important developing country issues, such as the relationship between the Convention on Biological Diversity (CBD) and the WTO Agreement on Trade-Related Aspects of Intellectual Property Rights (TRIPs) likely will not go forward for the time being. This issue has been seen by some as outside of the mandate of this round of talks.

A number of developing countries, including Brazil, China and India, have proposed that the TRIPs agreement be changed to make it mandatory to include disclosure of the origin of genetic resources in patent applications (*IPW*, Genetic Resources, 7 June 2006).

Who Is To Be Blamed?

All of the ministers who spoke today expressed disappointment about the situation, many pointing to the European Union and the United States.

European External Trade Commissioner Peter Mendelson expressed "profound disappointment and sadness," adding that the suspension of talks was neither "desirable nor inevitable." He said everybody showed flexibilities at the 23 July meeting except the United States. India's Minister Kamal Nath also said that in this meeting, the EU made a move off of its previous position and that everyone put something on the table except one country (seen as a reference to the United States), which said it could not see anything change in others' positions.

Geneva talks on WTO fail to reach agreement. Bridge Initiative International, World Trade Organization, Geneva, 2006.

Terminator Technology

IPRs on seeds are criminalizing their duty to the earth and to each other by making seed saving and seeds exchange illegal. The attempt to prevent farmers from saving seeds is not just being made through new IPR laws; it is also being made through the new genetic engineering technologies. Delta and Pine Land (now owned by Monsanto) and the U.S. Department of Agriculture (USDA) have established a new partnership through a jointly held patent (No. 5723785) to seed which has been genetically engineered to ensure that it does not germinate on harvest, thus forcing farmers to buy seed at each planting season. Termination of germination is a means for capital accumulation and market expansion. However, abundance in nature and for farmers shrinks as markets grow for Monsanto. When we sow seed, we pray, "May this seed be exhaustless." Monsanto and the USDA on the other hand are stating, "Let this seed be terminated so that our profits and monopoly are exhaustless."

There can be no partnership between the terminator logic which destroys nature's renewability and regeneration and the commitment to continuity of life held by women farmers of the third world. The two worldviews do not merely clash — they are mutually exclusive. There can be no partnership between a logic of death on which Monsanto bases its expanding empire and the logic of life on which women farmers in the third world base their partnership with the earth to provide food security for their families and communities.

Gene Patenting: Pros and Cons

Several claims that have been filed in biotechnology in recent years have not met the required criteria. A case in point is the patenting of genes in the human genome. A gene may be discovered, but hardly invented! The process of discovery could be invented. Gene discovery could lead to the invention of therapeutic measures for treating or preventing a specific disease by developing appropriate screening tests. Such inventions are patentable. However, the patenting of the discovery of a gene itself is highly contentious. A frequently mentioned example of this kind is the patent on the gene for Canavan disease. It is one of the most common cerebral degenerative diseases of infancy, caused by mutations in the gene for an enzyme called aspartoacylase. Symptoms of Canavan disease, which appear in early infancy and progress rapidly, may include mental retardation, loss of previously acquired motor skills, feeding difficulties, abnormal muscle tone, and an abnormally large head. Paralysis, blindness, or hearing loss may also occur. Patient organizations played a big part in raising funds to find the gene, collect DNA samples, and encourage scientists to conduct research. After the gene was discovered, the discoverer and his employer, Miami Children's Hospital, obtained a patent, charging royalties for the genetic testing.

Even drugs already on the market can face challenges based on newly patented genes. For instance, the University of Rochester won a patent on the DNA sequence that involves the use of Celebrex, among other painkillers. The university filed a suit against the pharmaceutical company Searle to stop selling the drug, which has been on the market since 1999.

Another complication now is the increasing involvement of the patients and their families in the patent claims. Traditionally, research has been based on free access to the biological samples taken from the patients. Additional claims and lawsuits would definitely impede such research. In recent years, patient organizations and lobbies have become extremely assertive and vocal in claiming their rights to owning their biological samples.

Pacific Region

The Pacific region is no exception. An increasing number of cases of biodiversity piracy have been reported in the islands of the Pacific. The following account was published in the *Genetics News* dated March 23, 2007:

Unethical Bio-Research And Patenting of Genes in Pacific By International Companies

A new book documents 16 'acrimonious' encounters between scientific researchers and indigenous communities and calls for Pacific states to take a united approach to gaining control over such patents in the region.

The book, *Pacific Genes & Life Patents*, is published by international indigenous activist group Call of the Earth and the United Nations University in Tokyo.

"Researchers are harvesting and patenting the Pacific region's genetic resources by simply gathering and taking ownership over almost everything in their path," says Aroha Mead, Senior Lecturer at Victoria University, Wellington, New Zealand, and co-editor of the book, *Pacific Genes and Life Patents*, launched at the university on March 20. She says lack of regulation and a lack of knowledge about the latest genetic technologies and intellectual patent law has made the region a major target for commercial gene hunters.

"In South Pacific cultures a plant is a living ancestor — and even a drop of human blood retains its life spirit after it has been collected for medical research or synthesized and specific DNA qualities isolated," says A.H. Zakri, Director of the United Nations University's Yokohama-based Institute of Advanced Studies. "The authors chronicle many actions over the years by the scientific and private sector communities that offend these deeply-held values. We hope this book helps advance international understanding."

Tribal Genes for Sale: $216

The Report continues:

"One of the earliest offences involved the US government, which filed patents on DNA cells taken from the Hagahai tribe in Papua New Guinea and the Solomon Islands in the early 1990's. Neither the individuals, their communities nor governments were informed; the US government

(*Continued*)

(Continued)

rejected their later objections as inconsequential. Hagahai T cells can be purchased today from American Type Culture Collection for $216."

Another scientific proposal to patent and license the Hawaiian genome as the intellectual property of the Hawaiian people was also rejected even though it might be worth hundreds of millions of dollars. Roche Pharmaceuticals paid US$200 million for the rights to the Icelandic genome. Cultural leaders of the Kanaka Maoli (native Hawaiian people) consider their human genetic material "sacred and inalienable," writes Kanehe.

"These are not research questions driven by Indigenous peoples of the Pacific."

Genetics and biotechnology are not going to solve the fundamental problems facing the Pacific Region, says Mead. "Climate change, waste management, regional security, over fishing through illegal fishing and bottom trawling, continue to threaten the Region's resources."

Recommendations

Among the book's recommendations is a Regional Pacific Intellectual Property Office to assess patent and trademark applications, informed by Pacific model laws and responses. Such an office could enable patent application assessments to be carried out in a more critical manner with regard to Pacific cultural heritage.

Equally important is the enactment by Pacific states of laws that eliminate or significantly reduce patents on life. While this might contravene existing international patent agreements where nearly anything is patentable, growing sectors of society around the world believe patents are out of control.

Culture clash

Scientific research and patenting can often offend deeply held cultural values, says co-editor Dr Steven Ratuva, of the University of the South Pacific in Fiji.

(Continued)

(Continued)

He says patents on genes in medicinal plants conflict with the traditional view that such plants are common property, available for everyone.

While fair compensation for exploiting indigenous knowledge can be important, there are other issues at stake, says Ratuva.

"It's not only a matter of money," he says. "There are certain aspects of the culture which a lot of communities think cannot be bought or sold."

"Plants and animals are not seen as mere physical or biological entities but also as embodiment of ancestral spirits," says Ratuva.

He says recognition of local people's world view, even if it appears absurd to outsiders, must be part of the process in working out any patent or bioprospecting agreements.

She says lack of regulation and a lack of knowledge about the latest genetic technologies and intellectual patent law has made the region a major target for commercial gene hunters.

Vetting patents

The book renews calls for a Regional Pacific Intellectual Property Office to vet patent applications and make sure they conform with Pacific Island cultural values.

Leaders at the intergovernmental Pacific Islands Forum say they also want such a regional office.

Mead says Pacific states should also pass laws to either prevent or significantly reduce patents on life.

But Professor Brad Sherman, director of the Australian Centre for Intellectual Property in Agriculture, says such laws would contravene current World Trade Organization (WTO) rules on intellectual property relating to plants and animals.

"It's at odds with the trend on intellectual property across the world over the last decade which has seen any prohibitions on patenting of life being removed from the laws," he says.

(*Continued*)

(Continued)

Economic sanctions?

Any country that contravenes the WTO rules would run the risk of economic sanctions, says Sherman, based at the University of Queensland.

He says in his experience bioprospectors are often public sector researchers, who are under pressure to generate income from patenting.

And he says tough gene patent laws would not stop such researchers from taking material out of the Pacific region and patenting it elsewhere.

Mead is aware going against the tide won't be easy but is committed. "Patents are out of control and a growing number of sectors of society are indicating that limits do need to be drawn," she says.

She says lack of regulation and a lack of knowledge about the latest genetic technologies and intellectual patent law has made the region a major target for commercial gene hunters.

Call of the Earth Llamado de la Tierra

Call of the Earth, Llamado de la Tierra is a global initiative on indigenous intellectual property policy hosted at UNU-IAS. The project brings together leading indigenous experts in cultural and intellectual property from around the world, among other things, to develop responses at local, national, regional and international levels to policy and legal developments that adversely impact on indigenous peoples' traditions of preserving their cultural heritage for future generations.

Gene Patents Jeopardize Gene Testing

In an article in GEN (May 1, 2007), Roger Klein discussed the impact of gene patenting on genetic testing.

As the use of genetic testing has become more frequent, holders of patents on genes, genetic variants, and their biological correlations

(Continued)

(Continued)

are using the threat of litigation to force laboratories to stop performing genetic tests. Some laboratories have already discontinued genetic testing for muscular dystrophy and other severe neurodegenerative illnesses. However, legal precedent on patent protection suggests that their threats may lack substance.

The 1970s and early 1980s in the United States were plagued by high unemployment, high inflation, and a decline in economic confidence. In response, Congress took a number of steps to encourage the growth of domestic technology industries. Among the most significant of these were changes to the U.S. patent system. Because we did not appear to be deriving maximum economic value from our substantial federal investments in basic science research, Congress in 1980 enacted the Bayh–Dole Act.

Bayh–Dole encouraged universities to patent, and thereby commercialize, inventions arising out of government-sponsored research grants. Accompanying the enactment of Bayh–Dole, and in the years subsequent to its passage, federal financial commitments dedicated to biomedical research dramatically increased. As a consequence of these governmental actions, the number of patents assigned to universities increased from 264 in 1979 to 3,259 in 2003.

Other important events took place on the legal front. In the landmark 1980 case of Diamond v. Chakrabarty, the U.S. Supreme Court made it clear that man-made, living organisms could be patented. Chakrabarty, a research scientist, applied for a patent on a Pseudomonas bacterium that was bioengineered to carry multiple plasmids.

The Patent Office (PTO) rejected Chakrabarty's patent application on the grounds that living organisms were not patentable. Chakrabarty appealed the PTO's decision, and the case ultimately made its way to the United States Supreme Court. Urging a broad interpretation of patent eligibility, the Supreme Court recited a now famous quote, stating that "anything under the sun that is made by man," including living organisms, could be patented.

In order to provide national uniformity and add greater certainty and expertise to the application of patent law, Congress in 1982 created

(*Continued*)

(Continued)

a national appeals court with exclusive jurisdiction for patent cases. This court is referred to as the Court of Appeals for the Federal Circuit. Since the Court's inception, Federal Circuit decisions have largely been viewed as expanding patent-eligible subject matter and strengthening the rights of patent holders relative to potential infringers. On no scientific field has the impact of this Court's decisions been more acutely felt than biotechnology. Thousands of patents have been issued on biotechnology inventions from transgenic mice and leukemia-derived cell lines to recombinant drugs and vaccines.

The coalescence of these events set the stage for enormous growth in the United States biotechnology industry. FDA approvals for biotechnologically produced drugs and vaccines grew from 2 in 1982 to 35 in 2002. The number of U.S. biotechnology companies expanded from 225 in 1977 to 1,457 in 2001.

The PTO and the Court of Appeals for the Federal Circuit have largely followed the Supreme Court's mandate to interpret patent eligibility broadly. Despite its dual roles as a physical substance and a store of biological information, our legal system has treated DNA as a chemical. In the frequently cited Federal Circuit case of Amgen v. Chugai, the Court wrote, "A gene is a chemical compound, albeit a complex one."

Prior precedents in chemical law have been applied to isolated DNA sequences, including those that permitted the patenting of purified chemical compounds like aspirin, Vitamin B12, and prostaglandins. The chemical analogy has allowed patents on isolated, purified human genes to circumvent the "product or nature" doctrine, a longstanding rule that prohibits the patenting of natural materials.

When DNA is used as a chemical compound to produce recombinant drugs and vaccines the chemical analogy is valid and useful. For example, recombinant drug production involves creation of a nucleic acid that didn't previously exist in nature (cDNA), which is then used to produce another, medically valuable chemical.

(*Continued*)

(Continued)

Enormous investments are required to develop, test, and obtain regulatory approval for pharmaceutical products. Patents on the human genes used to manufacture new drugs are central to the process of obtaining the necessary risk capital to introduce these important therapeutic agents into medical practice.

By contrast, genetic testing simply involves examining patients' DNA sequences for variants that may predispose them to disease, predict response to drug therapy, or signal a possible susceptibility to pharmacologic side effects. Technical advancements have made the process of setting up genetic tests straightforward, inexpensive, and routine using existing, justifiably patented instruments and techniques. The Human Genome Project has made reference sequences, with which patients' DNA sequences are compared, freely available. Consequently, gene patents in the genetic testing context decrease innovation in test development and limit the number of test providers, thereby raising healthcare costs and reducing or eliminating patient opportunities to confirm results.

Current U.S. law does not appear to permit patents on human genes, or patents on correlations between genetic variants and clinical phenotypes, to be used to restrict diagnostic testing for genetic disorders. Dating back to litigation over Samuel Morse's patents on the telegraph, U.S. law has not allowed the patenting of natural phenomena. The "natural phenomenon" doctrine, which is distinct from the "product or nature" doctrine discussed in the context of drug production, prohibits the patenting of laws of nature, like gravity or relativity, and biological relationships, like genotype–phenotype correlations.

If I want to perform a genetic test on myself, my sole aim is to extract the information contained within my genes and correlate that information with the likelihood that I will be afflicted with a clinical disease. I am not making or selling a drug. I am merely reading my genetic code in search of natural relationships, or phenomena, that long predate my existence.

(*Continued*)

(Continued)

Patent holders on genes, genetic variants, and their biological correlations are already using the threat of litigation to force laboratories to stop performing genetic tests for predisposition to, as well as likelihood of treatment response for, various forms of cancer.

As genetic testing moves into mainstream medicine, this legal scenario will begin to have much wider, more detrimental effects on healthcare in the U.S. Gene patents for drug production have been useful and societally beneficial. But it is in all of our interests that the scope of these patents not be extended to include genetic testing.

IPR and Developing Countries

Mainly because of the Chakrabarty patent, beginning in the 1980s, issues about intellectual property in biotechnology have become contentious and frequent. Such contentions have arisen at several levels: intramurally within institutions; between academia, business, and government; among various NGO factions in countries; and finally between countries. Those between developed and developing countries are the most contentious, involving technology sharing and technology transfer. It has become a dominant political and economic issue in the trade negotiations of the WTO.

A few years ago, an International Commission on Intellectual Property Rights in London declared that the internationally mandated expansion of IPR is unlikely to generate significant benefits for most developing countries while imposing additional costs such as higher prices for medicines and agriculture. It will make poverty reduction more difficult. The Commission urged developed countries, the World Trade Organization (WTO), and the World Intellectual Property Organization (WIPO) to consider the circumstances of poor countries when drafting the international IP systems.

The chairman of the Commission, John Barton, commented that it is not necessarily in the best interests of the developing countries to have more and stronger IP protection. Developing countries should not be coerced into adopting stronger IPR while ignoring their impact on their development and poverty reduction. Appropriate IP regimes should be allowed to develop for developing countries without external pressures.

Patenting Life

The following is an excerpt from *The New York Times* article by Michael Crichton (February 13, 2007):

> Gene patents are now used to halt research, prevent medical testing and keep vital information from you and your doctor. Gene patents slow the pace of medical advance on deadly diseases. And they raise costs exorbitantly: a test for breast cancer that could be done for $1,000 now costs $3,000.
>
> The results have been disastrous. Ordinarily, we imagine patents promote innovation, but that's because most patents are granted for human inventions. Genes aren't human inventions, they are features of the natural world. As a result these patents can be used to block innovation, and hurt patient care.
>
> When the gene for Canavan disease was identified in 1993, the families got the commitment of a New York hospital to offer a free test to anyone who wanted it. But the researcher's employer, Miami Children's Hospital Research Institute, patented the gene and refused to allow any health care provider to offer the test without paying a royalty. The parents did not believe genes should be patented and so did not put their names on the patent. Consequently, they had no control over the outcome.
>
> But forget the costs: why should people or companies own a disease in the first place? They didn't invent it. Yet today, more than 20 human pathogens are privately owned, including haemophilus influenza and Hepatitis C. And we've already mentioned that tests for the BRCA genes for breast cancer cost $3,000.
>
> Fortunately, two congressmen want to make the full benefit of the decoded genome available to us all. Last Friday, Xavier

Becerra, a Democrat of California, and Dave Weldon, a Republican of Florida, sponsored the Genomic Research and Accessibility Act, to ban the practice of patenting genes found in nature. Mr. Becerra has been careful to say the bill does not hamper invention, but rather promotes it. He's right. This bill will fuel innovation, and return our common genetic heritage to us. It deserves our support.

6

Traditional Knowledge and Intellectual Property Rights

The exploitation and conservation of traditional knowledge of our planet's biological wealth are among the major issues of concern whenever the interaction between developing and developed countries is discussed. Although, in recent years, the Convention on Biological Diversity (CBD), the World Intellectual Property Organization (WIPO), and the World Trade Organization (WTO) have focused increasing attention to resolve the legal, ethical, and moral issues involved, inappropriate exploitation of traditional knowledge and materials still continues today. Article 1 of the CBD has already defined the three major objectives: the conservation of biological diversity, the sustainable use of biological diversity's components, and the fair and equitable share of profits arising from their utilization.

World Intellectual Property Organization (WIPO)

The World Intellectual Property Organization (WIPO) (*Organisation mondiale de la propriété intellectuelle* or OMPI) is one of the specialized agencies of the United Nations. WIPO was created in 1967 with the stated purpose "to encourage creative activity, [and] to promote the protection of intellectual property throughout the world" (Convention establishing WIPO, Stockholm, July 14, 1967).

WIPO currently has 184 member states, administers 23 international treaties, and is headquartered in Geneva, Switzerland. The Vatican City and almost all UN members are members of the WIPO. Nonparticipating members are the states of Kiribati, Marshall Islands, Micronesia, Nauru, Palau, the entities of Palestinian Authority, Sahrawi Republic, Solomon Islands, Taiwan, Timor Leste, Tuvalu, and Vanuatu.

WIPO was formally created by the Convention Establishing the World Intellectual Property Organization (signed at Stockholm on July 14, 1967, and as amended on September 28, 1979). Under Article 3 of this Convention, WIPO seeks to "promote the protection of intellectual property throughout the world." WIPO became a specialized agency of the UN in 1974, as above-mentioned.

WIPO Members

Unlike other branches of the United Nations, WIPO has significant financial resources independent of the contributions from its member states. In 2006, over 90% of its income of around CHF500 million was expected to be generated from the collection of fees by the International Bureau (IB) under the intellectual property application and registration systems which it administers (the Patent Cooperation Treaty, the Madrid system for trademarks, and the Hague system for industrial designs).

Objectives and Functions

Articles 3 and 4 of the WIPO constitution define the objectives and functions of the organization.

Article 3
Objectives of the Organization

The objectives of the Organization are:

(i) to promote the protection of intellectual property throughout the world through cooperation among States and, where appropriate, in collaboration with any other international organization,
(ii) to ensure administrative cooperation among the Unions.

Article 4
Function

In order to attain the objectives described in Article 3, the Organization, through its appropriate organs, and subject to the competence of each of the Unions:

(i) shall promote the development of measures designed to facilitate the efficient protection of intellectual property throughout the world and to harmonize national legislation in this field;
(ii) shall perform the administrative tasks of the Paris Union, the Special Unions established in relation with that Union, and the Berne Union;
(iii) may agree to assume, or participate in, the administration of any other international agreement designed to promote the protection of intellectual property;
(iv) shall encourage the conclusion of international agreements designed to promote the protection of intellectual property;
(v) shall offer its cooperation to States requesting legal-technical assistance in the field of intellectual property;
(vi) shall assemble and disseminate information concerning the protection of intellectual property, carry out and promote studies in this field, and publish the results of such studies;

(*Continued*)

(Continued)

(vii) shall maintain services facilitating the international protection of intellectual property and, where appropriate, provide for registration in this field and the publication of the data concerning the registrations;

(viii) shall take all other appropriate action.

Traditional Cultural Expressions/Folklore and Traditional Knowledge: IGC Review of Policy Issues

At its tenth session, the Intergovernmental Committee (IGC) of WIPO discussed the draft provisions for the protection of traditional cultural expressions/expressions of folklore (TCEs/EoF) and for the protection of traditional knowledge (TK) against misappropriation and misuse.

Issues on Traditional Cultural Expressions/ Expressions of Folklore

1. Definition of traditional cultural expressions (TCEs)/expressions of folklore (EoF) that should be protected.
2. Who should benefit from any such protection or who hold the rights to protectable TCEs/EoF?
3. What objective is sought to be achieved through according intellectual property protection (economic rights, moral rights)?
4. What forms of behavior in relation to the protectable TCEs/ EoF should be considered unacceptable/illegal?
5. Should there be any exceptions or limitations to rights attaching to protectable TCEs/EoF?
6. For how long should protection be accorded?
7. To what extent do existing IPRs already afford protection? What gaps need to be filled?

8. What sanctions or penalties should apply to behavior or acts considered to be unacceptable/illegal?
9. Which issues should be dealt with internationally and which nationally, or what division should be made between international regulation and national regulation?
10. How should foreign rights holders/beneficiaries be treated?

Issues on Traditional Knowledge

1. Definition of traditional knowledge (TK) that should be protected.
2. Who should benefit from any such protection or who hold the rights to protectable TK?
3. What objective is sought to be achieved through according intellectual property protection (economic rights, moral rights)?
4. What forms of behavior in relation to the protectable TK should be considered unacceptable/ illegal?
5. Should there be any exceptions or limitations to rights attaching to protectable TK?
6. For how long should protection be accorded?
7. To what extent do existing IPRs already afford protection? What gaps need to be filled?
8. What sanctions or penalties should apply to behavior or acts considered to be unacceptable/illegal?
9. Which issues should be dealt with ternationally and which nationally, or what division should be made between international regulation and national regulation?
10. How should foreign rights holders/beneficiaries be treated?

One can only hope that the WIPO, through its deliberations, will lead to an equitable and fair protection of traditional knowledge. However, progress is slow and the future outcome is uncertain.

Mandate to Continue Negotiations on Core Issues

A key committee of the World Intellectual Property Organization (WIPO), meeting in Geneva from July 3 to 12, 2007, recommended that the WIPO General Assembly should renew its mandate to continue work on intellectual property and traditional knowledge (TK), traditional cultural expressions (TCEs) (also termed 'expressions of folklore'), and genetic resources. The current mandate of the Intergovernmental Committee on Intellectual Property and Genetic Resources, Traditional Knowledge and Folklore (IGC) expires in December 2007.

Member states noted that the IGC had made progress on its substantive work, and agreed to work towards further convergence on the questions under its mandate, with a view to making recommendations to the WIPO General Assembly. Delegates also affirmed that the Committee's work had greatly benefited from the enhanced participation of representatives of indigenous and local communities which had been made possible by various initiatives, including the successful launch of the WIPO Voluntary Fund (see PR/2007/499). The IGC session commenced with an Indigenous Panel, chaired by Mr. Greg Young-Ing of the Opsakwayak Cree Nation, at which seven representatives of indigenous and local communities explained to the Committee their communities' experiences, concerns and expectations regarding intellectual property and TK, TCEs, and Genetic Resources (GR).

WIPO Director General, Dr. Kamil Idris, applauded the positive spirit in which member states, intergovernmental organizations and non-governmental organizations had addressed the far-reaching and challenging issues that were before them. He said that "the open and cooperative manner in which discussions took place in the IGC augurs well for its renewed mandate. There is now a deeper understanding of the diversity of views held and stronger mutual respect for the different perspectives that are brought to the table. This is a solid foundation on which to base future work of these fundamentally important issues."

This session concentrated on core issues for the protection of TK and TCEs, focusing on the fundamental policy challenges that are

(*Continued*)

(Continued)

central to the quest for more effective protection against misuse and misappropriation. These issues cover such questions as definitions of traditional knowledge and traditional cultural expressions, the form and scope of protection, and the nature of the beneficiaries. This process has constituted the first systematic multilateral review of these fundamental intellectual property policy questions, building on a rich base of work in the Committee that has drawn on the experience of over 80 countries and many indigenous and local communities. The Committee requested the secretariat to prepare new working documents consolidating this exploration of the issues.

On genetic resources, the Committee reviewed the full range of options for its work in this area, on the basis also of an overview and reports from other UN agencies working in this field, including the Convention on Biological Diversity, the Food and Agricultural Organization, and the United Nations secretariat. The work of the Committee will continue with the options discussed and an update of work in the other forums. The options reviewed include patent disclosure requirements, with the European Community and Switzerland both proposing specific reforms to the patent system to provide for specific disclosure relating to genetic resources and traditional knowledge, and alternative proposals for dealing with the relationship between intellectual property and genetic resources; the interface between the patent system and genetic resources; the intellectual property aspects of access and benefit-sharing contracts; and a factual update of international developments relevant to the genetic resources agenda item (see documents WIPO/GRTK/IC/11/8(a) and (b)). Peru tabled a further analysis of its national initiatives against biopiracy, and Japan updated and extended its proposal for a database to ensure that information on genetic resources is better taken into account in patent examination.

Subject to the decision of WIPO General Assembly to renew the IGC's mandate, the next session of the IGC is anticipated to take place in February, 2008.

(*Continued*)

(Continued)

The IGC's work was enlivened by three performances of traditional music and dance — a troupe of traditional Mongolian musicians and dancers, who also introduced their musical culture to participants in an informal seminar, a group of Geneva-based traditional Indonesian musicians and dancers, who included the Chair of the IGC, Ambassador I Gusti Agung Wesaka Puja of Indonesia, and an impromptu performance by one of the indigenous panelists, Ms. Chukhman of the Kamchatka region in Russia. These performances gave a vivid, immediate insight into the vitality and deep cultural significance of the diverse forms of traditional cultural expression that the Committee was addressing.

There was general agreement that the work of the IGC, which formally ended with this session, should be extended. But there was disagreement as to how to characterise (the IGC's) future work, and what documents the work should be based on. For example, the African Group wanted to state clearly that the IGC should work towards an international agreement, while many developed countries opposed this. Discussions of the issues had showed that developed countries had a lot of limitations.

Sources said the United States, Canada and Japan opposed the language in an 11 July proposal from the African Group which said the IGC should work towards a legally binding instrument, which stated that it "should work towards producing concrete, substantive outcomes by the end of 2009." Western countries including the US and Canada took issue with a later reference to the mandate. The United States has resisted language that would commit the group to a negotiation for an international instrument.

There were 10 issues addressed in the meeting, including future work. The adopted text on the IGC's future work now reads as follows:

The Intergovernmental Committee reviewed the progress made on its substantive agenda items at the current and previous sessions of its current mandate, and

(i) Agreed that progress had been made on its substantive work to date;

(Continued)

(Continued)

(ii) Agreed that its work had greatly benefited from the enhanced participation of representatives of indigenous and local communities made possible by various initiatives including the successful launch of the WIPO Voluntary Fund, and also from the participation of intergovernmental organisations;

(iii) Agreed to recommend to the WIPO General Assembly that the current mandate of the Committee be renewed as set out in document WO/GA/30/8, paragraphs 93 to 95, namely that;

- the Committee "will continue its work for the next budgetary biennium on questions included in its previous mandate";
- "its work will focus, in particular, on a consideration of the international dimension of those questions, without prejudice to the work pursued in other fora", and
- "no outcome of its work is excluded, including the possible development of an international instrument or instruments";
- the IGC would be urged "to accelerate its work and to present a progress report to the session of the General Assembly" in September 2008;
- the General Assembly would further request "the International Bureau to continue to assist the IGC by providing Member States with necessary expertise and documentation."

(iv) With respect to the content of paragraph (iii), the Committee agreed to work towards further convergence of views on the questions included in its previous mandates, in particular, within the areas of TCEs and TK, on the Lists of Issues agreed at its Tenth Session, with a view to making appropriate recommendations to the General Assembly.

(v) Agreed concerning its substantive working document on item 7 (TCEs/EoF) that:

- the Secretariat should prepare a factual extraction, with attribution, consolidating the view points and questions of Members and Observers on the List of Issues considered

(*Continued*)

(Continued)

during the Eleventh Session including their comments submitted in writing for the Eleventh Session, subject to review of Member States and observers and without prejudice to any position taken on these issues", and

— As agreed at the Tenth session, document WIPO/GRTKF/IC/11/4(c) remains on the table in its existing form and comments made in relation to it are noted.

(vi) Agreed concerning its substantive working document on item 8 (TK) that:

— the Secretariat should prepare a factual extraction, with attribution, consolidating the view points and questions of Members and Observers on the List of Issues considered during the Eleventh Session including their comments submitted in writing for the Eleventh Session, subject to review of Member States and observers and without prejudice to any position taken on these issues, and

— as agreed at the Tenth session, document WIPO/GRTKF/IC/11/5(c) remains on the table in its existing form and comments made in relation to it are noted.

(vii) Agreed concerning its substantive working documents on item 9 (genetic resources) that: the Secretariat should prepare a further update of international developments based on document 11/8(b) which would include omissions identified in the current session, more recent developments, and any other relevant developments reported to the Committee.

WIPO Meeting on Intellectual Property and Genetic Resources, and Traditional Knowledge and Folklore, Geneva, July 3–12, 2007.

Shaman Pharmaceuticals

The problems involved have been succinctly discussed by King (1996), who cited two specific partnerships involving Belize and Tanzania and a midstage pharmaceutical company from California, Shaman Pharmacuticals. In the Senate Foreign Relations Committee Hearing on the US ratification of the Convention on Biological

Diversity (CBD), Shaman's CEO and founder, Lisa Conte, testified, emphasizing both biological diversity and cultural rights. She explained Shaman's philosophy: "to equitably compensate indigenous societies for their intellectual contributions to the identification of useful products in the drug discovery process."

As King (1996) wrote, Shaman was committed to several layers of reciprocity, in their program of ethnomedical collaboration with indigenous communications.

Does Shaman's approach serve the indigenous communities fairly and equitably? The foremost question one might ask is whether one can adequately compensate or reward traditional knowledge? How does one evaluate the value of an intellectual property which has contributed to the survival of humanity for millenia? I refer to the selection of various varieties of food crops and preservation of their germplasm, as well as the successful farming practices which farmers in the developing countries have practiced for several millennia. This applies to the conservation of medicinal plants also, which has traditionally passed on both the germplasm and the specific knowledge of medicinal value of various plant species to new generations.

Shaman's policy involved reciprocity of three kinds, short, medium and long term. Short term reciprocity involves devoting 10 to 15 percent of the research funds to immediate community needs, as defined by the community. These may involve either services or supplies or both, as decided by the community, not Shaman.

Medium term involves projects that might take several months or years, including major segments of the CBD, technology transfer and sustainable development goals. This may often involve collaboration with local governments, universities, Traditional Healer's Associations, and other groups. Long term benefits are those that are made available after a product reaches the market. A non-profit organization called "The Healing Forest Conservancy" (HFC) was established by Shaman to develop and deliver long-term compensation to the indigenous community.

These three categories of benefit sharing are created by Shaman because natural product discovery and development usually takes a long time. Shaman wanted to assure the community that benefit sharing starts right from the beginning and continues regardless of the final outcome of product discovery, development and marketing.

Belize

Many of the books about traditional Mayan medicine were destroyed when Spanish Conquistadors burnt them and banned the practice of traditional medicine. The practice of written knowledge was then replaced by oral communication, which was passed on through generations until the 20th century. In recent times, it was generally considered unsuitable for a modern world and was replaced by modern Western medicine; however, some healers, ethnobotanists, and younger Belizeans have recorded the remaining elements of Belizean traditional medicine before it too disappears forever.

Fig. 1. A Toda tribal village in South India. Courtesy of Michele Wambaugh.

India

Traditional Knowledge Digital Library (TKDL)

An ambitious project of the Indian Government, "Traditional Knowledge Digital Library", will produce an encyclopaedia of the country's traditional medicine in five languages — English, French, German, Japanese and Spanish — in an effort to stop people and corporations from other countries claiming ancient knowledge as their own and patenting them.

The electronic encyclopaedia will be made available next year. It will contain information on the traditional medicines, including exhaustive references, photographs of the plants and scans from the original texts and a discussion of their medicinal properties.

Ayurvedic texts are in Sanskrit and Hindi, unani texts are in Arabic and Persian and siddha material is in Tamil language. Material from these texts is being translated into five international languages, using sophisticated software coding to make it accessible and understandable to readers all over the world.

There is enormous material — there are at least 54 authoritative 'text books' on ayurveda alone, many of them are thousands of years old.

There are also 150,000 recorded ayurvedic, unani and siddha medicines; and 1,500 asanas (physical exercises and postures) in yoga, which originated in India more than 5,000 years ago.

India has suffered even though its traditional knowledge, as in China, has been documented extensively. But it has not been in an easily accessible form for western readers. It has never been translated and put out in the public domain.

(Sen 2001)

Neem and Turmeric

India has been embroiled in some lengthy litigation in the past decade — the government spent at least $6m in fighting legal battles against the patenting of turmeric and neem-based medicines. In 1995, the US Patent Office granted a patent on the wound-healing

(*Continued*)

(Continued)

properties of turmeric. India protested and fought a two-year-long legal battle to get the patent revoked. Also, India won a 10-year-long battle at the European Patent Office against a patent granted on an anti-fungal product, derived from neem, by successfully arguing that the medicinal neem tree is part of traditional Indian knowledge.

Dr Vinod Kumar Gupta, who is leading the traditional wealth encyclopaedia project and heads India's National Institute of Science Communication and Information Resources (Niscair), estimates that of the nearly 5,000 patents given out by the US Patent Office on various medical plants by the year 2000, some 80 percent were plants of Indian origin. Practitioners of traditional medicines say their importance cannot be denied — according to the WHO, 70 percent of the people living in India use traditional medicine for primary health care.

It is estimated that drugs containing one or more plant-derived active ingredients represented about 25 percent of all prescriptions dispensed from community pharmacies in the United States. In Eastern Europe, the corresponding figure is 60 percent because traditional medicine and modern medicine are amalgamated, and in China it is 80 percent. Both the Indian and the Chinese traditional systems use more than 200 plant species in their pharmacopoeias.

(Kurien 2002)

7

Impact of GMOs in Developing Countries

Biosafety Regulations and Biodiversity Protection in the Developing Countries

The main object of biotech regulation is to set standards to guarantee the safety and efficacy of new transgenic products that protect human, animal, and environmental health. Regulations also make sure that international quality and safety standards are maintained to facilitate trade. Four steps are involved to assure the safety of agricultural products:

1. To determine when a risk assessment is needed;
2. To accomplish a risk assessment in order to identify and measure potential risks;
3. To manage the identified risks in order to minimize them by setting conditions on the use; and
4. To fulfil regulation to ensure that products meet the safety criteria (quality, labeling, and purity).

In the case of GMOs for agriculture with novel traits, the following requirements are important: a case-by-case safety assessment, information on risks and tests, controlled release to minimize gene escape (field testing), and finally unconfined release into the environment once the previous requirements have been met.

In the developing countries, an increasing number of genetically modified (GM) crops are being released into centers of biodiversity. Since these centers are important for the future of the world's major food crops, there is great concern about the potential ecological impact of these releases. Therefore, developing countries are required to set national biosafety procedures to protect their crops and to avoid environmental disasters, while at the same time trading in safe GM products.

The accidental transfer of living modified organisms resulting from modern biotechnology may have an adverse effect on the conservation and sustainable use of biological diversity. The United Nation's (UN's) Convention on Biological Diversity (CBD) has carried out international negotiations to develop an international protocol on biosafety. After several years of negotiations, the Protocol, known as the Cartagena Protocol on Biosafety to the Convention on Biological Diversity, was finalized and adopted in Montreal in 2000. The Cartagena Protocol is in force since September 11, 2003.

Adverse Impact

Public education is urgently required to explain the possible risks of planting GM crops and consuming food products. Decisions can and should be made in a democratic manner, involving the public. Such is not the case so far in several countries, where major agribusinesses have expedited the planting of GM crops and mixed GM-derived food products with non-GM foods. This has been the case so far in the United States, where the largest share of the land is planted with GM crops in the world; yet, the public is not aware of this fact. There is a general increase in allergic problems, cancer rates, headaches, and numerous other health problems, but to what extent any of these might have been caused by consuming GM foods is not known. In other words, compared to the many millions of people who are exposed to the risks in the country, there is hardly any testing done at all. It is obvious that the so-called precautionary principle has no place in this situation! Developing countries are strongly urged not to follow this path.

Some international environmentalist groups such as Greenpeace have performed a valuable public service by appealing to the public and governments to take a critical look at the use of GM foods and related food products. Although at first these activities appeared to be simply alarming with no evidence, recent experiments in the U.K. on the harmful effects of GM pollen on the pollinating insects and the compelling evidence obtained by two scientists indicate a definite adverse impact of GM products on health and environment.

Professor Gilles-Eric Seralini (2005), President of CRIIGEN (Committee for Independent Research and Information on Genetic Engineering) at the University of Caen in France, reanalyzed the results of experiments testing the effect of Monsanto's GM maize (MON863) on rats. It must be noted that Monsanto did not voluntarily disclose any harmful effects in the first place. Indeed, Monsanto attempted to conceal the data until a lawsuit by Greenpeace forced their release to the public. The following report appeared in *The Independent* (UK) on May 22, 2005:

Judges Order Disclosure of Secret Study on GM Risks

Monsanto 'which dismisses the differences between the rats as pure chance' supplied the study to safety authorities on condition it was kept confidential. It has consistently refused to make the study public, saying it 'contains confidential business information which could be of commercial use to our competitors'. Last week the dispute ended in a German court, where Greenpeace argued the study should be published under a European Union law, namely the public should have access to documents.

French Professor Gilles-Eric Seralini, a molecular endocrinologist at the University of Caen, agrees that the results indicate a toxic reaction. Seralini is a member of two French government commissions that evaluate GM food, one of which originally rejected a request for approval of the corn variety in October, 2003 due to the adverse findings of the study. Seralini won a French lawsuit allowing him to express his concerns in public, and now Greenpeace has won a German court battle that makes public the data that is the source of his concerns.

Professor Seralini said: "These revelations are profoundly disturbing from a health point of view. They are certainly sufficient to require new and more carefully conducted feeding studies and an immediate ban from human or animal consumption of GM maize MON 863 and all its hybrids. This maize cannot now be considered safe to eat. We are now calling urgently for a moratorium on other approved GMOs while the efficacy of current health testing methods is reassessed."

The reanalysis by Prof. Seralini (2005) produced clear evidence that the GM maize caused hepatorenal toxicity in rats. Seralini's group also confirmed the toxic effects of the commonly used herbicide Roundup on human embryonic cells in culture. Another investigation by Dr. Irina Ermakova (2006) at the Institute of Higher Nervous Activity and Neurophysiology of the Russian Academy of Sciences showed that rats fed on GM soy died in significantly larger numbers than those fed on non-GM soy.[a] These results have raised serious concerns which must be studied carefully by all developing countries before initiating any biotech programs.

Public Responsibility

Thus, the public has the right to be informed and the responsibility to learn about biotechnology and how to deal with it. The public and the private sectors should jointly undertake to address public concerns on the safety of biotechnological products and applications.

---

[a] Irina Ermakova's research was critically examined in the following paper: Marshall, A. (2007) GM soybeans and health safety — a controversy reexamined. *Nat Biotechnol* **25**: 981–987.

Publications should give simple and precise impartial facts about biotechnology to the interested public. Education programs in schools should provide easy-to-understand information about biotechnology. It is only through education that we can demystify biotechnology and avoid misunderstandings or extreme positions.

Economics of GM Cotton in Developing Countries

The farm-level economic impact of transgenic crops has been studied by Terri Raney (2006) of the Food and Agriculture Organization (FAO) in Rome, among others. The most extensive studies of this kind in developing countries have been conducted in Argentina, China, India, Mexico, and South Africa. The data discussed here for transgenic (IR = insect-resistant) cotton are based on 2 or 3 years of commercial farm production. The following figures reflect the average percentage difference between IR and conventional cotton for all farmers over all seasons:

Performance advantage of IR over conventional cotton (%)

	Argentina	China	India	Mexico	South Africa
Yield	33	19	34	11	65
Revenue	34	23	33	9	65
Pesticide costs	–47	–67	–41	–77	–58
Seed costs	530	95	17	165	89
Profit	31	340	69	12	299

Modified from Raney (2006).

In yield, South Africa is first, followed by India in second place. In revenue, India exceeded China by ten points. In profit, China outperformed all of the others. South Africa performed uniformly well in all of the categories.

India also showed large net gains from IR cotton adoption at the national level, although some significant variation was noted across states; one state especially, Andhra Pradesh, experienced negative results. By 2005, India had approved 20 IR cotton varieties for use all over the country. The area of the country planted with IR cotton almost tripled from that of the previous year.

China's success has been attributed to its highly developed public agricultural research system which has independently produced two IR transgenic entities, incorporating them into several locally adapted cotton varieties. These are available at much lower prices, competing locally with Monsanto's high-priced IR cotton varieties. The public sector's active role in China in developing and distributing IR cotton varieties is mainly responsible for the low cost, eventually leading to large net profit gains in China.

In Argentina, Monsanto strictly enforced its intellectual property rights (IPRs) in the case of IR cotton varieties while failing to patent its soybean innovation. Consequently, the Argentine farmers did not adopt the IR cotton widely, while enthusiastically embracing the herbicide-tolerant (HT) soybeans.

Several studies have reported economic success for South African farmers adopting IR cotton, especially in the Natal province where a local cooperative provided seed on credit as well as technical advice. The benefits were widely shared by all farm types. Pesticide use also declined significantly. A comparison of IR and non-IR maize varieties indicated that large commercial farmers benefited from yield, pesticide, and income advantages; whereas smallholders experienced higher yields. Gouse *et al.* (2005) examined the economic impact of Bt white maize, the first transgenic staple food crop to be adopted anywhere in the world. IR varieties of yellow maize have been grown by large commercial farmers in South Africa since 1998, but it is mainly used in animal feed. White maize is preferred for human consumption. IR varieties were first made available to farmers in 2001. The study recommended that IR maize varieties be made more accessible to small farmers.

India

Even though the planting of IR cotton proceeded very rapidly in India, it is less successful in some important respects in comparison with the performances in China and South Africa. The first economic studies were based on farm-level field trials. Morse *et al.* (2005) evaluated the relative merits of two official and two unofficial Bt cultivars with a popular non-Bt counterpart in Gujarat. The Indian Government has not authorized the unofficial varieties. They are less costly because no royalties are paid, and they may be less reliable than the official varieties because they have not gone through a lengthy approval process. The two official Bt varieties outperformed the unofficial Bt cultivars as well as the non-Bt varieties in yield and net profit. The unofficial locally produced Bt hybrid also outperformed the non-Bt hybrids.

The following table gives performance advantage figures for IR over non-IR cotton in four major states of India:

Performance advantage of IR over non-IR cotton in India

State	Yield	Revenue	Chemical Costs	Profits
Maharashtra	32	29	–44	56
Karnataka	73	67	–49	172
Tamil Nadu	43	44	–73	229
Andhra Pradesh	–3	–3	–19	–40
National Average	34	33	–41	69

Source: Qaim *et al.* (2006) *Rev Agric Econ* **28**: 48–58.

Countries Growing Genetically Modified Organisms (GMOs)

The world's leading producers of GM crops are the United States, Argentina, Brazil, Canada, India and China. In 2006, GM crop production also reached noteworthy levels in Paraguay, South Africa, Uruguay and Australia. In the EU, GM crops have remained uncommon. Appreciable GM maize production in the EU only took place in Spain on an area of nearly 60,000 hectares. In Portugal, Germany, France and the Czech Republic, transgenic crops were primarily grown for small-scale field trials.

In 2005, Iran and the Czech Republic were added to the list of countries commercially growing transgenic crops. As of 2006, 38 percent of GM crops are grown in developing countries.

(Continued)

(Continued)

Global Area of Genetically Engineered Crops, 1996 to 2006: By Country (Million Hectares)

Country	USA	Argentina	Brazil	Canada	China	Paraguay
1996	1.5	0.1	—	0.1	—	—
1997	8.1	1.4	—	1.3	0.0	—
1998	20.5	4.3	—	2.8	<0.1	—
1999	28.7	65.7	1.4*	4.0	0.3	—
2000	30.3	10.0	3.6*	3.0	0.5	—
2001	35.7	11.8	5.7*	3.2	1.5	—
2002	39.0	13.5	6.3*	3.5	2.1	—
2003	42.8	13.9	3.0	4.4	2.8	—
2004	47.6	16.2	5.0	5.4	3.7	1.2
2005	49.8	17.1	9.0	5.8	3.3	1.8
2006	54.6	18.0	11.5	6.1	3.5	2.0

*Illegal cultivation of GMOs: calculated area.

Global Area of Genetically Engineered Crops, 1996 to 2006: By Country (Million Hectares)

Country	India	South Africa	Uruguay	Australia	Mexico	Romania
1996	—	—	—	<0.1	<0.1	—
1997	—	—	—	0.1	<0.1	—
1998	—	<0.1	—	0.1	—	—
1999	—	0.1	—	0.1	<0.1	<0.1
2000	—	0.2	<0.1	0.2	<0.1	<0.1
2001	—	0.2	<0.1	0.2	<0.1	<0.1
2002	<0.1	0.3	<0.1	0.1	<0.1	<0.1
2003	0.1	0.4	0.1	0.1	<0.1	<0.1
2004	0.5	0.5	0.3	0.2	0.1	0.1
2005	1.3	0.5	0.3	0.3	0.1	0.1
2006	3.8	1.4	0.4	0.2	0.1	0.1

(*Continued*)

(Continued)

Global Area of Genetically Engineered Crops, 1996 to 2006: By Country (Million Hectares)

Country	Philippines	Honduras	Colombia	Iran	Spain	Portugal	Germany
1996	–	–	–	–	–	–	–
1997	–	–	–	–	–	–	–
1998	–	–	–	–	<0.1	–	–
1999	–	–	–	–	<0.1	<0.1	–
2000	–	–	–	–	<0.1	–	<0.1
2001	–	–	–	–	<0.1	–	<0.1
2002	–	<0.1	<0.1	<0.1	<0.1	–	<0.1
2003	<0.1	<0.1	<0.1	0.1	<0.1	–	<0.1
2004	0.1	<0.1	<0.1	0.5	0.1	–	<0.1
2005	0.1	<0.1	<0.1	1.3	0.1	<0.1	<0.1
2006	0.2	<0.1	<0.1	<0.1	0.1	<0.1	<0.1

Source: (with the kind permission of ISAAA, Clive James, 2006)

FAO Report

Several developing countries now have well-developed biotechnology programmes; they are approaching the leading edge of biotechnology applications and have significant research capacity, according to a new FAO assessment on the status of research and application of crop biotechnologies in developing countries (2005).

Based on a review of the information in the FAO database on Biotechnology in Developing Countries (FAO-BioDeC), which covers both genetically modified (GM) crops and non-GM biotechnologies, the assessment suggests that developing countries will soon have new GM crops available such as virus-resistant papaya, sweet potato and cassava as well as rice tolerant to abiotic stresses (salinity and drought).

(*Continued*)

(Continued)

Focus on Food Security

Most of the GMOs commercialized so far in developing countries have been acquired from developed countries and focus on a limited number of traits (mainly herbicide tolerance and insect pest resistance) and crops (commodities such as cotton, soybean and maize).

However, the FAO assessment reveals that several developing countries have been conducting research on a wider range of crops, such as banana, cassava, cowpea, plantain, rice and sorghum, and on traits relevant for food security, such as abiotic stress tolerance and quality.

Argentina, Brazil, China, Cuba, Egypt, India, Mexico and South Africa have taken the lead. A second group of countries has medium-scale agricultural biotechnology programmes, usually in a few key areas. Other developing nations have relatively limited research capacity, according to the FAO report.

"We hope that research activities in developing countries will increasingly focus on issues important for food security," said Andrea Sonnino, from FAO's Research and Technology Development Service.

Noticeable Gaps

There are, however, some noticeable gaps in research. For example, no research is reported in the field of nematode resistance despite the considerable losses caused by these plant parasites. Another fundamental but neglected research problem concerns post-harvest losses.

The study also notes that biosafety capacity building is needed to enable many countries in Africa, Eastern Europe, Latin America and the Near East to fully benefit from GMO technology.

Regarding non-GM biotechnologies, many are being used on a commercial scale but only a few studies have been carried out to assess their socio-economic impacts. The report highlights that this is an area needing urgent attention as it is likely to help guide research

(*Continued*)

(Continued)

and technology policies and investments towards wider and efficient utilization of all biotechnologies.

FAO-BioDeC

Launched in 2003 as an online searchable database, FAO-BioDeC currently has about 2000 entries from 71 developing countries, including countries with economies in transition.

It is regularly updated and has recently been expanded to include extensive data from the forestry sector and some initial data on livestock.

The assessment presents a first analysis of the information contained in the database as of 31 August 2004.

Impact of GMOs in Developing Countries

Agrobiodiversity

The MS Swaminathan Research Foundation has been promoting a great deal of discussion of the impact of GMOs on the food security and economy of developing countries, especially India.

Agrobiodiversity in Southern India

Area covered by different crops at Kudankulam

Crops	**Area (ha)**
Pulses	12
Peanut (groundnut)	7
Jatropha	4
Fodder grass	30
Fruit crops	14
Total	67

Source: Swaminathan and Jana (1992).

Elite high-oil-yielding accessions of Jatropha in Tamil Nadu

District	No. of Accessions
Coimbatore	57
Erode	68
Salem	22
Namakkal	17
Karur	62
Trichy	41
Dindigul	13
Madurai	11
Virudunagar	24
Total	315

Source: Swaminathan and Jana (1992).

Climate Change Likely to Increase Risk of Hunger

Climate change is likely to undermine food production in the developing world, while industrialized countries could gain in production potential, FAO Director-General Jacques Diouf stated in a speech at the M.S. Swaminathan Foundation Conference in Chennai, India (August 7, 2007).

Crop yield potential is likely to increase at higher latitudes for global average temperature increases of up to 1°C to 3°C depending on the crop, and then decrease beyond that. On the contrary, at lower latitudes, especially in the seasonally dry tropics, crop yield potential is likely to decline for even small global temperature rises, which would increase the risk of hunger.

Greater frequency of droughts and floods would affect local production negatively, especially in subsistence sectors at low latitudes.

Rainfed agriculture in marginal areas in semi-arid and sub-humid regions is mostly at risk. According to Diouf, India could lose 125 million tons of its rainfed cereal production — equivalent to 18 percent of its total production.

The impacts of climate change on forests and on forest dependent people are already evident in increased incidences of forest fires and outbreaks of forest pests and diseases. Climate change adaptation

(*Continued*)

(Continued)

will be needed in a variety of ecosystems, including agro-ecosystems (crops, livestock and grasslands), forests and woodlands, inland waters and coastal and marine ecosystems.

Using New Biotechnologies

Science and technology must spearhead agricultural production in the next 30 years at a pace faster than the Green Revolution did during the past three decades. Exploiting the new biotechnologies, including in particular *in vitro* culture, embryo transfer and the use of DNA markers, can supplement conventional breeding approaches, thus enhancing yield levels, increasing input use efficiency, reducing risk, and enhancing nutritional quality.

But, most genetically modified (GM) crops cultivated today are focused on herbicide tolerance and resistance to pests. Development of GM crops with traits valuable for poor farmers, especially within the context of climate change — such as resistance to drought, extreme temperatures, soil acidity and salinity — is not yet a reality.

Commodification

According to the social activist of India, Vandana Shiva, the G8 countries are mainly responsible for the increasing commodification and privatization of the world's resources, especially the biodiversity which mainly resides in the developing countries. Indeed, Vandana Shiva argues that the policies of the G8 countries have increased the cost of seeds to such an extent that the farmers of the developing countries can hardly afford them. The GM technology made it necessary for the farmers to buy fresh seeds every year. The international seed market is controlled by large multinationals such as Monsanto which control the seed prices of GM crops. Consequently, many Indian farmers who planted Bt cotton went into bankruptcy and several committed suicide rather than face their debtors. In conclusion, one might say that the G8 countries have indirectly caused these deaths.

(Shiva 2007)

BT Cotton Farmers Rake In

New Delhi — Cotton farmers have earned an additional income of Rs 7,039 crore in 2006 after a 50% increase in yield due to use of Bt cotton seed, a study conducted jointly by the Associated Chamber of Commerce & Industry (Assocham) and IMRB International has revealed.

Introduction of two-stacked genes Bollguard II Bt cotton has benefited farmers by making a saving on pesticide use to the tune of Rs 1,600 per acre. Bollguard II Bt cotton was allowed for commercial cultivation in central and western India in 2006 and according to Assocham-IMRB study, farmers growing conventional cotton spend Rs 2,900 per acre on pesticide use, while those growing Bt cotton (with one gene cry 1 Ac) spend Rs 2,000 per acre and farmers growing Bollguard II Bt cotton spend Rs 1,300 per acre.

Thus the farmers growing Bollguard II Bt cotton have the advantage of saving Rs 1,600 per acre on pesticide use over those growing conventional cotton. Bollguard II Bt cotton has the advantage of controlling both Bollworm and the sucking pest, Spodopetra, while Bt cotton (with one gene cry 1 Ac) controls only Bollworm. The Bt technology does not totally eliminate pesticide use, it curtails the number of sprays, said the study. The number of sprays was about 4.6 times less per acre for control of Bollworm on Bt cotton (with one gene cry1Ac). The number of sprays was two times less per acre for control of Spodopetra on Bollguard II Bt cotton.

Bollguard II farmers earned a profit of Rs 15,136 per acre, while farmers growing Bt cotton (with one gene cry 1 Ac) earned a profit of Rs 12,541 per acre. Farmers growing conventional cotton earned a profit of only Rs 4,784 per acre, the study said, and added "this is despite the fact that Bt seeds are 2.5 times costlier than conventional seeds" and increased used of water and fertilisers. Another study conducted by Assocham in collaboration with Indicus Analytics showed that the area under cotton increased to over 8 million acres with 2 million farmers cultivating it.

New green revolution. *Outlook Business*. New Delhi, December 5, 2006.

Bt Cotton Consumption Not Cause of Animal Death

Mr R.K. Sinha, Executive Director of The All-India Crop Biotechnology Association (AICBA) asserted that the deaths of sheep, goat and cattle were not related to consumption of Bt cotton leaves and plants (July 10, 2007). He was responding to a report by the Centre for Animal Disease Research and Diagnosis of IVRI (Indian Veterinary Research Institute) to the Genetic Engineering Approval Committee (GEAC).

He added further that none of the reports or analyses conclude that Bt toxin was responsible for sheep mortality in Adilabad and Warangal districts of Andhra Pradesh. The GEAC reviewed reports published in international journals and views expressed by the Punjab State Agriculture University (Ludhiana). However, the compounds such as nitrates and nitrites found in the viscera of sheep were in no way connected to Bt cotton.

Cattle, sheep death not related to Bt cotton consumption. *The Hindu Business Line*. Chennai, India, July 10, 2007.

Kerala — "GM Free Zone"

TRIVANDRUM — There are regional differences in the acceptance of GM technology. After Uttarakhand, Kerala has said a firm no to Genetically Modified (GM) seeds test in the state. For instance, in the State of Kerala in India, Agriculture Minister Mullakkara Ratnakaran has conveyed the government stand in this regard in a letter to Federal Agriculture Minister Sharad Pawar. He has urged the federal minister to declare the state as 'GM-free zone'.

He said that the government would not allow the current move to permit a Maharashtra-based firm to do experimental cultivation of GM seeds in the northern district of Palghat, which is known as the rice bowl of Kerala.

The letter is in the wake of reports that the Federal government was in the process of approving a proposal by an agency under the Monsanto Corporation, a multi-national seed monopoly company, to

(*Continued*)

(Continued)

conduct field trials of GM Rice after the field trials on Bt-Cotton in Maharashtra. Ratnakaran has urged Pawar to stop the move as it would endanger the biodiversity of the state. Half of the geographical area of the state is part of the Western Ghats, which is a biodiversity hotspot.

The minister told a private television channel earlier that the planting of Bt-Cotton was mainly responsible for suicide by a large number of farmers in Vidharbha and Andhra Pradesh and added that the state government will not allow such things to happen in the state.

Chief Minister V.S. Achuthanandan has also made known his opposition to GM seed trials in the state. He said that the state government will not allow it under any circumstances as it would be catastrophic to the local farmers.

Kerala: GM-free status sought for State. *The Hindu Business Line*. Chennai, India, July 13, 2007.

'India's GM Regulation isn't Equipped for the Coming Flood of New Applications'

Dr. C Kameswara Rao pointed out in the *Indian Express* (October 4, 2007) that India is poised to become the world's second largest cotton producer, as reported in this newspaper, thanks largely to higher yields from genetically modified cotton. The most important follow-up question to this good news is why are the chances of more breakthroughs being held back by bureaucratic infighting. The department of science and technology and the environment ministry have been battling for months over whose nominee should head the proposed national biotechnology regulatory authority. This body, recommended by the M.S. Swaminathan task force, is to function as a single window for preliminary approval, research evaluation and final clearance, taking care of inter-ministerial wrangles that characterise the current three-stage process. It is surprising there has been no top-of-the-government intervention as yet to sort out this bureaucratic turf battle.

(*Continued*)

(Continued)

The surprise is greater because India, unlike, say Europe, has a fairly rational GM policy. Of course there have been examples of unnecessary obfuscation by government regulators, which is also one of the reasons Swaminathan advocated a single-window system.

But overall, India's GM policy has been a reasonably good mixture of positive attitude to new technology and abundant caution while testing it. That the government has set up a special committee that will evaluate independent assessments of field trials for Bt brinjal — the reason is that Bt brinjal, if cleared, will be India's first GM food crop — is one example of cautious policy. On the other hand, the system of event based clearance — this means once a GM crop from one party has been cleared, other parties planning to employ the same variety need not seek approval — shows policymakers have learnt flexibility.

But this government, which says farming is a high priority, should consider itself warned: its current GM regulation is equipped neither to efficiently handle the flood of applications coming India's way — GM tomato and golden rice, among others — nor to manage the resultant high intensity NGO activism.

Indian Express, Banglore, India, October 4, 2007.

Latin American and Caribbean Region

Projections of agriculture and food supply are not encouraging. Since 1950, 25% of the globally world's topsoil has been lost and continued erosion at the present rate will result in the further irreversible loss of at least 30% by the year 2050. A similar percent may be lost to land degradation. FAO has projected that over the next 20 years: arable land in developing countries could be expanded by 12% at acceptable economic and environmental costs (although such expansion would inflict damage to the remaining biodiversity). The increase in food demand that is expected to occur in these countries during the same period is 61%. In Latin American countries, the only large

(Continued)

(Continued)

tracts of land that have the potential for conversion to arable land are the Brazilian Cerrados and the Llanos of Colombia and Venezuela.

The Andean region comprises five countries: Bolivia, Colombia, Ecuador, Peru and Venezuela, with a total area of 4,104,816 km^2, and a total population of 103 million. Numerous plants that have provided food for humanity came from the Andean Region including: Potato (*Solanum tuberosum*), Sweet Potato (*Ipomoea batatas*), Maize (*Zea mays*), Tomato (*Lycopersicon esculentum*), Beans (*Phaseolus vulgaris*), Cassava (*Manihot esculenta*), Peanut (*Arachis hypogea*), Pineapple (*Ananas comosus*), Cocoa (*Theobroma cacao*), Peppers (*Capsicum annum, C. pubescens & C. frutescens*), Papaya (*Carica papaya*), la Mora de Castilla (*Rubus glaucus*), Cotton (*Gossypium hirsutum & G. barbadense*) and Tobacco (*Nicotiana tabacum*).

The full nutritional and medicinal value of many other plants of LAC origin has not yet been determined. These include grains such as Quinua (*Chenopodium quinoa*); Kañiwa (*Chenopodium pallidicaule*) and Amaranthus (*Amaranthus caudatus*). Tubers such as Bitter Potato (*Solanum juzepczukii*); Oca or Ibia (*Oxalis tuberosa*); Ulluco (*Ullucus tuberosus*); Mashwa or Cubio (*Tropaeolum tuberosum*). Roots such as Arracacha (*Arracacia xanthorrhiza*); Achira (*Canna edulis*); Jicama (*Pachyrhisus tuberosus*); Yacón (*Polymnia sonchifolia*); Mauca or Chago (*Mirabilis expansa*); Maca (*Lepidium meyenii*) and Ajipa (*Pachyrhisus ahipa*). Legumes such as Cacha (*Phaseolus polyanthus*); Tarwi (*Lupinus mutabilis*); Torta (*Phaseolus lunatus*); Pajuro (*Eritrina edulis*) and Pacay (*Inga feuillei*). Vegetables such as: Zapallo (*Cucúrbita máxima*) and Achokcha (*Cyclanthera pedata*). Fruits such as: Pitaya (*Acanthocereus sp*); Pepino (*Solanun variegatum*); Uchuva (*Physalis peruviana*); Tomato tree (*Solanum betacea*); Granadilla (*Passiflora ligularis*); Curuba (*Passiflora mollisima*); curuba de Indio (*Passiflora mixta*); Tin-Tin (*Passiflora pinnastistipula*); Curuba Antioqueña (*Passiflora antioquiensis*); Badea (*Passiflora quadrangularis*); Cherimoya (*Annona cherimolia*); and Ciruela de Fraile (*Bunchosia armeniaca*).

(*Continued*)

(Continued)

Biotechnology

During the last ten years, the number of field tests of transgenic crops estimated to be near 870 in this region. However, most transgenic crops tested in agro-ecosystems of Latin American countries have been developed in northern industrialized countries.

Cotton and some types of corn are of great economical importance in the region. In certain regions of Mexico, transgenic varieties of cotton resistant to insects or herbicides are grown commercially. Since 1995, more than 100,000 hectares have been planted in Mexican cotton regions. This experience is of great importance to the other countries in the region for analyzing the level of monitoring needed, especially with respect to the behavior of insect populations and to the development of resistance by insect pests.

Argentina and Mexico are noted in their adoption rate of transgenic soybean resistant to glyphosate (Round Up).

Biosafety

While some countries have Biosafety regulations, the majority do not. What is even more critical is that many do not have the inter-disciplinary personnel needed to carry out risk analyses and risk management within a methodological framework as stipulated by modern regulations. Hence their potential advantages can not be utilized to guarantee necessary Biosafety requirements to protect the environment, human health, the agricultural production, and the equitable distribution of the benefits for the welfare of its inhabitants.

Latin American region represents about 6% of the total field trials now taking place in developing countries worldwide. Argentina, Brazil, Chile, and Mexico have had the largest number of trials and these have been increasing steadily. Bolivia, Brazil, Colombia, Cuba, and Peru established Biosafety legislation following the mandate of the Convention of Biological Diversity. Chile, Costa Rica and Uruguay

(*Continued*)

(Continued)

adapted existing legislation for seeds and plant health inspection services.

Countries without specific legislation in the region include Dominican Republic, Ecuador, El Salvador, Guatemala, Honduras, Nicaragua, Panama, Venezuela, and the majority of the Caribbean countries.

(Modified from a report by Rodrigo Artunduaga-Salas, UNDP)

The Significance and Importance of the Biosafety Protocol for Developing Countries

The Biosafety Protocol is very important, particularly for developing countries, for many reasons. It is an international law that regulates genetically modified organisms (GMOs). This is a recognition of the fact that GMOs are inherently different and carry special risks and hazards, and hence need to be regulated internationally. Countries always had the sovereign right to regulate GMOs and their products at the national level; the Protocol now establishes an internationally binding framework of minimum standards.

The Protocol establishes the foundations of international law on the transboundary movement of GMOs. While many aspects of biosafety regulation are best addressed by national biosafety legislation, many aspects relating to the transboundary movement of GMOs are difficult to regulate domestically. An international law is therefore necessary.

Developing countries are and will continue to be the prime importers of GMOs and products derived from GMOs, which are exported primarily from the North. Public rejection in the North of GMOs and their products means that increasingly, markets are being sought for GMOs and their products in developing countries. And most developing countries do not yet have national biosafety laws or regulations. Developing countries also face an even greater environmental risk than countries of the North because most of the global centers of crop origin and diversification are located in the South.

The Precautionary Principle has been reaffirmed and operationalized in the decision-making procedures in the Protocol. This means that in the absence of scientific certainty, Parties should err on the side of caution and ban or restrict the import of the GMOs on account of its potential adverse effects. The sound reaffirmation of the Precautionary Principle in the Protocol also further establishes this Principle as a principle of international law.

Some Key Weaknesses in the Biosafety Protocol

However, the Protocol, though significant, was a heavily negotiated text and is riddled with many deficiencies. Many categories of GMOs have been excluded from the general scope of the Protocol and from the prior informed consent procedures. But the risks from all GMOs are the same, whether they are used in agriculture, medicine, or research, and regardless of whether they are classified as commodities or pharmaceuticals.

Products derived from GMOs (e.g. soy proteins, a product of transgenic soya beans) are excluded entirely from the general scope of the Protocol. As such, products derived from GMOs remain unregulated internationally. But naked DNA, which is the genetic material inserted into the recipient organism, subsists in the products derived from GMOs and has been shown to survive passage through the gut and can enter the blood stream.

The Protocol also does not apply to the transboundary movement of genetically engineered pharmaceuticals for humans that are addressed by other "relevant international agreements or organizations".

The bulk of GMOs are excluded from the advance informed agreement (AIA) procedure. The AIA procedure requires that potential importers of GMOs are first notified and furnished with relevant information. This triggers a process of decision making based on risk assessment and the Precautionary Principle. While GMOs for food, animal feed, or processing are clearly within the general scope of the Protocol, they are excluded from the AIA procedure. These form the bulk of traded GMOs — around 90% of the main GMO exporters' exports (e.g. soya, canola, maize).

Likewise, GMOs that are destined for contained use (the Protocol defines contained use as specific measures that limit the contact and impact of GMOs on the external environment) and GMOs in transit (i.e. that are passing through the territory of a third party) are also excluded from the AIA procedure.

National Biosafety Implementation

The Protocol only sets the international framework for the regulation of GMOs, in particular the transboundary movement of GMOs. It must be implemented at the national level; indeed, Parties have an obligation under the Protocol to do so. In order to properly and effectively implement the Protocol, a few key requirements are critical. A potential importer must be aware of the fact that a particular GMO is likely to enter into its territory, and must also be able to assess its potential risks and hazards in order to make a decision.

In other words, to effectively implement the Biosafety Protocol in line with its objectives and provisions, a country must have the following:

1. Full knowledge that GMOs will be crossing its national boundaries; and
2. Ability to assess the safety or otherwise of GMOs, and make a decision either to ban or import the GMOs with or without conditions.

Full Knowledge of Pending Import of GMOs

The "backbone" of the Protocol is the advance informed agreement (AIA) procedure which regulates the transboundary movement of GMOs. It requires the exporting Party to obtain the prior informed consent of the importing Party before GMOs can cross national boundaries. The exporting Party must first notify the importing Party that it intends to export GMOs to the other Party. The obligation to notify thus lies with the exporting Party. This should have ensured that countries would be fully aware of any GMOs that will be entering into its territory.

However, many categories of GMOs are excluded from the AIA procedure, and some categories of GMOs are even excluded entirely from the general scope of the Protocol. Only the first intentional transboundary movement of GMOs for intentional introduction into the environment of the Party of import (e.g. for planting, field testing) will be subject to the AIA procedure. This means that subsequent exports will not be subject to the AIA procedure. This is severely lacking as on account of the uncontrollable, random nature of the genetic engineering process, each transgenic line will be distinct, even though the same materials, gene constructs, and vector systems are used. Also, because of the inherent instability of the transgenic lines, further changes may occur during cultivation, so that, in effect, the properties will become quite different from the originally approved line.

GMOs intended for direct use for food, feed, or for processing, which form the bulk of traded GMOs, are excluded from the AIA procedure, and a different mechanism applies. Once domestic approval has been given for any GMOs for food, feed, or for processing that may be traded internationally, the approving Party must make this information available to the Biosafety Clearing House. (The Biosafety Clearing House is basically a website administered by the Secretariat to the Convention on Biological Diversity.)

This is basically the extent of the obligation of the potential exporting Party. A country cannot even be sure if the GMOs will be shipped into its territory, but will have to initiate procedures for assessing whether or not the GMOs should be admitted into its territory. In other words, a domestic approval by one Party shifts the onus onto all other countries to decide whether or not they will accept the GMOs, when countries do not even know whether the GMOs will even be exported to their country.

Ability to Assess the Safety of GMOs

To decide on whether or not to import a GMO, Parties will base their decision on risk assessment and the Precautionary Principle. Socioeconomic considerations can also be taken into account, but this provision is qualified. This consideration is, however, of particular importance to developing countries.

The information that must be provided as a minimum, along with the notification by the exporting Party that it intends to export the GMO, is listed in an annex to the Protocol. This information includes a risk assessment. This risk assessment will be furnished by the exporting Party. The importing Party must evaluate the risk assessment in order to make an informed, scientific decision.

One of the difficulties here lies with the issue of conflict of interest — the risk assessment would have been conducted by or commissioned by the exporter. Another difficulty lies with whether or not the information provided by the exporting Party is sufficient for the importing Party to assess the safety of the GMOs. The Protocol only specifies the minimum information required. The exporting Party is therefore under no obligation to provide more than the minimum information required. For imports of GMOs for food, feed, or processing, the minimum information that the exporting Party has to supply is even less than that required for the other categories of GMOs.

The information required to be supplied as a minimum and the specifications for risk assessment that are provided for in the Protocol do not properly take into account the molecular genetic characterization data that indicates the stability of the transgenic line. This information is crucial for a number of reasons. Failure to supply informative data on these characterizations will mean in practice that unstable lines may be approved, which will change its characteristics in successive generations of growth; or multiple transgenic lines, all with different characteristics may be released after a single cell line has been approved. These data are also crucial for risk assessment and risk management, and for identification and traceability when considering liability.

Thus, these two considerations form the backdrop and the context for considering the main areas for capacity building in developing countries.

Key Areas for Capacity Building in Developing Countries

Translated into the national operative framework, countries and developing countries in particular need to build up their capacity on three key fronts: biosafety regulation, scientific capacity, and monitoring and enforcement capabilities.

Biosafety Regulation

Clearly, the Biosafety Protocol has its shortcomings. However, all these can and should be made good by national biosafety legislation. The Protocol sets down minimum standards. This is explicitly recognized in the Protocol. Parties may take action that is "more protective of the conservation and sustainable use of biological diversity than that called for" in the Protocol.

The best way to ensure that obligations under the Protocol are effectively implemented is to enact national biosafety legislation. This will have the force of law and its implementation can be backed by punitive measures. There are also many aspects of biosafety regulation that are not addressed by the Protocol and which can be best addressed by domestic regulation. Comprehensive national legislation will also ensure that the unique risks and hazards of GMOs are fully taken into account and regulated specifically and appropriately.

Parties are obliged to implement their obligations under the Protocol. In order to do so effectively, countries require comprehensive national biosafety legislation. While the strengthening of the Protocol and rectification of its deficiencies should be the long-term goal, developing countries should work towards strengthening the biosafety regime within the national framework.

The Protocol is not comprehensive in scope, does not address all aspects of biosafety regulation, and covers only some aspects of transboundary movement of GMOs. National biosafety legislation must strive to fill the gaps in the Protocol to ensure the highest standards of biosafety, based on the Precautionary Principle.

Key elements in national biosafety legislation which would plug the gaps in the Protocol and ensure tighter biosafety regulation would include requiring approvals for all activities relating to all GMOs and their derived products on a case by case basis. Approvals should also be required for every stage of GMO development — from research in contained conditions to field trials, and to full-scale releases into the environment. In addition to risk assessment, a cost-benefit analysis should also be conducted to determine if there is even a need for the GMOs, and if there are sustainable or safer alternative technologies.

The formulation of national biosafety legislation must benefit from an open and participatory process. Given the volume and strength of worldwide public disquiet and consumer opposition towards genetic engineering biotechnology, a national process that is transparent, accountable and which involves all levels of public participation is crucial.

Scientific Capacity

Once a biosafety regulatory system is in place, countries will need to be able to make decisions on applications to export GMOs to its country. The decision-making body must have the backing of local scientific capacity to screen applications and make decisions on import. The training and capacity building of local scientists in biosafety assessment is therefore critical. Government scientists, university scientists, scientists from research institutions and scientists from civil society organizations should all be part of the local scientific pool of expertise.

The scientific body may have to do a number of things. Most importantly, it will have to evaluate the risk assessment submitted by the exporting Party as part of the information supplied along with the notification to the importing Party. It could also conduct its own risk assessment or instruct the exporting Party to undertake another risk assessment if it is not fully satisfied with the risk assessment supplied by the exporting Party.

In light of new scientific information about potential adverse effects, the importing Party can review its decision. Where a change in circumstances has occurred that may affect the outcome of the risk assessment, or additional scientific information has become available, the exporting Party may request the importing Party to review its decision. All this requires scientific biosafety capacity.

It is also advisable that the scientific body should not comprise scientists from industry because of the inevitable conflict of interest.

Monitoring and Enforcement Capability

There has to be an effective national monitoring and enforcement capacity. The best laws, regulatory mechanisms, and scientific expertise will be of little use if there is no effective monitoring and enforcement capability to ensure sound biosafety regulation.

Monitoring and enforcement has to take place on a few levels. Firstly, scientific developments globally have to be tracked and monitored from day to day. New scientific evidence of actual and potential risks of genetic engineering biotechnology and of the food, crops, pharmaceuticals, etc. is constantly emerging. These developments in the scientific arena must be closely followed and considered in risk assessment and risk management. New scientific information is also a basis for the decision-making body to review its decision.

Secondly, countries must be on the alert for GMOs that may slip through the regulatory process and enter into the country inadvertently. This will entail an extremely vigilant customs at all entry points into the country.

Countries also need to be aware of other means by which GMOs could inadvertently pass through their borders. This could be through foreign agencies providing bilateral aid and food aid. A substantial part of food aid could comprise genetically engineered seeds and food. The growing consumer rejection of genetically engineered foods in Europe, and which has now spread to North America, is giving rise to concerns that excess and unwanted genetically engineered seeds and food would be dumped on developing countries. This could bypass the procedures and mechanisms for regulating the transboundary movement of GMOs.

An article on March 30, 2000 in the *London Independent* reported that the US Department of Agriculture is exporting hundreds of thousands of tons of genetically engineered maize to the Third World through the United Nations and American aid agencies. According to the article, American farmers who are facing a rejection of their genetically engineered crops are left with one unquestioning market — emergency aid, which is the last unregulated export market open to the US farmers.

This is unfairly taking advantage of a country's urgent needs, and violates the spirit and tenor of the Protocol. It undermines all that developing countries had fought so hard for in the Protocol negotiations. Countries in crisis should not have to be faced with the dilemma between allowing their people to starve to death and allowing their genetic pool to be contaminated and potentially hazardous food to be fed to their populations. Aid programs and international agencies should be prohibited from including genetically engineered food, seeds, and crops in their aid or projects.

All this implies that countries will also require some means of testing or access to testing facilities to find out whether or not an organism is genetically engineered. This is particularly relevant in the light of the GMO scandal that hit Europe in May 2000, where 11,600 acres of farmland were found to be planted with GM-contaminated oil seed rape.

Proposals for Action

1. Governments must sign and ratify the Cartagena Biosafety Protocol as soon as possible in order for it to come into force.
2. A national committee must be set up to formulate national biosafety legislation that meets the minimum requirements of the Biosafety Protocol and goes beyond that to provide for

comprehensive biosafety regulation and a stricter national biosafety regime.

3. An approvals committee must be set up that will have the backing of local scientific expertise, and should also comprise representatives from civil society organizations and exclude industry representatives.
4. Effective monitoring and enforcement mechanisms must be established.
5. Capacity for further negotiations must still be built up. The interpretation and implementation of the Biosafety Protocol will be an ongoing process. There are also provisions in the Protocol — e.g. liability and redress, and segregation and identification of GMOs for food, feed, or processing — which still have to be finalized.
6. Monitor developments in other fora that could undermine the Protocol (e.g. World Trade Organization agreements) or which could be used to strengthen the Protocol.

8

Agricultural Biodiversity

> The future of global food security depends on the success of our efforts in the conservation and enhancement of agrobiodiversity.... It is now widely acknowledged that tribal and rural women have not only conserved biodiversity of great economic, cultural and social value, but have also enhanced them through selection and value addition through knowledge and information. Most of the domesticated plants and animals of today were identified centuries ago. The medicinal plants we now use were identified by our ancestors among the wild flora for specific uses such as food, fibre, oil and medicine. Articles 8 (F), 10 (C and D), and 15 (7) of the CBD all recognize the role of indigenous and local communities in the conservation and improvement of genetic resources. (Excerpt from Swaminathan 1996, preface).

Biodiversity is the total variety of all living entities on this planet, including all living organisms as well as their surrounding habitats, or ecosystems, and the genetic material of which they are made. Agrobiodiversity is that component of biodiversity that contributes to food and agriculture production. The term "agrobiodiversity" encompasses within-species, species and ecosystem diversity.

Agricultural biodiversity of all food species is a vital subset of general biodiversity. It is highly threatened by the globalization of food markets and tastes, intellectual property systems, and the spread of unsustainable industrial food production; but it provides the basis for the food security and livelihood security of billions of people and the development of all food production, including for industrial agriculture and for the biotechnology (life) industries. It is the first link in the food chain, developed and safeguarded by farmers, herders, and fishermen throughout the world.

Modern agricultural and development practices have persistently neglected and eroded agrobiodiversity and indigenous knowledge. As human population growth and habitat destruction have accelerated beyond all expectations, they have accelerated the damage at the grassroots level. The narrow focus on market-oriented agricultural components, especially genetically modified (GM) crops, and cash crop farming have severely damaged the conservation of agrobiodiversity. Fueled by rapidly increasing population pressure, this trend will continue for a long time in the foreseeable future, eliminating many options for food security and nutrition among the rural poor. Consequently, the erosion and neglect of agrobiodiversity and the associated indigenous knowledge will diminish the capacity of farmers to control their subsistence systems, including their food security and nutrition. Their fate will be increasingly controlled by large multinational agribusiness corporations and government beaurocrats.

The variability of animals, plants, and microorganisms used directly or indirectly for food and agriculture (including, according to the Food and Agriculture Organization (FAO) definition, crops, livestock, forestry, and fisheries) comprises the diversity of genetic resources (varieties, breeds, etc.) and species used for food, fodder, fiber, fuel, and pharmaceuticals. It also includes the diversity of nonharvested species that support production (e.g. soil microorganisms, predators, pollinators) and those in the wider environment that support agroecosystems (agricultural, pastoral, forest, and aquatic), as well as the diversity of the agroecosystems themselves.

Although the term "agricultural biodiversity" is relatively new, the concept itself is quite old. It is the result of the careful selection and inventive developments of farmers, herders, and fishermen over millennia. Agricultural biodiversity is a vital subset of biodiversity. It is a creation of humankind whose food and livelihood security depend on the sustained management of those diverse biological resources that are important for food and agriculture.

Agricultural biodiversity, also known as agrobiodiversity or the genetic resources for food and agriculture, includes the following:

- Crop varieties, breeds of livestock, fish species, and various wild species within field, forest, rangeland, and aquatic ecosystems; and
- Nonharvested species such as soil microbiota and pollinators, and other species in the wider environment.

Agricultural biodiversity results from the interaction between the environment, genetic resources, and management systems and practices used by culturally diverse peoples, resulting in the different ways land and water resources are used for production. It thus encompasses the variety and variability of animals, plants, and microorganisms which are necessary to sustain key functions of the agroecosystem as well as its structure and processes for, and in support of, food production and food security (FAO, 1999).

Agricultural biodiversity has spatial, temporal, and scale dimensions, especially at agroecosystem levels. These agroecosystems — ecosystems that are used for agriculture — are determined by three sets of factors: the genetic resources, the physical environment, and the human management practices. There are virtually no ecosystems in the world that are "natural" in the sense of having escaped human influence. Most ecosystems have to some extent been modified or cultivated by human activity for the production of food and income and for livelihood security.

The interaction between the environment, genetic resources, and management practices determines the evolutionary process, which may involve, for instance, introgression from wild relatives, hybridization between cultivars, mutations, or natural and human selections. This process eventually results in genetic material (farmers' crop varieties or animal breeds) that is well adapted to local abiotic and biotic environmental variation.

Agroecosystems may be identified at different levels or scales, for instance, a field/crop/herd/pond, a farming system, a land-use system, or a watershed. These can be aggregated to form a hierarchy of agroecosystems. Ecological processes can also be identified at different levels and scales. Valuable ecological processes that result from the interactions among species and between species and the environment include, *inter alia*, biochemical recycling, the maintenance of soil fertility and water quality, and climate regulation (e.g. microclimates caused by different types and densities of vegetation).

Classification of the Main Usage Categories of Wild Plants

- Food plants, including beverages for humans (seeds, fruits, leaves, stems, petioles, roots, tubers, etc.)
- Food additives, including processing agents and additive ingredients used in food preparations
- Animal food, including forage and fodder for vertebrates
- Bee plants, including pollen or nectar sources for honey production
- Invertebrate foods, including plants eaten by invertebrates useful to humans (e.g. silkworms)
- Materials, including woods, fibers, tannins, latex, resins, essential oils, waxes, oils
- Fuels, including fuelwood, charcoal, fuel alcohol
- Social uses, including masticatories, smoking materials, hallucinogens, psychoactive drugs, contraceptives, abortifacients, plants used for ritual or religious purposes
- Vertebrate poisons, including both accidental or useful poisonous plants (e.g. those used in hunting, fishing)
- Nonvertebrate poisons, including accidental and useful poisons (e.g. molluscicides, herbicides, insecticides, bactericides, fungicides)
- Medicines, including human and veterinary uses
- Environmental uses, including ornamentals; barrier hedges; windbreaks; soil improvers; erosion control; and indicators of heavy metals, pollution, or underground water
- Cosmetic and perfumery plants
- Genetic resources, including wild relatives of crops

What role can precision breeding play in bringing the ecofarming movement forward? The following are a few of the major scientific issues needing particular attention:

- Soil health care: Maintenance of soil health requires attention to the physical, chemical, microbiological, and erodibility characteristics of the soil.
- Water quality: The quality of irrigation water, with particular reference to salt concentration, is important in relation to crop growth.
- Plant health care: Steps will have to be taken to protect crops from the triple alliance of weeds, pests, and pathogens. Pest pressure is particularly high in tropical and subtropical agriculture, since crops as well as alternate hosts are available in the field, particularly throughout the year.
- Genetic homogeneity: Experience has shown that genetic homogeneity enhances genetic vulnerability to pests and diseases. Monoculture of transgenic crop varieties over large areas will enhance prospects for both the breakdown of resistance and the outbreak of pest epidemics.
- Abiotic stresses: With intensive agriculture, problems of salinization, waterlogging, and pollution are increasing in intensity. Bioremediation techniques will hence become increasingly important. Droughts, floods, cyclones, and other natural calamities pose additional threats to crop security. The consequences of potential changes in climate as a result of global warming are yet to be understood fully, but it is clear that anticipatory research should be initiated to meet potential adverse changes in temperature, precipitation, sea level, and ultraviolet B radiation.
- Postharvest management: Uniform ripening, uniform skin color, processing and keeping quality, and capacity to withstand transportation over long distances are all becoming important in the market, particularly in vegetables, fruits, and flowers. Globalization of trade is opening up new markets for agricultural produce, but markets are also becoming very choosy in terms of the quality of the produce.

Thus, there will be a need for genetic material which can help to reduce or eliminate dependence on market-purchased chemicals on the one hand and enhance adaptation to market preferences on the other. Research on bioremediation techniques will have to be stepped up to clean up problems arising from soil and water pollution. It is not surprising that the very first patent given to any living organism was for a microorganism developed via genetic engineering by Dr Ananda Chakrabarty for cleaning up pollution caused by oil spills (Chakrabarty 1981).

Blending of Recombinant DNA Technology and Organic Farming Methods

For every problem, there is a solution. Methods are being developed to find substitutes for antibiotic markers in recombinant DNA experiments. Also, techniques which help to be as precise as possible in the transfer of alien DNA are being standardized. It is likely that most of the current biosafety and environmental concerns associated with GM crops will be satisfactorily addressed scientifically during the next few years, so that precision breeding will become an important component of an economically and ecologically efficient precision farming system. The following examples from the work and approach of the scientists of the M.S. Swaminathan Research Foundation will help to illustrate the power of blending traditional practices with frontier technologies.

Prebreeding and Participatory Breeding

An integrated prebreeding procedure leading to the production of novel genetic combinations and designer genotypes, and participatory breeding involving the development of location-specific varieties jointly with farming families, would help to combine genetic efficiency and diversity in a mutually reinforcing manner. This will help to avoid the danger inherent in spreading single genotypes over large areas. Also, sustainable agriculture needs for its sustenance location-specific varieties.

At the M.S. Swaminathan Research Foundation (MSSRF), a group of scientists headed by Dr Ajay Parida have been working for the past 8 years on the transfer of salinity-tolerance genes from mangrove tree species to annual crops. This program was initiated with support from the Department of Biotechnology, Government of India, to prepare genetic material which will prove to be useful if sea levels rise, thereby safeguarding coastal agriculture.

One such gene, *betaine aldehyde dehydrogenase* (*BADH*), cloned from a highly salt tolerant mangrove species, *Avicennia marina*, is currently being evaluated in transgenic tobacco and Brassica systems for its efficacy. *BADH* converts betaine aldehyde to glycine betaine. Glycine betaine is an effective, compatible solute and its accumulation confers salinity tolerance in plants. The transgenic tobacco and Brassica, overexpressing the *BADH* from Avicennia, conferred salinity tolerance up to 250 mM NaCl. Work on the isolation of a gene that can convert the ubiquitous choline into betaine aldehyde is being actively pursued. Other genes relating to stress resistance isolated and characterized from the mangrove species include *catalase* (*CAT*), *superoxide dismutase* (*SOD*), *glyoxalase*, and *sodium hydrogen antiporter*; these genes are being evaluated for their expression in transgenic systems. Also, transformation work is in progress in rice and Vigna (Ajay Parida, personal communication).

Once transgenic plants with the desired salinity tolerance are developed, they will be used, after obtaining the necessary clearance from the regulatory authority, in breeding programs undertaken jointly with farm families. The aim will be to transfer to numerous locally adopted varieties the salinity tolerance character in both coastal and inland areas, where salinity is a problem.

Bioremediation: Sequestration of Salt

Salinity is responsible for major crop losses, particularly in semiarid and irrigated agriculture. High salinity in soil may result from excessive irrigation or the excessive application of chemical fertilizer. Usually, sulfates, chlorides, and bicarbonates of Na^{+}, Mg^{2+}, and Ca^{2+}

contribute to the salinity of soil. Sodium (Na^+) is the predominant soluble cation in most saline soil water, particularly in coastal areas. An alternate approach for practicing crop cultivation in saline environments is the amelioration of soil salinity in agricultural habitats. Conventionally, it is done by the addition of gypsum, followed by leaching out of excess salts by flooding. A biological approach to solving this problem will be preferable. *Anabaena torulosa*, a blue-green alga, was found to grow and enrich the nitrogen status of moderately saline "Kharland" soils. *A. torulosa* was found not to intracellularly accumulate Na^+ but rather to entrap the cation in its extracellular mucopolysaccharide sheath, thereby reducing the availability of this deleterious cation to the crops. Research on the biological sequestration of salt from the soil may be rewarding.

Under a collaborative research program between the Bhabha Atomic Research Centre and MSSRF, a salt-tolerant culture of *A. torulosa*, along with AL31 were given for testing in the southern coastal region. Several field trials have been conducted so far. The trials have shown that *A. torulosa* established very well along with local *Ananaena* sp. in the field condition. Enhanced nitrogenase activity was observed in the field after transplantation. Up to 64% salt sequestration was observed when 1000 mL of innoculum was added (Sudha Nair, personal communication).

Biotechnological Applications in Organic Farming

MSSRF scientists are integrating a wide variety of biotechnological applications to improve the productivity, profitability, stability, and sustainability of major cropping systems. Among the techniques of particular value are vermiculture, biopesticides, biofertilizers including stem-nodulating green manure crops, azolla, blue-green algae, and improved rhizobial strains. Such biopesticides and biofertilizers are best produced by village-level self-help groups. In fact, there are good opportunities for gainful employment in the area of producing such biological software for sustainable agriculture.

The era of precision breeding opened up by advances in genomics and genetic engineering has become an ally in the movement for environmentally sustainable advances in agriculture, a phenomenon I have christened as the "evergreen revolution" (see Swaminathan 1999). Knowledge is a continuum. The 20th century was marked by spectacular advances in crop productivity, triggered by Mendelian breeding. The 21st century will witness even more spectacular progress from an intelligent integration of Mendelian and molecular breeding.

The enormous power which transgenic technology has conferred on human kind imposes an ethical obligation, which should be discharged by developing a transparent and multistakeholder method of risk-benefit analysis, capable of inspiring public confidence and trust. At the same time, however, the tendency to decry all advances in the breeding of transgenic crops will not be in the interest of sustainable food and nutrition security. India's population exceeds one billion, and there is no option in the future except to produce more crop per unit of land and per drop of water.

Linking Food and Ecological Security

Current government stocks of wheat, rice, and other grains in India exceed 45 million tons. The Indian Government may have to purchase another 15–20 million tons of wheat and rice in the near future. A considerable proportion of these stocks remain in gunny bags and temporary storage structures. The Government of India has announced a scheme for the construction of large numbers of rural godowns. Severe drought in several parts of Rajasthan, Gujarat, Madhya Pradesh, and other states is compounding the problems of poverty-induced endemic hunger and drinking water scarcity.

Community Grain Banks

Professor M.S. Swaminathan (see Appendix) suggested that the time is therefore opportune to launch an imaginative Community Grain Bank movement. On average, one ton of wheat or rice supports the food needs of five individuals in India. Community grain banks each with 200 tons of wheat or rice — or other locally acceptable staples like ragi, jowar, bajra, and maize — could be established, to begin with, in "hunger hotspot" villages. Remote areas with poor communication, such as the desert areas of Rajasthan and hill-, tribal-, and drought-affected areas, can be given priority in starting the Community Grain Bank movement. About 25 000 grain banks can be established very soon if the Government of India will immediately approve the release of five million tons of grain for this purpose. Because large quantities of Government stocks are in gunny bags, it is easy to move them to the community grain banks, where they can be stored using the low-cost technology standardized by the Food Corporation of India. It will be sad if the Government sits over 60 million tons of food grain, allowing some of them to rot rather than taking them to places where, in Gandhiji's words, "God is Bread."

Based on the experience of the initial 25 000 village-level grain banks, another 25 000 can be established later in the year, thus using 10 million tons of the surplus stock in a socially meaningful manner. Let the first year of the new millennium be a year of decisive action in our resolve to provide every individual in the country an opportunity for a productive and healthy life.

The community grain banks can be sustained with locally procured grains, wherever feasible. They should be linked to the rural godowns scheme. The banks could function under the overall umbrella of the Gram Sabha, and can be operated by local self-help groups of women and men. This will ensure their relevance to local conditions, in addition to involving low transaction costs. The community grain banks could be used for initiating at the local level food for work, food for nutrition (i.e. distribution of food among pregnant and nursing mothers, infants, and old or infirm persons), waste land and watershed development, ecological restoration of common property resources, and establishment of community water banks (see Swaminathan, *Sunday Hindu*, 15 October 2000). They can also be the vehicles for operating the targeted public distribution, Antyodaya Anna Yojana, and other Central and State Government schemes. Thus, the community grain banks can become instruments of ecorestoration, water harvesting, and hunger elimination.

We should link conservation, cultivation, and consumption in a mutually reinforcing manner. For this, it will be useful to foster the establishment of community gene, seed, water, and grain banks in every village. Future agricultural production programs will have to be based on a three-pronged strategy designed to foster an "ever-green revolution", which leads to increased production without associated ecological or social harm. The following are the major elements of this strategy for producing more in an environment-friendly manner:

1. Defending the gains already made. This calls for the conservation and enhancement of soil and water resources as well as forests and biodiversity through an integrated package of government regulation, education, and social mobilization (through Panchayats and local bodies). The traditional "green revolution" areas are in urgent need of such an integrated natural resource management strategy, so that the pattern of present production does not erode future prospects. The Punjab, which is India's "granary" today, will become food insecure in 15 to 20 years from now if the current unsustainable land and water use practices continue. Defending the gains already achieved will also require stepping up maintenance research to ensure that new strains of pests and pathogens do not cause crop losses. Special steps are needed to prevent the introduction of invasive alien species, which are coming into the country along with imported food and agricultural commodities. These invasive alien species — like new and aggressive weeds, nematodes, etc. — can cause incalculable harm to the future of Indian agriculture.
2. Conserving and enhancing land and water resources. Water harvesting, watershed development, and economic and

efficient water use can help to enhance productivity and income considerably. Conjunctive use of different water sources should become the rule rather than the exception. Unless there is equity in water sharing, there will be no cooperation in water saving. Therefore, equitable methods of water sharing should be promoted. Where water is scarce, high-value but low-water-requiring crops should be grown. In this context, the organization of Pulses and Oilseed Villages should become a national movement. This should be a major aim of the Pulses and Oilseeds Technology Missions. Solving internal shortages of pulses and oilseeds through imports will only add to the economic woes of dryland farming communities. Pulses and oilseeds are important income-earning and soil-enriching crops in dryland areas. Various estimates of land degradation exist. The Ministry of Rural Development has also published a Wasteland Atlas of India. The following kinds of soil degradation have been quantified:

- Wind erosion: 19.7 million ha.
- Salinization: 4.1 million ha.
- Water logging: 3.1 million ha.
- Water erosion: 69.6 million ha.
- Soil fertility decline: 13.7 million ha.

Thus, there are vast opportunities for local self-help groups to launch wasteland development enterprises by identifying the precise nature of soil degradation and developing scientific restoration measures. Based on agroecological conditions, choose tree species which can help to initiate suitable enterprises. For example, a plant pesticide model of wasteland development could involve the planting of neem and melia. Appropriate species can be chosen and planted, depending on soil and water conditions, for undertaking the preparation of furniture, doors, windows, etc. or paper, fiber, or fruit packaging industries.

Ecoagriculture

Although food production systems for the world's rural poor have typically had devastating effects on the planet's wealth of genes, species, and ecosystems, that need not be the case in the future. In ecoagriculture, agricultural landscapes can be designed more creatively to take the needs of human populations into account while also protecting, or even enhancing, biodiversity. The innovative concept of "ecoagriculture" underlies the management of landscapes for both the production of food and the conservation of wild biodiversity.

Key areas of interest are the following:

- assessing the global impact of agriculture on wild biodiversity;
- reconciling biodiversity conservation and agricultural goals; and
- identifying how policies, markets, and institutions can be reshaped to support ecoagriculture.

In the tropical regions of the developing world, increased agricultural productivity is most vital for food security, poverty reduction, and sustainable development. When much of the world's wild biodiversity is threatened, it also draws on lessons learned in developed countries. Dozens of examples from around the world present proven strategies for small-scale, low-income farmers involved in commercial production. Ecoagriculture thus offers new approaches to agricultural production that complement natural environments, enhance ecosystem function, and improve rural livelihoods.

Women's Contributions

It may not be widely known to the Western world that women, especially tribal women, have historically played a very important role in conserving biodiversity in the developing countries. This is particularly true of the conservation of agricultural biodiversity in the developing countries. No one has stated this more eloquently than the eminent agricultural scientist of India, Prof. M.S. Swaminathan (1996):

Historians have long recognised the critical role of tribal and rural families, particularly women, in the selection of plants and animals for domestication. More than 10,000 years ago, when men were hunting and gathering food, women started selecting plants from the wild for cultivation. For example, at the Donyi-Polo Temple at Along in Arunachal Pradesh, there is a portrait of a woman credited with the introduction of rice into cultivation. This marked the transition from food gathering to growing. It is a mark of the acuteness of the early selectors that not a single plant has been added to the list of major domesticated crops over the past many centuries. This is why even the World Intellectual Property Rights Organisation (WIPO) has agreed to recognise the IPR rights of communities that have not only conserved biodiversity but also added value through selection and identification of the properties of economic value.

From the beginning of the 20th century, industrialised countries have enacted legislation to safeguard and reward the IPR contributions of commercial plant breeders. Since 1961 such legislation has been coordinated by the Union for the Protection of New Varieties of Plants (UPOV), which has its headquarters in Geneva along with the WIPO. The Director General of the WIPO is the Secretary General of the UPOV, which is an inter-governmental organisation currently with 63 countries as members. The International Convention for the Protection of New Varieties of Plants (the UPOV Convention) came into force in 1968 and has since been revised in 1972, 1978, and 1991. The 1991 Act, which entered into force on April 24, 1998, has further strengthened the IPR rights of plant breeders.

Chairing Commission II of the general conference of the United Nations' Food and Agriculture Organisation (FAO) in 1979, I urged that the irony of the poverty of primary conservers co-existing with the prosperity of commercial breeders should be ended by developing procedures for recognising and rewarding the contributions of the

(*Continued*)

(Continued)

former. This was followed by a resolution moved by India, Mexico, and several other countries at the FAO General Conference in 1981 emphasising the need for equity in sharing benefits from a nation's agro-biodiversity heritage. In 1983, a Commission on Plant Genetic Resources was set up by a resolution of the FAO Council, of which I was then the Independent Chairman. Deliberations in this Commission led to the birth of the concept of farmers' rights. There was a general consensus that farmers' rights should be defined as collective rights of communities involved in the conservation of agricultural plant diversity. This concept got enshrined in Article 9 of the International Treaty on Plant Genetic Resources for Food and Agriculture of the FAO (FAO Treaty), which came into force on July 29, 2004.

Much of the crop genetic diversity occurs mainly in developing countries, where farm families have been identifying and conserving economically valuable plants. Centres of civilisation have also been major centres of domestication of crop plants and conservation of agro-biodiversity, thereby indicating the intimate relationships between cultural and biological diversity. The concept of farmers' rights, however, did not find ready acceptance among most developed countries and multinational seed companies. To resolve this conflict, the Keystone Centre in Colorado, U.S., convened a series of multi-stakeholder dialogues under my Chairmanship during 1989–91. The second dialogue in this series was hosted by the M.S. Swaminathan Research Foundation (MSSRF) at Chennai in January 1990. Major multinational seed companies participated in this dialogue and issued the following consensus statement on farmers' rights:

> Farmers' rights, a concept which has been developed and adopted in FAO recognises the fact that farmers and rural communities have greatly contributed to the creation, conservation, exchange and knowledge of genetic and species utilization of genetic diversity; that this contribution is on-going and not simply something of the past; and that this diversity is extremely

(Continued)

(Continued)

valuable. Local communities bear much of the burden of protecting germplasm and the rest of the world has an obligation to help them carry out this task and help them in utilizing the material. Yet neither the marketplace nor current intellectual property systems have any way of assigning a value to this valuable material. No compensation or reward mechanisms exist.

The Chennai consensus paved the way for the benefit sharing provisions of the Convention on Biological Diversity (CBD) adopted at the Earth Summit at Rio de Janeiro in 1992, which gives explicit recognition to the rights of the primary conservers and incorporates the principles of prior informed consent and equity in benefit sharing. The MSSRF convened in January 1994 a multi-stakeholder consultation to prepare a draft Act for conferring concurrent recognition to breeders' and farmers' rights. At a follow-up consultation convened by the MSSRF in 1996, a revised draft legislation was prepared, titled "Plant Variety Protection and Farmers' Rights Act," to emphasise that the rights of breeders and farmers, who are allies in the struggle for sustainable food security, should be mutually reinforcing.

The 1996 Chennai draft provided the basic text for the final Act adopted by Parliament in 2001, under the title, "Plant Varieties Protection and Farmers' Rights Act" (PPVFR Act). This is the first legislation of its kind in the world that simultaneously recognises and rewards the contributions of breeders and farmers to the development of new crop varieties.This concept is in conformity with the provisions of both the CBD and the FAO Treaty. It satisfies the need for enacting a *sui generis* legislation for protecting the IPR of plant breeders stipulated under the WTO Agreement in Agriculture.

The Indian Act recognises the multiple roles of farmers — as conservers, cultivators, and breeders. As cultivators, farmers are entitled to keep and plant their own seeds (plant back rights). As breeders, they are entitled to the same rights as commercial breeders. As conservers, they are entitled to recognition and reward from a National Gene Fund.

(*Continued*)

(Continued)

To implement these provisions, the Government of India set up a Protection of Plant Varieties and Farmers' Rights Authority in 2005 in New Delhi under the leadership of the distinguished agricultural scientist, S. Nagarajan. The Authority and the MSSRF organised in 2006 a consultation at Koraput, an important centre of genetic diversity in rice, to develop a methodology for according recognition to tribal and rural communities for their selfless contributions to sustainable food security. The Koraput Declaration urged:

> The preamble to the PPVFR Act calls for recognition to the contributions of farm families to crop improvement made at any time. Agro-biodiversity centres like the Koraput region in Orissa, where tribal families have preserved and improved rice genetic material over many centuries, need to be protected from genetic erosion. Tribal families who have conserved important genetic material for public good at personal cost deserve recognition and reward.

The report further recommended that the National Gene Fund should be activated since severe genetic erosion is occurring in major agrobiodiversity rich areas. There is need to create an economic stake in conservation. This can come only by according economic benefits and social prestige to the primary conservers, who are often women.

The M.S. Swaminathan Foundation proposed that tribal and rural communities living in penury but conserving valuable genetic diversity should be recognised as "Genome Saviours." This is what wlll become a reality when the first Genome Saviour Awards are presented. This step would lead not only to financial support for the revitalisation of the on farm conservation traditions of local communities, but also help to spread genetic literacy.

The loss of every gene and species limits our options for the future, particularly in the context of potential adverse changes in climate. This is why genome and gene saviours are invaluable; their past and current efforts provide the tools to meet the food security challenges of today and tomorrow.

Rural Women's Role in Conserving Agrobiodiversity

The traditional role played by the rural poor, especially low-caste and tribal women, has been recognized increasingly by social activists. One such example is described below. Poorer sections of rural society and the dalits (low-caste groups) are basically subsistence farmers. For them, traditional crop varieties are very important, since these crops have harmonized over a long period of time with the environment and hence are easier to grow. It is easier for the women to manage traditional varieties of crops as they have done for centuries. Diminishing farm biodiversity has made the livelihood systems extremely vulnerable. In the event of one single pest attack, the entire crop may disappear, creating in its wake hunger and famine. The hardest hit by this phenomenon are women, who have to constantly worry about food availability for their families. This was clearly demonstrated in Africa many years ago, when the shortsighted British colonial administration planted a single clone of bananas which was wiped out by a pest in one stroke, causing much hardship to the local communities.

Mono-cropping and promotion of hybrid varieties on a large scale have accelerated this process. Under market conditions, the seeds are available only to the rich farmers. The poor will not be able to afford them or follow the farming practices they require, which can only be followed by resource-rich farmers. The shortage of local varieties of seeds which require fewer resources adversely affects the sustainability of poorer farmers.

Conserving and Promoting Agricultural Diversity

Seed fairs are increasingly popular events for promoting diversity. African interest in these was rekindled by exchange visits in the 1990s between Zimbabwe and Peru, where seed fairs are a traditional, spiritual, and cultural mechanism for keeping seed diversity alive.

Zimbabwean seed fairs are now annual events in many villages, and the word has spread to many countries throughout the continent.

(*Continued*)

(Continued)

This has been achieved through informal information exchange, publications, and some formal NGO networks such as PELUM. In Tharaka, Kenya, for example, they are called "seed shows" and have been held annually since 1996, when they were initiated by ITDGPractical Action.

In 1998, 29 women and 47 men as well as some community groups mounted displays. A panel of judges evaluated the displays, and the most diverse ones were awarded prizes. The total number of crop varieties displayed increased in 1998 to 149 from 134 in 1997. In 2001, 46 farmers displayed 206 varieties.

Participants like seed shows for many reasons: farmers can obtain rare crop varieties; they identify seed sources; it is a good forum for the exchange of ideas on farming and exchange of seeds; farmers are exposed to national agricultural research work; the spirit of competition boosts farmers' morale and motivates farmers to diversify their crops, indirectly enhancing food security; and it is a venue for interaction between farmers, students, researchers, extension staff, and other development agents.

Conserving biodiversity is a business opportunity. Global Economics. IUCN Report. Geneva, March 27, 2008.

Agrobiodiversity and Indigenous Knowledge

Another author who discussed, succinctly, the relationship between agrobiodiversity and indigenous knowledge is Joseph Gari of the FAO's Population and Development Service. The following account is based, in part, on his publications (kindly made available by the author):

> Agrobiodiversity comprises the whole plant resource diversity that human societies use and manage for agriculture, food, healthcare, and livelihood. It includes the enormous diversity of crops and crop varieties that small-scale farmers conserve and cultivate, representing both the basis for their subsistence and a source of income. To some extent, it also embraces wild food and medicinal plants that rural populations use for nutrition, healthcare and livelihood purposes. The maintenance and use of agrobiodiversity

> relies on extensive indigenous knowledge systems, which address aspects such as cultivation practices, uses, and genetic resource management of such plant species.
>
> The fundamental roles of agrobiodiversity and the associated indigenous knowledge in sustaining the agricultural dynamics of rural communities throughout Sub-Saharan Africa have been absent from policies and programmes related to agriculture, natural resource conservation, and rural development. However, agrobiodiversity and indigenous knowledge represent locally available resources with enormous value and potential for food security and rural development. All too often, the only assets which remain in poor rural communities for livelihood and even survival are the local biodiversity and indigenous knowledge; these assume increasing significance as other resources dwindle or disappear. (Gari and Villarreal 2003)

According to Gari (2004), agrobiodiversity represents the foundation of food security, livelihood options, and well-being for millions of poor farmers. The interaction between agrobiodiversity and indigenous knowledge provides the rural poor with numerous benefits and opportunities, such as the capacity to address environmental conditions, provision of food and nutritional supplies, access to local market opportunities, and options to cope with evolving needs. Although the way rural people conserve, use, and manage agrobiodiversity define their food security, livelihood, cultural dynamics, and development opportunities, such considerations have not been accordingly recognized and supported in policies and program relevant to rural development that could benefit these indigenous communities.

The AIDS epidemic is generating an additional paradox regarding agrobiodiversity and indigenous knowledge. As AIDS disrupts customary agricultural systems, sociodemographic structures, and community dynamics, it further impairs the maintenance of agrobiodiversity and indigenous knowledge. At the same time, as poor households and communities become severely impacted and impoverished, agrobiodiversity and the associated indigenous knowledge become increasingly important for achieving food security, whilst coping with the specific needs and changes owing to HIV/AIDS.

Overall, agrobiodiversity and indigenous knowledge represent locally available resources for mitigating the impact of AIDS and developing agroecological strategies aimed at household food security and nutrition. At the same time, they represent grassroots components for sustainable agricultural and rural development, capable of enhancing the autonomy and dignity of rural people throughout their development process. Therefore, they constitute a fundamental component in the agricultural sector response to the concurrence of food insecurity and HIV/AIDS impact, especially in heavily affected areas in sub-Saharan Africa.

The consideration of agrobiodiversity and indigenous knowledge leads one to new modes of agricultural and rural development that directly address household food security and nutrition aspects, while providing an integral approach to the entire agriculture-food-nutrition-health complex at the rural grassroots. Next, there follows a set of strategic components that can empower the rural poor in the face of food insecurity and AIDS impact, provided adequate changes in policies, agricultural program, and practices at the farmer level are accomplished.

Remedial Measures

The above considerations lead one to believe that a change in agricultural practices is urgently required to aid the rural poor. Several remedial measures proposed below are, in fact, a part of the traditional practices that were followed by the rural poor in the past.

1. *Traditional, neglected, and underutilized crops.* Rural people are often custodians of a rich diversity of crops and crop varieties. Most of these crops are neglected and underutilized. Some of these crops are downgraded as "traditional" or "indigenous", in opposition to the handful of economically valuable crops that are preferred by large multinational corporations (MNCs) such as Monsanto. The traditional neglected crops are eliminated from

agricultural and development programs, and their use has declined among rural people themselves. However, they do represent strategic crop genetic resources in household food security and nutrition, whilst providing many options for improving rural livelihoods and addressing evolving needs. These underutilized crops include rich nutritional sources such as many neglected legume crops which provide high protein supplies, and leafy vegetables which are an excellent source of micronutrients.

2. *Agricultural diversification*. Agricultural diversification represents a key strategy to combat food insecurity and the impact of AIDS. It is the basis for providing and enhancing a balanced nutritional supply among poor rural families, particularly in the context of subsistence agriculture and socioeconomic marginalization. Agricultural diversification also provides a local mechanism to manage agroecological risks. The promotion of agricultural diversification opposes narrow crop and farming approaches, as supported by modern agricultural development paradigms, and enhances the local control over food production and nutrition. In addition, agricultural diversification can represent a mechanism to alleviate labor shortages, as it allows for diffusing labor loads through time; this is particularly critical for AIDS-affected households, which suffer severe labor shortages.

   Overall, agricultural diversification provides a resilient and reliable strategy for food security and balanced nutrition, with specific advantages and benefits in the context of HIV/AIDS. Agrobiodiversity and indigenous knowledge are critically relevant for agricultural diversification efforts; in fact, they have traditionally supported diversified agricultural systems.

3. *Low-input agriculture*. Given the view that the AIDS epidemic generates an economic crisis in poor rural households, low-input agriculture becomes a pressing need. Modern agricultural models have promoted high material inputs at the farmer level, such as the regular purchase of chemical products and improved seeds; however, this has impaired the economic development of small-scale farmers, as the investment efforts have not often resulted in increased gains, but in a downward

cycle of financial debits, development dependency, environmental degradation, and health impoverishment. Low-input agriculture aims at increasing the net economic gains of farming households, aside from arresting the related ecological and health impact. This represents a relevant strategy for poor farmers, more so in the context of AIDS.

Overall, the success of low-input agriculture is highly dependent on local crop genetic resources (as they have co-evolved with local agroecological conditions), on the optimal use of indigenous knowledge (which contains advanced understanding of the farming constraints and potential in a particular agroecological area), and on the enhancement of the seed autonomy of farmers (so to ensure seed availability and access). In consequence, the focus on agrobiodiversity and indigenous knowledge may support more economically viable agricultural systems, which become particularly useful in view of the economic impact of HIV/AIDS among the rural poor.

4. *Wild food plants*. As rural communities confront hunger and malnutrition, the increased use of wild food plants represents a strategy for food security and nutrition. In fact, the role of wild food plants in emergency situations is well extensible to the context of AIDS. The value and potential of wild food plants in the nutrition of rural people is, however, neglected in agricultural and environmental programs. The promotion of wild food plants requires simultaneous support to community ecological conservation and management systems in order to facilitate equitable access and ensure sustainable use. In some cases, highly valuable wild food plants may be considered for cultivation, for instance in home gardens. In fact, the domestication of many crops evolved from an increased value and use among people.
5. *Medicinal plants*. Medicinal plants constitute a fundamental component of traditional healthcare systems in rural communities throughout Africa. In fact, they constitute the fundamental basis of health care for the largest part of the world's population. Medicinal plant diversity and the associated indigenous healthcare knowledge represent affordable and locally available

resources capable of addressing many AIDS-related diseases and health problems. Their role is particularly significant in view of the fact that current economic and trade inequalities limit access to modern drugs and treatments among the poor.

In developing countries, plants are the main source of medicine. According to the World Health Organization, as many as 80% of the world's people rely on traditional medicine for their primary health care, mostly using remedies made from plants. The use of traditional medicine in developing countries is increasing because populations are increasing, governments want to encourage indigenous forms of medicine rather than rely on imported drugs, and there are strong moves to revive traditional cultures.

The world trade in plant medicines is worth billions of dollars. In 1994, China exported US$2 billion of plant drugs. Germany imports around US$100 million of plant drugs. The number of medicinal plants in trade is also astonishing. Germany imports at least 1560 plant species for medicinal purposes, but experts there estimate that only 50–100 of them are cultivated on a large scale; the remainder are thought to be collected from the wild.

Two of the largest users of medicinal plants are China and India. Traditional Chinese medicine (TCM) uses over 5000 plant species; India uses some 7000. In China, sales of traditional medicines have more than doubled in the last 5 years, while India's booming export trade in medicinal plants has risen almost threefold during the last decade. In 1990, Chinese doctors used 700 000 tons of plant material. China has about 250 000 doctors trained in traditional medicine; India, 460 000. All use plant drugs.

This boom in local use and export trade is depleting many species from the wild, bringing some to the edge of extinction. Few are cultivated; 80% of the species used in China and 95% of those in India are wild-collected. Data on threatened species are rare, but national studies show that 120 medicinal plants are rare or endangered in India, at least 77 in China, and 75 in Morocco. It is reasonable to assume that at least 1000 plant species used in medicine today are threatened with extinction.

Threatened species include Tetu lakha (*Nothatodytes foetida*), a small tree from rainforests in South India and Sri Lanka that is used for anticancer drugs in Europe; *Saussurea lappa* from India whose root is used for chronic skin disorders; and *Fritillaria cirrhosa* from Sichuan, China, used for respiratory infections.

India

Around 1500 B.C., Ayurveda was delineated into eight specific branches of medicine. There were two main schools of Ayurveda at that time: Atreya, the school of physicians; and Dhanvantari, the school of surgeons. These two schools made Ayurveda a more scientifically verifiable and classifiable medical system. Through research and testing, they dispelled the doubts of the more practical and scientific minded, removing the aura of mystery that surrounded the concept of divine revelation. Consequently, Ayurveda grew into a respected and widely used system of healing in India. People from numerous countries came to Indian Ayurvedic schools to learn about this world medicine in its completeness. Chinese, Tibetans, Greeks, Romans, Egyptians, Afghanistanis, Persians, and others traveled to learn the complete wisdom and bring it back to their own countries.

According to the World Health Organization (WHO), more than 1 billion people rely on herbal medicines to some extent. The WHO has listed 21000 plants that have reported medicinal uses around the world. India has a rich medicinal plant flora of some 2500 species; of these, at least 150 species are used commercially on a fairly large scale. Foreign researchers have always appreciated the traditional Indian healers.

Herbs by Botanical Names

- Abelmoschus moschatus
- Abrus precatorius
- Acorus calamus
- Adhatoda vasica
- Aegle marmelos
- Aesculus hippocastanum
- Aloe barbadensis
- Aloe vera
- Alpinia galanga
- Andrographis paniculata
- Annona squamosa
- Argyreia nervosa

(Continued)

(Continued)

- ¤ Artocarpus heterophyllus
- ¤ Asparagus racemosus
- ¤ Azadirachta indica
- ¤ Bacopa monniera
- ¤ Bambusa arundinacea
- ¤ Bambusa vulgaris
- ¤ Bauhinia variegata
- ¤ Boerhaavia diffusa
- ¤ Boswellia serrata
- ¤ Calendula officinalis
- ¤ Carica papaya
- ¤ Cassia augustifolia
- ¤ Cassia fistula
- ¤ Cassia tora
- ¤ Cedrus deodara
- ¤ Celosia cristata
- ¤ Centella asiatica
- ¤ Chlorophytum borivilianum
- ¤ Cichorium intybus
- ¤ Cichorium intybus Linn
- ¤ Cinchona officinalis
- ¤ Citrus aurantium
- ¤ Coleous forskohlii
- ¤ Coleus forskohlii
- ¤ Commiphora mukul
- ¤ Convolvulus alsinoides
- ¤ Curculigo orchioides
- ¤ Curcuma longa
- ¤ Curcuma zedoaria
- ¤ Cydonia oblonga
- ¤ Cymbopogon citrates
- ¤ Cymbopogon flexuosus
- ¤ Cyperus rotundus
- ¤ Datura stramonium
- ¤ Emblica officinalis
- ¤ Ephedra vulgaris
- ¤ Evolvulus alsinoideses
- ¤ Fagopyrum esculentum
- ¤ Ferula foetida
- ¤ Ficus benghalensis
- ¤ Ficus racemosa
- ¤ Ficus religiosa
- ¤ Fumaria officinalis
- ¤ Garcinia cambogia
- ¤ Glycyrrhiza glabra
- ¤ Gymnema sylvestre
- ¤ Hedychium spicatum
- ¤ Hemidesmus Indicus
- ¤ Hibiscus rosa-sinensis
- ¤ Holarrhena antidysenterica
- ¤ Hyocyamus niger
- ¤ Ipomea carnea
- ¤ Jatropha curcas
- ¤ Juglans regia
- ¤ Lantana camara
- ¤ Lawsonia inermis
- ¤ Linum usitatissimum
- ¤ Luffa cylindrical
- ¤ Matricaria chamomilla
- ¤ Mentha arvensis/pipertia
- ¤ Mesua ferrea
- ¤ Momordica charantia
- ¤ Morchella conica
- ¤ Morinda citrifolia
- ¤ Moringa oleifera
- ¤ Morus alba

(Continued)

(Continued)

- Mucuna pruriens
- Nardostachys jatamansi
- Nelumbo nucifera
- Nyctanthes arbortristis
- Nymphaea lotus
- Nymphaea nouchali
- Ocimum sanctum
- Operculina turpethum
- Papaver somniferum
- Peganum harmala
- Phyllanthus niruri
- Picrorhiza kurroa
- Piper cubeba
- Piper longum
- Piper nigrum
- Plantago ovata husk
- Pluchea lanceolata
- Plumbago indica
- Plumbago zeylanica
- Polygala senega
- Polygala vulgaris
- Psoralea corylifolia
- Pterocarpus marsupium
- Pterocarpus santalinus
- Pueraria tuberosa
- Punica granatum
- Raphanus sativus
- Rauwolfia serpentina
- Rheum emodi
- Ricinus communis
- Rosa centrifolia
- Rosa damascena
- Rubia cordifolia
- Salacia reticulata Wight
- Sapindus mukorossi
- Sapindus trifoliatus
- Saraca asoca
- Solanum indicum
- Solanum nigrum
- Solanum xanthocarpum
- Strychnos nux-vomica
- Swertia chirata
- Syzygium cumini
- Tamarindus indica
- Tephrosia purpurea
- Terminalia arjuna
- Terminalia belerica
- Terminalia chebula
- Thymus vulgaris
- Tinospora cordifolia
- Tribulus terrestris
- Trichosanthes dioica
- Trigonella foenum-gracum
- Triticum sativum
- Tylophora asthmatica
- Tylophora indica
- Valeriana wallichi
- Vetiveria zizanoides
- Vinca rosea
- Viola odorata
- Vitex negundo
- Withania somnifera
- Woodfordia fruticosa
- Wrightia tinctoria
- Zingiber officinalis

(*Continued*)

(Continued)

Herbs by Common Names

- Achicoria amarga
- African eggplant
- Akhrot
- Al-henna
- Al-khanna
- Alholva
- Ameirao
- Aloe
- Amaltas
- Ambrette seeds
- Amla
- Anantamool
- Anar
- Annual hibiscus
- Anon
- Anona blanca
- Aphium opium poppy
- Apple thorn
- Apple wood
- Arjuna
- Arjuna herb
- Arjuna root
- Artayniya-e hindi
- Ashok
- Ashwagandha
- Asogam
- Asok
- Asparagus
- Asparagus root
- Assafetida
- Ates
- Ati
- Atteeka
- Aute
- Bach
- Baehra
- Bael
- Balsam apple
- Balsam pear
- Balsamodendrom mukul
- Bamboo
- Bamia moschata
- Banyan
- Barbados aloe
- Barbe de capucin
- Basil
- Beggary
- Beleric
- Belliric myrobalan
- Bigarade orange
- Bindu
- Bird feed
- Bird's foot
- Bitter gourd
- Bitter melon
- Bitter oleander
- Bitter orange
- Bitter stick
- Black caraway
- Black catnip
- Black cumin
- Black henbane
- Black morels
- Black night shade
- Black onion seeds
- Black pepper

(Continued)

(Continued)

- Black plum
- Black sesame seeds
- Blaue wegwarte
- Blue sailors
- Blue
- Sailorschicory
- Bodhi tree
- Bosbeesklou
- Boyotu
- Brahmi
- Bread wheat
- Brindall berry
- Buckwheat
- Bush morning glory
- Bush tomato
- Cain ghe
- Calamus
- Caricaceae
- Carry me seed
- Cassia senna
- Castor
- Cedarwood
- Cedarwood oil
- Ceylon lead root
- Chakvad
- Canca piedra
- Chaste tree
- Chebulic myrobalan
- Chicory
- Chicory
- Chikana red mango
- Child pick-a-back
- China rose
- Chinagreye
- Chirette indien
- Chuo
- Cichorei
- Cicoria cikoria
- Citrus dulcis
- Clavo
- Clearing nut tree
- Cluster fig
- Cockscomb
- Coco grass
- Coffee pod
- Coffeeweed
- Coleous
- Coleus
- Commiphora
- Common chicory endive
- Common cowitch
- Common curculigo
- Common fumitory
- Common henbane
- Common wheat
- Connessi bark
- Coral jasmine
- Corn mint
- Cot chu
- Cowitch
- Crab's eye
- Crattock
- Creat
- Creole senna
- Crested celosia
- Cubeb
- Curacao aloe
- Curcuma

(*Continued*)

(Continued)

- ¤ Curcumae domestica
- ¤ Curcumin 95%
- ¤ Custard apple
- ¤ Da liang
- ¤ Daikon
- ¤ Datura
- ¤ Datura seeds
- ¤ Daun marisan
- ¤ Deadly nightshade
- ¤ Devil's dung
- ¤ Dhatura tatula
- ¤ Dishcloth gourd
- ¤ Djarak
- ¤ Dowa i pechish
- ¤ Drumstick tree
- ¤ Dwarf morning-glory
- ¤ Dyers's oleander
- ¤ Dysentery rose bay
- ¤ Early morels
- ¤ Earth smoke
- ¤ East indian figtree
- ¤ Egyptian privet
- ¤ El adkham
- ¤ El galangal
- ¤ Elephant creeper seeds. Original argyreia seeds
- ¤ Emblica
- ¤ Emetic swallow wory
- ¤ Endive
- ¤ English violet
- ¤ Ephedra
- ¤ Eugenia jambolana
- ¤ European horse chestnut
- ¤ European walnut
- ¤ Evulgaris
- ¤ Fenegriek
- ¤ Fenugreek
- ¤ Field mint
- ¤ Figuier des pagodes
- ¤ Fire morel
- ¤ Fire plant
- ¤ Fire-flame bush
- ¤ Fish poison
- ¤ Fistula
- ¤ Flax seeds
- ¤ Fleam
- ¤ Fodder radish
- ¤ Foenum graecum
- ¤ Foetid cassia
- ¤ Food of the gods
- ¤ Foxtail amaranth
- ¤ Fumiterry
- ¤ Fumus
- ¤ Fun li chi
- ¤ Galanga maior
- ¤ Galanga root
- ¤ Gale of wind
- ¤ Galgant
- ¤ Galu gasturi
- ¤ Garden thyme
- ¤ Garden violet
- ¤ Gattilier incise
- ¤ Geelbeesklou
- ¤ Ghi kunvar
- ¤ Giloy
- ¤ Ginger grass
- ¤ Ginger root
- ¤ Gloria de la manana

(Continued)

(Continued)

- Gokhru
- Golden eye-grass
- Golden shower
- Gond part used: bark
- Gooseberry
- Gourd towel
- Graines reglisse
- Grani neri
- Greater galanga
- Greek hayes
- Grosser galgant
- Gudmar
- Guggul
- Gulancha tinospara
- Gulanshe tinospara
- Gulf leafflower
- Gum resin
- Hang dou kou
- Hardad
- Hawaiian baby woodrose seeds
- Hedichium
- Hei zhong cao
- Henbane
- Henna
- Herpestis monnieria
- Hibiscus abelmoschus
- Hibisucus flower
- Himalayan rhubarb
- Hint meyankoku
- Hog weed
- Holy basil
- Holy fruit tree
- Horse chestnut
- Horse purslane
- Horse radish tree
- Huang ching
- Huang ping
- Hung tou
- Indian aloe
- Indian banyan
- Indian bdellium tree
- Indian camphorweed
- Indian echinacea
- Indian fig
- Indian fleabane
- Indian gentian
- Indian ginseng
- Indian gooseberry
- Indian ipecac
- Indian jalap
- Indian kino tree
- Indian kudzu
- Indian madder
- Indian mulberry
- Indian nard
- Indian nightshade
- Indian olibanum tree
- Indian pennywort
- Indian rhubarb
- Indian rose chestnut
- Indian sarasaparilla
- Indian senna
- Indian snake root
- Indian spikenard
- Indian valerian
- Indian winter cherry
- Ingo

(Continued)

(Continued)

- Iron wood tree
- Isabgol
- Isithibathibana
- Ispaghula
- Jal brahmi
- Jamaica mignonette
- Jaman
- Jambolan
- Japanese mint
- Japanese radish
- Jatamansi
- Jatropha seeds
- Java galangal
- Java pepper
- Jequerit
- Jequerity
- Jimson weed
- Kaner'apra
- Kapur kachri
- Kariyat
- Khas-khas
- Knaros
- Kino
- Konch
- Kotalahimbatu
- Krapata
- Ku tzu malaysia
- Kurchi bark
- Kutki
- Laburnum
- Land caltrops
- Lantana weed
- Laos
- Latai
- Latay
- Lautafifi
- Leadwort
- Leafy daikon
- Lemongrass
- Lenggundi
- Lenkuas
- Liane reglisse
- Lilac
- Linsed
- Lipstick flower
- Loofah gourd
- Luban
- Lukelangafu
- Lukera batuzi
- Ma liao tou
- Ma qian zi
- Mad apple
- Madagascar periwinkle
- Madderwort
- Makoy
- Malabar kino tree
- Malabar nut
- Malabar tamarind
- Malay tea
- Malaysian scurfpea
- Man ching
- Mandyju'ra
- Mangosteen
- Margosa tree
- Marigold
- Marking nut tree
- Mehndi
- Melega saga

(Continued)

(Continued)

- Mendee
- Meniran
- Mesua
- Moql
- Moqle-arzagi
- Moraceae
- Morels
- Moringa
- Mukul
- Mulethi
- Musk mallow
- musk okra
- Musk root
- Musta siemen
- Myrobalan
- Naukyo
- Neem
- Neem chal
- Negundo chastetree
- Nidor
- Niger seed
- Night jasmine
- Noni
- Ntunfulu
- Nutgrass
- Nux-vomica
- Nymphaeaceae
- Olibanum
- Orchid
- Oregano
- Ornamental okra
- Osaka
- P'o ku chih
- Pala indigo plant
- Palma Christi
- Paratella
- Paternoster
- Peepal tree
- Pennywort
- Peppercorns
- Periploca of the woods
- Periwinkle
- Persian walnut
- Peruvian bark
- Pha cot chi
- Phyllanthus emblica
- Phyllanthus plant
- Picrorhiza
- Pig weed
- Pink rose
- Piperaceae
- Plantago
- Pluchea
- Plum
- Podina
- Pointed gourd
- Poison berry
- Poison nut
- Pom
- Pomegranate
- pomme canelle
- Poppy seeds
- Portugal orange
- Posta
- Pot marigold
- Precatory bean
- Psoralea seed
- Psyllium husk

(*Continued*)

(Continued)

- ¤ Psyllium seeds
- ¤ Psysic nut
- ¤ Puncture vine
- ¤ Purging fistula
- ¤ Purging nut
- ¤ Quaker buttons
- ¤ Quebra pedra
- ¤ Quince
- ¤ Quince seeds
- ¤ Quinine
- ¤ Quinine bark
- ¤ Radish
- ¤ Rakat
- ¤ Red cockscomb
- ¤ Red Indian kudzu
- ¤ Red rose
- ¤ Red sandal wood
- ¤ Red saunders
- ¤ Redwood fig
- ¤ Reer
- ¤ Revat chinni
- ¤ Rhubarb
- ¤ Ricin
- ¤ Rosary pea
- ¤ Rosy-flowered leadwort
- ¤ Ruby wood
- ¤ Rumbodo
- ¤ Sacred basil
- ¤ Sacred tree
- ¤ Sadabahar
- ¤ Safed moosli
- ¤ Safran boubou
- ¤ Safran de malabar
- ¤ Safran des indes
- ¤ Sahijan
- ¤ Salacia
- ¤ Sampalok
- ¤ Saptarangi
- ¤ Sarasaparilla
- ¤ Satavari
- ¤ Scurf-pea
- ¤ Seneca snakeroot
- ¤ Senega roots
- ¤ Senna
- ¤ Serpentine root
- ¤ Seville orange (sweet)
- ¤ Shankhahuli
- ¤ Shankhpushpi
- ¤ Shatterstone
- ¤ Shinaji tea
- ¤ Shinajitea
- ¤ Shoeback plant
- ¤ Shrub verbena
- ¤ Siamese galanga
- ¤ Siamese ginger
- ¤ Sickle pod
- ¤ Sickle senna
- ¤ Small soapnuts
- ¤ Smooth lawsonia
- ¤ Snake root
- ¤ Snake-wood
- ¤ Soapnut in printed cotton bags with reversible strings along with wash bags
- ¤ Soapnut laurifolia
- ¤ Soapnut powder

(Continued)

(Continued)

- Soapnut shells
- Soapnut whole
- Soapnuts powder
- Somlatha
- Sonth (dried)
- Spikenard
- Spogel seeds
- Sponge gourds
- St. Thomas tree
- St. Thomas lidpod
- Stink weed
- Stone breaker
- Strychnine tree
- Succory
- Suganda
- Sugar apple
- Sweet ginger
- Sweet liquorice
- Sweet root
- Sweet violet
- Sweet-flag
- Sweetsop
- Sweetwood
- Taggar
- Tailed cubebs
- Tailed pepper
- Tamarin
- Tamarind
- Tamarindo
- Tapotapo
- Tar vine
- Tellicherry bark
- Tento muido
- Terra merita
- Terre-mérite
- Thorn apple stramonium
- Thuzna
- Thyme
- Thyme leaved gratiola
- Tinnervelly senna
- Tinospara
- Tora
- Torongil de limon
- Tovara
- Tulsi
- Turmeric
- Turpeth
- Vada tree
- Valerian
- Valerian jatamansi
- Vapor
- Vegetable sponge
- Vetiver
- Walnut
- Wanaithile
- Wash sponge
- Wax dolls
- Weesboontje
- Wheat
- White indian kudzu
- White poppy
- White turmeric
- Wild asparagus
- Wild indigo
- Wild leadwort
- Wild sage

(Continued)

(Continued)

- ¤ Winter cherry
- ¤ Withania
- ¤ Wonder tree
- ¤ Woodfordia
- ¤ Woodfordia
- ¤ Wu you hua
- ¤ Yellow toreador
- ¤ Yellow bauhinia
- ¤ Yellow bell
- ¤ Yellow sage
- ¤ Yellow-berried nightshade
- ¤ Yorka okra
- ¤ Zedoary root
- ¤ Zygophyllaceae

Mother Herbs Ltd., Delhi, Indra, 1994.

6. *Indigenous agroecological knowledge and practices.* As illustrated above, indigenous agroecological knowledge and practices are inseparable from agrobiodiversity, and hence critical in agricultural strategies relevant to household food security and nutrition. They represent the connecting axis between rural people and ecosystems, being important for providing local food security whilst maintaining sustainable agroecological systems. As a consequence, the indigenous agroecological knowledge and practices deserve adequate recognition, support, and enhancement to strengthen the agricultural systems of small-scale farmers. Through cross-cultural exchange and cooperation, they can better improve and adapt to the evolving needs of farmers.
7. *Community seed systems.* The agrobiodiversity dynamic among small-scale farmers relies significantly on local seed systems. In fact, community seed systems represent the basis for the conservation, access, and exchange of plant genetic resources, and hence the foundation for the effective use of agrobiodiversity in poor rural communities. Community seed banks and farmer seed fairs are relevant practices to strengthen and improve community seed systems.

In conclusion, these strategic components represent valuable forces to enhance the local capacity to cope with chronic food insecurity and the AIDS impact. They pose significant changes in agricultural development policies and programs, emphasizing a priority focus on the rural grassroots.

Social Dimensions

Using agrobiodiversity and indigenous knowledge to combat food insecurity and AIDS impact requires a simultaneous attention to the social matrix. Four relevant concerns are suggested:

1. *Enhancing appropriate coping mechanisms.* Agrobiodiversity and indigenous knowledge are relevant ingredients in local coping mechanisms to food insecurity and AIDS impact, since they represent the locally available resource base for agriculture, food production, and healthcare. This grassroots dynamic demands adequate recognition and support from agricultural policies and programs.
2. *Exploring agrobiodiversity-based livelihood options.* Agrobiodiversity encompasses a number of options for livelihood and economic development in rural communities, but they are severely constrained in the existing socioeconomic framework. The effective use of agrobiodiversity to enhance rural economies requires active support at all levels, including policies, institutional mechanisms, and targeted interventions; for instance, in research projects, extension programs, rural microcredit schemes, and market policies.
3. *Gender equity.* Gender inequalities aggravate food insecurity and HIV/AIDS impact among rural women, whilst constraining their roles and capacity to combat such constraints at household and community levels. Rural women play numerous and vital roles throughout the agriculture-food-nutrition-health complex, including particularly the conservation and use of agrobiodiversity, as well as the transmission of indigenous knowledge. All of these elements become of utmost importance in the context of AIDS, but the persistent gender inequalities constrain their effective use to cope with food insecurity and HIV/AIDS.

The interrelated roles of rural women comprise active conservation and management of biodiversity (e.g. seed conservation, selection of neglected but nutritious crops), holding and transmitting of indigenous and local knowledge (e.g. farming practices, knowledge on wild food plants), specific agricultural tasks (e.g. home gardens, subsistence agriculture), postharvest practices (e.g. food processing and preservation), and provision of household well-being (e.g. nutrition through meal preparation, care-giving for ill family members). A priority focus on gender equity is required to adequately mainstream agrobiodiversity and indigenous knowledge in order to combat food insecurity and AIDS impact at household and community levels.

Gender approaches are relevant also in the view that women are the dominant actors in subsistence agriculture, playing relevant roles in agrobiodiversity management and transmission of indigenous knowledge, yet they suffer unequal access to agricultural productive resources (e.g. land and credit schemes) and notable sociopolitical exclusion in many cases. Accordingly, a gender-oriented approach to agrobiodiversity and indigenous knowledge is highly required to ensure that men and women equally participate in, and benefit from, the use of agrobiodiversity for food security and HIV/AIDS mitigation.

4. *Community mobilization*. The scale of the AIDS impact in some rural areas is likely to undermine community dynamics, which are however increasingly important to support affected households, mitigate the impacts, and launch rural development. In particular, AIDS may erode the community-based conservation of agrobiodiversity as well as social reproduction systems that facilitate the exchange and transmission of indigenous knowledge. In this context, community-led initiatives that promote the conservation, use, and management of biodiversity and the associated indigenous knowledge are highly important.

   Such community mobilizations are relevant to enhance the local capacity to cope with food insecurity and AIDS impact, and are equally important to maintain biodiversity and indigenous knowledge as grassroots components for rural development. In particular, as AIDS generates a rural emergency situation, the active stance for conserving and making available agrobiodiversity and indigenous knowledge becomes increasingly important to maintaining the prospects of reconstruction, development, and autonomy among surviving community members, including the youngest generation.

Conserving Local Breeds: Role of Indigenous Communities

The Aseel is a chicken breed in India. For centuries, Adivasi communities living in the East Godavari District have reared and selectively shaped this breed especially for its meat. Today, infectious diseases, high production losses and government policies promoting non-local breeds threaten its existence.

In 1996, a group of organisations studied the local production system in 24 villages. A number of improvements were initiated: promotion of local fodder crops to improve feeding; training of village animal health workers and introduction of basic healthcare practices such as vaccinations and regular deworming; and education of women — who are responsible for the poultry — in improved animal husbandry.

A follow-up survey conducted a year later revealed that overall mortality had fallen from 70% to 17%. The following year (1998–99) the mortality was down to 6% and the number of Aseel poultry had trebled. A further mechanism to enlarge the population was the revival of 'vaata', a traditional system of sharing and asset building. Initially, 196 women in 20 villages received 200 hens and 67 cocks.

Within one year, the birds had produced 1,414 chicks and the initial investment of Rs 60,000 could be recovered. The main problems faced by the project were the difficulty to obtain vaccines in small quantities, difficult access to markets and policies that favour cross-breeding.

Anthra, Yakshi, Girijana Deepika, and Womens Gottis of East Godavari Adivasi Areas, Andra Pradesh, India.

India is one of the megadiversity zones of the world, showing wide agroclimatic, geophysical, and ethnic variation. At least 167 crop species and 320 wild relatives of crops have their primary, secondary, or regional centers of diversity here. In particular, the tribal-inhabited belt is the center of domestication and remarkable genetic diversity in food crops.

Home Gardens and Agrobiodiversity

The tribal-inhabited belt is more often the center of domestication and genetic diversity of food crops (cereals and pseudocereals, millets, grain legumes, vegetables, spices and condiments, oil plants, etc.), being maintained by peasants and subsistence farmers. These areas hold unique and important genetic materials, which should be strictly protected against heavy grazing; intensive farming; commercial logging; construction of highways, dams, and hydroelectric stations; resettlement projects; mining operations; etc. The genetic diversity is held by the tribals in their dooryard gardens, Baris (land attached to their houses and huts), kitchen gardens, and fields. Some examples of such cultivated crops are *Piper peepuloides, Parkia roxburghii, Moghania vestita, Vigna umbellata, Inula racemosa, Coix lacrymajobi, Digitaria cruciata* var. *esculanta, Hodgsonia heteroclita,* and several species of Alocasia, Colacasia, Amorphophallus, and Dioscorea cultivated by the tribes in northeast India.

The primitive cultivars grown by farmers are valuable sources of genetic material for modern plant breeding. IR-72, a modern variety of rice (*Oryza sativa*), was developed by cross-breeding 22 landraces from 7 nations: India, Indonesia, China, Philippines, Vietnam, Thailand, and Malaysia. In 1969, *O. nivara*, a wild rice from India, was discovered to resist grassy stunt virus (GSV). It was crossed and back-crossed with IR-24 three times to produce a variety resistant to GSV and with IR-24 grain qualities.

As an example, field studies of home gardens in southern Vietnam illustrate the wealth of genetic resource diversity available for home gardening. They contain a vast variety of vegetarian sources, including leaves, tender stems, fruits, and even flowers as food sources. Some leafy and fruit vegetables are important as cash crops. Others contribute to family nutrition, diet diversification, and local culinary preferences. Asia has a very rich diversity of species and many varieties of eggplant, gourds, and fruit trees. Dark green vegetables and certain fruits are usually a rich source of vitamin A and other micronutrients. Home gardens are often orchards because of the abundance of fruit trees, such as numerous varieties of citrus trees, banana, coconut, kumquat, cashew nut, pineapple, etc. Home gardens also boast a rich diversity of root and tuber crops as food sources, such as sweet potato, cassava, taro, tannia, as well as minor root crops which are used as spices or medicinal plants (e.g. ginger, turmeric, mustard, galanga, etc.). Many species in home gardens serve multiple purposes: as food, medicinal sources, religious materials, and/or ornamentals.

Pedigree of IR-72: A Modern Variety of Rice

Twenty-two ultimate landraces	
India	**China**
Oryza nivara	Cina
Arikarai	DGWG
Eravapandi	Pa Chiam
Gowdalu	Tsai Yuan Chan
Kitchili Samba	
Latisail	**Philippines**
Mudgo	Marong Paroc
Thekkan	Sinawpagh
Vellai Kar	Tadukan
Unknown Variety	
Unknown Variety	**Vietnam**
	Tetep
Indonesia	
Benong	**Thailand**
	Germ Pai
Malaysia	
Seroup Besai	

(Maheshwari 1988)

Harshberger (1896) outlined, for the first time, the purpose of ethnobotanical gardens. Plants of ethnobotanical importance like maize, sunflower, tomato, potato, tobacco, rice, pumpkin, yam, taro, arrowroot, cassava, sweet potato, and amaranths associated with tribal people were grown over the ages as part of the life-support system for the survival, subsistence, and livelihood of tribals. In modern times, these gardens — also referred to as indigenous gardens, home gardens, forest gardens, or heritage farms — should play a major role in the conservation and maintenance of "heirloom" varieties of economic plants. People in rural communities could be encouraged to construct ethnobotanical nurseries where locally useful species could be cultivated. This would not only provide a source of medicinal and edible plants, but would also serve to familiarize younger people with the herbs that were traditionally used in the community. The cultivated plots could serve as demonstration gardens, where overharvested wild species could be brought into cultivation and eventually integrated into home gardens or managed forests.

Home gardens (variously termed as kitchen gardens or forest gardens), are generally characteristic of the humid tropics. They represent intensive, multistoried combinations of crops, trees, and livestock, and are the dominant form of land use in Kerala and northeast India. These gardens have many variations, but all are designed to supply family requirements of food, fodder, fuel, and timber, and to generate additional income through the sale of surplus products. A typical example of a multistoried system might include coconut + black pepper + cocoa + pineapple, grown primarily for sale, in addition to family food crops. The most important crops are coconut and cassava in upland areas, and rice in the lowlands. Other tree crops include cocoa, jackfruit, cashew, arecanut (betel), nutmeg, and clove, as well as teak, *Pterocarpus marsupium*, *Erythrina variegeta*, *Artocarpus hirsuta*, *Bombax ceiba*, *Albizia falcataria*, *Ailanthus excelsa*, and bamboo grown for timber, fodder and fuel, and to support vines. Agricultural crops include sugarcane, sweet potato, colocasia (taro), yams, pulses, vegetables, sesame, ginger, and turmeric. Livestock form the third component, with cattle, goat, and poultry the most common domestic animals.

Home gardens are economically viable, ecologically sound, and biologically sustainable (Abdul Salam and Sreekumar 1991).

Tribal Markets — "Hats"

Tribal markets, as seen in tribal areas of the country, represent a distinct organizational structure in tribal societies. The "hat", as it is known locally, is a weekly marketing facility evolved by the tribals for the sale or barter of minor forest produce (MFP), seasonal vegetables, fruits, seeds, tubers, and other commodities. It is the only center of economic activity in the tribal area; and to attend the weekly markets, the tribals may have to cover a distance of over 25 km on foot. They collect wild and cultivated plants, forming the basis of several cottage and rural industries like those of herbal drugs, fibers and flosses, bamboo and canes, oils and fats, cordage, mats and basketry, oilseeds of forest origin (mahua, sal, neem, karanj, etc.), gums, resins, tanstuffs, guggul or incense materials, dyes, fermented drinks, soap and cosmetics, toys, drums, musical instruments, Kattha extraction, agricultural implements, brooms and brushes, perfumes (sandalwood oil, Khus oil), etc. Tendu (*Diospyros melanoxylon*), the leaves of which are used as wrappers for cigarettes (bidis) is another important nontimber product of the forests. These products are derived from over 1000 plant species (FAO 1994).

With a view toward helping the tribals in their economic development, tribal cooperative societies have been organized in different states. At the national level, the Tribal Cooperative Marketing Development Federation of India Limited (TRIFED) was set up in 1987 to handle items of tribal produce. It has also been declared as the central, nodal agency for organizing the collection, processing, storage, and development of oilseeds of tree and forest origin (Maheshwari 1990).

Conclusions

There is a growing recognition of the need for a crash program on agroethnobotanical studies in every agroecological zone of the country before land degradation leads to the permanent loss of genetic diversity of crops. The systematic study of the botanical knowledge of indigenous people and other ethnic groups, and of the use of locally available domesticated plants and their landraces or folk varieties, has been called "agroethnobotany"; this might cover realms of foods, medicines, clothing, or religious rituals.

Until recently, *in situ* conservation programs have focused on forest genetic resources, both at the national level and internationally under the leadership of the FAO; while there has been little attention to the *in situ* conservation of crops and their wild relatives. The landraces and other farmer varieties would not meet the criteria for Plant Breeder's Rights (PBR) protection under the Union for the Protection of New Varieties of Plants (UPOV) Convention that varieties must be distinct, uniform, and stable. Because the traditional varieties are often variable and, therefore, important sources of genetic diversity, they cannot be protected under existing PBR schemes. Hence, the domestic patent and intellectual property rights (IPR) legislation should include provisions to maintain the farmer's privilege of planting saved seed in successive seasons.

The role of small-scale farmers and their traditional varieties, farming systems, and knowledge in developing a truly sustainable agriculture may have been neglected by the formal research system over the past decades. Agroethnobotany provides a useful tool in determining the amount of agrobiodiversity present, its current status, and future strategies. India is inhabited by about 450 tribal communities, constituting about 8% of the total population of the country. Their knowledge about specific plant usage is transmitted largely through word of mouth and tradition. Much of this agroethnobotanical knowledge has, therefore, remained endemic to certain regions or tribes and needs to be systematically surveyed, documented, and utilized (Maheshwari 1988).

Agroethnobotanical information is key to preventing the loss of irreplacable genetic resources. There are still 74 primitive tribal groups in the Indian region, who were identified on the basis of their preagricultural level of technology, low level of literacy, and stagnant or diminishing population. They are the traditional conservators of biodiversity at the grassroots. A dynamic farmer-led approach to landrace conservation, enhancement, and utilization is thus recommended.

Agroforestry

There are four main types of agroforestry:

1. Agrosilviculture: This form combines trees and annual crops. Wild or semiwild species include those used in live fences and windbreaks, medicinal and aromatic plants, and shifting agriculture.
2. Silvipastoralism: This form has trees and livestock on wooded pasture or rangeland. Wild or semiwild species include those used in pasture, browsing and fodder trees, and regenerating forest.
3. Agrosilvipastoralism: This form combines trees, crops, and livestock. Wild or semiwild species include those used in pasture and rangeland, browsing and fodder trees, shrubs, multipurpose trees, and medicinal and aromatic plants.
4. Home gardening: This form combines various combinations of multispecies trees, shrubs, and perennial and annual herbs.

To protect biodiversity and the other benefits that biodiversity provides, agroforestry systems should continue and expand, especially in buffer zones near protected areas. There are two basic ways to encourage agroforestry systems:

1. Provide incentives to farmers who already have agroforestry systems.
2. Restore vegetation in monoculture systems with perennials and tree crops.

Although growing crops under existing forest canopies would easily create very diverse agroforests, it is problematic if farmers turn to existing forests and especially forest reserves to do so. However, it is important to provide farmers who already have agroforestry systems with economic incentives to prevent them from turning agroforests into intensive monocultural systems. Creating agroforestry systems by restoring degraded agricultural habitats is one way to avoid the problem of forest conversion while still maintaining and promoting the recovery of biodiversity. In such a restoration process, sun coffee, for example, slowly transforms to shade coffee. Furthermore, areas with sun-grown crops like vegetables can be planted with sun-tolerant trees such as cinnamon, allspice, and fruit trees, which eventually grow into shade-tolerant hardwoods. This process leads to the recovery of forested areas with diverse canopies from which people can still retrieve economic benefits.

To sum up, agroforests provide important habitats for biodiversity; ecologically sustainable buffer zones for protected areas; a high-quality matrix that promotes movement between forest fragments; and ecosystem services such as pest control, pollination, and erosion control. Furthermore, agroforests produce important sources of income for local people. It is possible to see agriculture as a diverse system and treat it as an extension of natural habitats that can be guided to grow our needs. So, agricultural biodiversity is not merely the result of human activity: human life is dependent on it not just for the immediate provision of food and other goods, but for the maintenance of areas of land that will sustain production and for the maintenance of the wider environment.

Erosion of Agrobiodiversity

These locally diverse food production systems are under threat and, with them, the accompanying local knowledge, culture, and skills of the food producers. With this decline, agricultural biodiversity is disappearing and the scale of loss is extensive. With the disappearance of harvested species, varieties, and breeds, goes a wide range of unharvested species.

- More than 90% of crop varieties have disappeared from farmers' fields.

- Half of the breeds of many domestic animals have been lost.
- In fisheries, the world's 17 main fishing grounds are now being fished at or above their sustainable limits, with many fish populations effectively becoming extinct.

The genetic erosion of agricultural biodiversity is also exacerbated by the loss of forest cover, coastal wetlands, and other wild uncultivated areas, and by the destruction of the aquatic environment. This leads to losses of wild relatives which are important for the development of biodiversity, and losses of wild foods essential for food provision, particularly in times of crisis.

For over 12 000 years, agricultural biotechnology played an important role in sustaining and strengthening food, nutrition, health, and livelihood security. However, even though crop productivity has increased significantly, the number of children and adults who are malnourished and undernourished has increased dramatically. The eminent agricultural scientist Prof. M.S. Swaminathan (personal communication, 2005) has cautioned that only 30 of the 7000 edible species are today contributing 95% of the world's dietary requirements; three of them — rice, wheat, and maize (corn) — are getting most of the attention, while other nutritious species and medicinal plants are neglected.

Causes for Declining Agrobiodiversity

There are many causes of this decline, which has been accelerating throughout the 20th century in parallel with the demands of an increasing population and greater competition for natural resources. The principal underlying causes include the following:

- The rapid expansion of industrial and Green Revolution agriculture, intensive livestock production, and industrial fisheries and aquaculture (some production systems using genetically modified varieties and breeds) that cultivate relatively few crop varieties in monocultures, rear a limited number of domestic animal breeds, and fish for, or cultivate, few aquatic species.

- Globalization of the food system and marketing, and the extension of industrial patenting and other intellectual property systems to living organisms, which have led to the widespread cultivation and rearing of fewer varieties and breeds for a more uniform, less diverse but more competitive global market.

The consequences are as follows:

- Marginalization of small-scale, diverse food production systems that conserve farmers' varieties of crops and breeds of domestic animals, which form the genetic pool for food and agriculture in the future.
- Reduced integration of livestock in arable production, which reduces the diversity of uses for which livestock are needed.
- Reduced use of 'nurture' fishery techniques that conserve and develop aquatic biodiversity.

Genetic erosion is the loss of genetic diversity, including the loss of individual genes and particular combinations of genes (i.e. gene complexes) such as those manifested in locally adapted landraces. The term "genetic erosion" may be used in a narrow sense, i.e. the loss of genes or alleles, as well as more broadly, referring to the loss of varieties. The main cause of genetic erosion in crops, as reported by almost all countries, is the replacement of local varieties by improved or exotic varieties and species. As old varieties in farmers' fields are replaced by newer ones, genetic erosion frequently occurs because the genes and gene complexes found in the diverse farmers' varieties are not contained in the modern variety. In addition, the sheer number of varieties is often reduced when commercial varieties are introduced into traditional farming systems. While some indicators of genetic erosion have been developed, according to the FAO (1996, 1998), there have been few systematic studies of the genetic erosion of crop genetic diversity that have provided quantifiable estimates of the actual rates of genotypic or allelic extinction in the Plant Genetic Resources for Food and Agriculture (PGRFA).

Variety replacement is the main cause of losses. The replacement of local varieties or landraces by improved and/or exotic varieties and species is reported to be the major cause of genetic erosion around the world. It is also cited as the major cause of genetic erosion in all regions except Africa.

For instance, a survey of farm households in the Republic of Korea showed that, of 14 crops cultivated in home gardens, an average of only 26% of the landraces cultivated there in 1985 were still present in 1993. The retention rate did not exceed 50% for any crop, and for two crops it was zero. These results are disturbing, as such home gardens have traditionally been important conservation sites, especially for vegetable crops.

In China, in 1949, nearly 10 000 wheat varieties were used in production. By the 1970s, only about 1000 varieties remained in use. Statistics from the 1950s show that local varieties accounted for 81% of production, locally produced improved varieties made up 15%, and introduced varieties 4%. By the 1970s, these figures had changed drastically; locally produced improved varieties accounted for 91% of production, introduced varieties 4%, and local varieties only 5% (FAO 1996, 1998).

Benefits of Agricultural Biodiversity

Prime Minister Manmohan Singh of India, in his message to the Chennai Consultations on Hunger and Agricultural Biodiversity on September 22, 2005, pointed out that, in the past, the community food tradition assured that a wide range of food crops rich in protein, iron, micronutrients, and vitamins were available to the people; however, commercial agriculture has narrowed the range of food crops available.[a]

Agricultural biodiversity can help in developing decentralized community food security systems which benefit local communities. They are also beneficial for long-term security by establishing gene banks, seed banks, and grain banks which can be

[a] This remark also appeared in *India Together*, April 1, 2008.

managed by local people. The diversity of crops could also reduce pesticide use. Furthermore, tropical fruits, sweet potato (with beta-carotene), and other vegetable crops can fight vitamin A deficiency in children.

Agricultural biodiversity provides the important raw material for improving the quality of crops, livestock, and fish. It will create opportunities for entrepreneurship by generating employment and additional income from a whole range of value-added foods, medicines, nutraceuticals, biofuel, and other sources. On a global scale, nearly 2.5 billion people depend directly on wild and traditionally cultivated plant species to meet their daily needs.

These benefits were emphasized at an international conference on agricultural biodiversity that was held at the M.S. Swaminathan Research Foundation in Chennai, India, in 2005. There is a close relationship between agricultural biodiversity and cultural diversity. Agricultural biodiversity is reflected in the cultural activities of the community, such as their songs, dances, dramas, and poetry. It is also closely tied to the culinary diversity and preferences of the community.

Chennai Recommendations

- Incorporate agricultural biodiversity conservation and sustainable use in national development.
- Enhance the use of agricultural biodiversity and its traditional knowledge for promoting nonfarm employment and income generation that would be consistent with traditional rights, cultural identity, ecosystem integrity, and gender equity.
- Strengthen the multilateral system of exchange provisions of the FAO International Treaty on Plant Genetic Resources for Food and Agriculture to expand its coverage of plant species important to food security and income generation.
- Ensure fair and equitable benefit sharing of commercial gains accrued from accessed genetic resources.

- Work towards a similar treaty on the multilateral exchange of animal genetic resources relevant to food and agriculture.
- Recognize and reward the invaluable contributions of rural and indigenous people, particularly women, in the conservation and enhancement of agricultural biodiversity.
- Confer social prestige and economic benefit to the primary conservers.
- Promote local markets and facilitate access to international markets for the products of agricultural biodiversity, especially traditional and functional foods.
- Advocate and strengthen national nutrition literacy through participatory knowledge management.
- Train agricultural extension workers and health and nutrition professionals in the importance of dietary diversity and evidence-based beneficial effects of traditional foods to re-establish the relevance of regional agricultural biodiversity in fighting hunger and poverty.
- Ensure that food and nutrition support safety net programs, especially food aid and school feeding programs as well as food banks, are fostering greater dietary diversity. Broadening the food basket with more indigenous crops as part of national nutrition policy can do this.
- Restructure research and development priorities to enhance the productivity, profitability, and value chain development of a wider range of agricultural biodiversity. By bringing in hitherto neglected species, a greater economic stake in their conservation can be created.
- Change the mindset so as to prevent the perennial loss of vanishing crops and dying wisdom through international initiatives to change the public image of underutilized and orphan crops. This could be done by steps such as redesigning "coarse serials" as "nutritious cereals", or classifying a wide range of leafy vegetables, tubers, grain legumes, and tropical fruits as "health foods". Saving plants for saving lives and livelihoods should become everybody's business.

Increased global cooperation in the conservation and sustainable use of agricultural biodiversity will go a long way towards the eradication of world hunger.

Corporate Responsibility

Some private industries are showing a keen interest in protecting biodiversity as part of their social responsibility. For instance, the company Spice Energy has declared the following goals:

- Biodiversity conservation
- Solid waste management
- Restoration of borrow areas and quarry sites
- Green belt
- Fisheries development
- Environmental monitoring
- Landscaping
- Health care
- Resettlement and rehabilitation
- Catchment area treatment
- Compensatory forestation

India: Destruction of Forest Biodiversity

Forest biodiversity in India is being destroyed at a rapid rate, mainly because of human population growth and expansion. India is one of the richest biodiversity regions of the world. It is home to almost 50,000 species of plants and 90,000 species of animals. It has about 170 crop species and 350 wild relatives. India has 2.4% of the world's surface area but has over 8.0% of the world's biodiversity, making it one of the 12 "hotspot" regions of biodiversity in the world. It is the center of origin of 50,000 varieties of rice, mango and other major crops, making it seventh among nations contributing to world agriculture.

However, India's biodiversity is under serious threat. Population expansion and human activities are the main culprits which are

(*Continued*)

(Continued)

causing habitat destruction. Climatic factors, invasive species and commercial forestry have also led to the destruction of the forest biodiversity.

As most of the minerals in India are in forest areas, there is tremendous pressure to cut forests for mining. If tribal areas have to be developed, roads, water and power have to be reached to them. But most of the tribal areas are in forests. Once roads enter forests, there are edge effects. With the ingress of population and vehicle, trees become the first casualty. Says a former forest official: "Forests are hardly ever seen as a symbol of development. But roads are. Mining in a forest is totally destructive. Government corporations often violate rules that lead to biodiversity destruction. As they are a part of the government setup, they do not take the required permissions."

Forest Replacement

Compensatory forestation has been in practice in India but former Director General of forests in India, S.K. Pande, pointed out that compensatory forestation does not replace fully the original forest. What was in existence before was the original forest area, a complex ecosystem, which had evolved over millennia. Planting one or two species to replace the original forest is not the answer. In fact, there is no alternative to conserving as much forest area as possible.

India has one of the largest networks of protected areas in the world. Its wildlife sanctuaries and national parks take up around 1,12,274 square kilometers. India has more than 50,000 varieties of rice, over 5,000 varieties of sorghum, 1000 of mango and as many as 500 varieties of pepper.

In biological diversity, India is one of the richest countries in the world. But widespread destruction has already taken place and this is continuing. Urgent measures to reverse the damage are both necessary and possible.

Ramesh Menon, documentary film maker in India, wrote: "After years of public participation, India's National Biodiversity Strategy

(Continued)

(Continued)

and Action Plan, reached its final stage in 2004. And then, things unravelled. Though numerous recommendations have been made by the NBSAP, New Delhi has gone silent on it."

India has lost over half of its forests, forty per cent of its mangroves and a large part of its wetlands, according to the National Biodiversity Strategy and Action Plan. Numerous species of plants and animals are extinct and several hundreds are under the threat of extinction.

With co-coordinated action, biodiversity loss can still be salvaged, at least to some extent. Some recommendations:

- Have a policy framework that constantly monitors dangers to biodiversity and acts on it.
- Ensure that commercial development is not allowed in areas that need ecological security.
- Expanding biosphere reserves may not be easy, but at least what exists must be zealously protected.
- Document traditional knowledge before it is lost and use it to save resources that are being eroded.
- Train people in natural resource management. Involve village communities in forest management.
- Research areas in immediate danger.
- Create wildlife protected areas with strict penalties.
- Return to diverse crop cycles. Food policies have to be so designed to promote practices that help conserve diversity.
- Avoid unsustainable farming methods that lead to wastage of water and excessive use of pesticides.
- Encourage appropriate and sustainable technologies.
- Build capacities of local communities to tackle biodiversity issues and sensitise them.
- Develop energy sources that will save natural resources like wood.

For being a part of a complex system that protects biodiversity, farmers have to be rewarded. The government could introduce nutritionally superior foods into the public distribution system like coarse millets

(*Continued*)

(Continued)

and also get it into programmes like Food for Work. This will keep diverse strains of the grain alive as there is a demand for it.

When a nation loses biodiversity, it stands to lose not only its wealth, but its future. There is little time for India to lose as its biodiversity is already in danger.

(Swaminathan and Jana 1992)[b]

Preserving the Web of Life

Agricultural Biodiversity and Sustainable Agriculture

NB: These pages were created for the Earth Summit in Johannesburg (WSSD) in 2002, and have not been updated since. They are included here for reference only.

Agricultural biodiversity, the vast number of locally-adapted seed varieties and animal breeds, underpins the food security of our planet. This interdependent life-support system helps sustain local eco-systems that provide, not just food to eat, but also clean water, healthy top-soils, living landscapes, clean air, and even a sink for excess carbon dioxide.

Agricultural biodiversity is disappearing rapidly, a loss that is contributing to poverty and environmental degradation. The effects of industrialised agricultural production threaten in particular, agricultural biodiversity. Mono-cropping, genetic modification and increasing restrictions on access to genetic diversity diminish agricultural biodiversity.

The ten international agreements* to preserve agricultural biodiversity since the first Earth Summit held in Rio de Janeiro ten

*Ten agricultural biodiversity agreements since 1992.
Agenda 21 highlights the importance of the sustainable use of agricultural biodiversity, and this is echoed in many other agreements developed in subsequent meetings organised through the Convention on Biological Diversity (CBD)

(Continued)

[b] See also Ramesh Menon in *India Together*, June 2, 2005.

(Continued)

years ago have not delivered reductions in losses of agricultural biodiversity, because their provisions have not been implemented effectively. The World Summit on Sustainable Development (WSSD) in Johannesburg presents an opportunity for governments to commit to their implementation.

It is also an opportunity to agree further action in key areas. Governments must:

- Take immediate action to ratify the Biosafety Protocol and the International Treaty on Plant Genetic Resources for Food and Agriculture, and implement these and other existing agreements that concern the conservation, sustainable use and equitable sharing of the benefits arising from the use of agricultural biodiversity.
- Ensure the 'free flow' of agricultural biodiversity without threats of privatisation through patents, and other intellectual property rights that restrict access to plant, animal and aquatic genetic resources.
- Prioritise agro ecological approaches in agricultural research, development and extension policies.
- Ensure that existing environmental and agricultural agreements which preserve agricultural biodiversity have precedence over trade agreements, where these conflict.

---

and the Food and Agriculture Organisation of the United Nations (FAO), for example:

- FAO Leipzig Global Plan of Action on the Sustainable Use and Conservation of Plant Genetic Resources for Food and Agriculture (1996)
- FAO World Food Summit's Commitment 3 to Sustainable Agriculture (1996)
- FAO Global Strategy on Farm Animal Genetic Resources (1997)
- FAO Code of Conduct for Responsible Fisheries (1995)
- Four CBD Decisions on Agricultural Biodiversity (III/11 (1996), IV/6 (1998), V/5 (2000), VI/5 and VI/6 (2002)) which mandate the Programme of Work on Agricultural Biodiversity, managed by FAO
- CBD Cartagena Protocol on Biosafety (2000)
- FAO International Seed Treaty (ITPGRFA) (2001)

(*Continued*)

(Continued)

- Agree a global moratorium on the release of GM crops, livestock, fish and other aquatic organisms in the form of grain, food, food aid, animal feed, seeds, embryos, live animals or living organisms, in accordance with the Precautionary Principle. In particular, implement an immediate ban on the release into the environment of GM crops in centres of origin and diversity of those crops, and prohibit the development and use of Genetic Use Restriction Technologies.

What Is Agricultural Biodiversity?

The taste, texture and diversity of the food we eat, as well as its nutritional qualities, depend on the genes of the plants and animals from which it comes. These plants and animals grow, thrive, resist pest and diseases and live in symbiosis with surrounding species. Collectively they comprise what we call agricultural biodiversity and are a vital part of what is termed 'biodiversity' — the variability among living organisms on the Earth. This agricultural biodiversity is the product of the application of the knowledge and skills used by women and men to develop agriculture, livestock production and aquaculture. Agricultural biodiversity is thus both a product of agriculture and an essential component of ecosystems and their sustainability.

Although some 7,000 species of plants and many hundreds of animal species and thousands of aquatic plants are edible, human societies have focused on a few to feed themselves. Only about 100 crops, a handful of grasses and a dozen animal species are considered by some to be essential for feeding the world. Just four crops provide more than half the dietary energy for the whole world's population — maize, potatoes, rice and wheat.

Such dependence on a few species is a potentially perilous strategy, at risk from pest or disease epidemics and climate change. Fortunately for mankind, resourceful indigenous peoples, women and men farmers, forest dwellers, pastoralists and fisher-folk have developed a myriad of varieties of every crop, breeds of livestock and sub-species of fish and

(*Continued*)

((Continued)

other aquatic organisms. These provide for every possible social, cultural and economic need and are suited to a kaleidoscope of different ecosystems, climates and pest and disease threats. By developing, selecting and improving local varieties and livestock breeds, swapping seeds and animals amongst themselves and sharing these with neighbours, agricultural biodiversity has been maintained.

The exchange of seeds and breeds across the world has resulted in a vast number of locally adapted varieties and breeds. Maize, which originated in what is now Oaxaca, Mexico, is a staple crop in Africa and Asia, as well as all of the Americas and much of Europe. Apples originated in the Himalayas but now there are varieties suited to every community in all temperate regions of the world. Rice came from S E Asia, wheat from the Fertile Crescent, potatoes from Peru, and the humble lettuce has its origin in Slovenia.

The biosphere — the Earth's surface environment and atmosphere — is dependent on agricultural biodiversity. For every crop variety, livestock breed or aquatic organism growing or being raised on a farm or on common pasture or in ponds, there are thousands of other species on which it depends — other plants, animals, insects, pollinators, predators and soil biota (fungi, bacteria, soil insects, worms). In one teaspoon of healthy soil there are estimated to be more than 100 million soil organisms of some 50,000 different species, each with its specific functions and niches within the soil structure. Pollinators, including bees, provide free services that have been valued as being worth more than $50 billion annually. All of these are interdependent life-support systems and sustain local ecosystems. In turn these ecosystems provide not just a productive environment but also clean water, healthy top-soils, living landscapes, clean air, even a sink for excess carbon dioxide.

Threats to Agricultural Biodiversity

For the majority of the 2.7 billion people who live on less than two dollars a day, survival depends on consuming locally-grown food, a

(*Continued*)

(Continued)

fact that highlights the importance of agriculture to poverty reduction. Threats to the agricultural biodiversity which underpins food security for these people are threats to sustainable development and poverty reduction. These threats are very real.

Changes in Production Systems

In the 1990s, the adoption of modern varieties of wheat, rice, and maize in developing countries reached around 90, 70, and 60 per cent respectively. In Latin America the take up of modern rice varieties leapt from 4 to 58 per cent in two decades, in Asia from 12 to 67 per cent. Access to, and use of, a wide range of agricultural biodiversity is threatened by this simplification of production systems.

As food production becomes increasingly industrialised, with fewer niches available for species other than those targeted for production, we are witnessing a rapid decline in the diversity of varieties used. The UN's Food and Agriculture Organisation (FAO) estimate that more than 90 per cent of crop varieties have disappeared from farmers' fields in the past 100 years. Agricultural plant varieties are continuing to disappear at 2 per cent a year. Livestock breeds are being lost at 5 per cent annually. The current extinction rate of species ranges from approximately 1,000 to 10,000 times higher than natural extinction rates.

These major changes in production lead to simplified and less resilient agro-ecosystems, reducing not only the number of niches but also the range of products and their distribution over time and space. Single crops are more vulnerable to the rapid spread of disease — this greatly heightens the vulnerability of resource-poor farmers. Through reduction of field margins, elimination of intercropping, destruction of soil biodiversity, pollution of water courses and so on, the essential niches for the species which support production — e.g. predators, pollinators, soil biota (fungi, bacteria, insects, worms) etc. — are removed. In the USA, for example, over 50 pollinator species are listed as threatened or endangered, and wild honeybee populations have dropped 25 per cent since 1990.

(Continued)

(Continued)

Sustainable agro ecology, the option that sustains agricultural biodiversity and food production, has been shown to be successful in restoring yields. Surveys of this type of production, in more than 10 million hectares in 51 countries, reveal that yields can increase by 200–300 per cent in the more degraded production systems. Even in modern smallholder production systems, yields increased by about 10 per cent, despite the sharp reduction in use of pesticides and added chemical fertilisers.

Genetic Engineering

While FAO records show that governments' most cited reason for biodiversity loss is variety and breed replacement on the farm, a further threat is presented by the adoption of genetic engineering in industrialised agricultural systems. During the six-year period 1996 to 2001, the global area under GM crops increased more than thirty-fold, from 1.7 million hectares in 1996 to 52.6 million hectares in 2001. The seven principal GM crops grown in 1998 were (in descending order of area) soybean, maize, cotton, canola (rapeseed), potato, squash, and papaya. GM crops have reached the field trial or commercial scale in over 40 countries, including those with high proportions of their populations dependent upon agricultural biodiversity, such as Nicaragua, Honduras, Swaziland and Vietnam.

Genetic modification is a threat to both the genetic integrity of agricultural biodiversity and its ownership. The resultant location of an inserted gene, the impact of modification on the structure of the genome and the impact and location of promoters is unknown in most cases and could have long-term deleterious effects. It is not just that the new genetically modified organisms (GMOs) may produce unexpected proteins that could cause allergies in humans, nor that they may behave in an erratic and unexpected way, it is also that the impacts they may have on other living organisms and the environment are unpredictable.

(*Continued*)

(Continued)

A concern is that the process of modification so alters the arrangement of genetic material in the nucleus — it "scrambles the genome" — that the GMO will behave very differently from other species that have been developed through normal sexual reproduction. The resultant gene constructs could then spread through the biosphere by way of horizontal gene transfer, through seeds, pollen, soil micro-organisms and so on, with unknown consequences. The focus on genetic engineering in agricultural research and development skews resource allocations away from the more sustainable option of agro ecology.

Genetic Patents

The insertion of patented genes into plants and animals, using genetic engineering technologies, transfers ownership of those plants and animals to the gene's patent holders. As these genes spread through the agro-ecosystem, so ownership of agricultural genetic resources will be further concentrated.

Agricultural biodiversity and its component genetic resources for food and agriculture are therefore under threat from privatisation through patents and other intellectual property rights. This results in moving knowledge and genetic resources from the informal sector into the formal sector, and from public domain to private ownership, reducing benefits for the originators of that knowledge. These are usually people and communities in the informal sector. Agricultural biodiversity was developed through the free exchange of seeds and other genetic resources and is better conserved and utilised through common access arrangements and the realisation of community, farmers' and traditional rights.

Terminator Technologies

Threats also arise from the development of one particular type of restriction of access. This is through the use of genetic modification to

(*Continued*)

(Continued)

produce a GMO including Genetic Use Restriction Technologies (GURTs). They include a range of use restriction technologies that limit a plant's ability to produce certain traits. The most dramatic of these is the variant that prevents germination of seeds produced by a plant. These technologies have been dubbed "Terminator Technologies".

The widespread use of GURTs will result, by definition, in reduced access to genetic resources. Farmers prefer a wide range of genetic materials available for local crop development. Increased use of GURTs may result in greater reliance on formal seed markets that are less efficient and accessible to cash-poor farmers. Finally the economic power of the corporations developing and marketing GURTs could induce a shift away from local germplasm sources and further erode local and traditional seed systems that lie at the heart of crop genetic diversity.

Restoring Marine Diversity

Construction of Artificial Reefs

In Kerala, SW India, local civil society organisations, supported by ITDGPractical Action, have worked with fishing communities to restore aquatic biodiversity in their fishing grounds. The solution was the construction of simple artificial reefs by village fishermen in response to loss of fishing grounds through destructive effects of trawling.

India is the world's seventh largest producer of fish products and one quarter of India's catch is from the fishermen of Kerala who use very simple craft and gear. Norwegian fishery advisers advocated the introduction of trawlers. The village fishermen survive at subsistence levels and did not have the capital to invest in this technology. They saw the market price of their catch collapse, fall in catches through overfishing and destruction of natural reefs. Militant actions were taken to keep trawlers away. Kerala fishing policy was changed, introducing a closed season for trawlers. But the fisherfolk took long-term actions themselves.

Artificial reefs were constructed using any available materials, rocks, coconut palm stumps, tyres, concrete well rings and later

(*Continued*)

(Continued)

triangular ferro-concrete units cast on the beach. These have restored aquatic ecology and fish breeding sites, provided inshore fishing locations, made the fishery more reliable and created a sense of ownership and stewardship for the resource. The unmarked reefs also protect the fishing grounds by erecting on the sea floor a significant disincentive to trawlers whose nets snag on the underwater obstructions.

International Collective in Support of Fishworkers

The International Collective in Support of Fishworkers (ICSF) is an international non-governmental organization that works towards the establishment of equitable, gender-just, self-reliant and sustainable fisheries, particularly in the small-scale, artisanal sector.

ICSF draws its mandate from the historic International Conference of Fishworkers and their Supporters (ICFWS), held in Rome in 1984, parallel to the World Conference on Fisheries Management and Development organized by the Food and Agriculture Organization of the United Nations (FAO).

Action Plan for WSSD

Ten years ago, at Rio, there was recognition that agricultural biodiversity is fast disappearing and that this was contributing to poverty and environmental degradation. International action to arrest this decline and restore agricultural biodiversity has resulted in a Treaty, a Protocol, a Code of Conduct, and action plans and programmes. In all, ten international agreements to preserve agricultural biodiversity have been negotiated since 1992, an indication of the importance attached by the UN to this issue. Together these agreements could go some way to arrest the decline in agricultural biodiversity. However, none of their measures have yet been effectively implemented and

(*Continued*)

(Continued)

they have, so far, failed to deliver reductions in losses of agricultural biodiversity.

The WSSD Plan of Implementation should call for immediate implementation of all these agricultural biodiversity instruments and programmes. Their combined impact could go some way to restoring the agricultural biodiversity that preserves the web of life on earth. It should specifically cite them in both the 'Agriculture' and 'Biodiversity' sections of the Plan of Implementation.

A considerable contribution to sustaining agricultural biodiversity can be made through continued support for work through civil society organisations and producer organisations with local small-scale producer groups to conserve, develop and use sustainably all genetic resources for food and agriculture including plant, animal and aquatic genetic resources.

Action is also required nationally and internationally. The draft Plan of Implementation for WSSD includes many agreed actions concerning the conservation, sustainable use, intellectual property, benefit sharing and biosafety of agricultural biodiversity. Actions for strengthening sustainable agriculture and food security, the protection of local natural resource management, soils, and the use of environmentally sound pest management practices have been agreed. The draft plan also calls for the involvement of local communities and especially women and the recognition of their resource rights. It 'invites' countries to ratify the International Seed Treaty (ITPGRFA) and the Biosafety Protocol.

Further actions governments must take are:

- The immediate ratification of the Biosafety Protocol and the International Treaty on Plant Genetic Resources for Food and Agriculture and implement these and other existing agreements that concern the conservation, sustainable use and equitable sharing of the benefits arising from the use of agricultural biodiversity.

(*Continued*)

(Continued)

- Prohibit patent on genetic resources for food and agriculture, to ensure the 'free flow' of agricultural biodiversity without threats of privatisation through patents, and other intellectual property rights that restrict access to plant, animal and aquatic genetic resources.
- Give priority to agro ecological approaches in agricultural research, development and extension policies. Changing the focus of agricultural, livestock, forestry and fisheries research away from industrial production systems and genetic engineering towards small-scale agro ecological approaches, in collaboration with producers, would sustain and develop agricultural biodiversity.
- Agree a global moratorium on the release of GM crops, livestock, fish and other aquatic organisms in the form of grain, food, food aid, animal feed, seeds, embryos, live animals or living organisms, in accordance with the Precautionary Principle. In particular, implement an immediate ban on the release into the environment of GM crops in centres of origin and diversity of those crops, and prohibit the development and use of Genetic Use Restriction Technologies (GURTs).
- Ensure that environmental and agricultural agreements that preserve agricultural biodiversity have precedence over trade agreements, where these conflict.

The fast-disappearing varieties of crops, livestock breeds and aquatic organisms threaten the planet's web of life. Urgent action is needed to restore this vital component of biodiversity so essential to food security and ecosystem integrity. WSSD must rise to the challenge of sustaining the agricultural biodiversity of the food crops, livestock breeds and aquatic organisms that feed us and sustain the biosphere.

Fig. 1. GM corn.

Fig. 2. Tribal dancers in Araku Valley, Andhra Pradesh, India. Courtesy of Michele Wambaugh.

Fig. 3. Haystack, Araku Valley, Andhra Pradesh, India. Courtesy of Michele Wambaugh.

Fig. 4. Winnowing grains, Araku Valley, Andhra Pradesh, India. Courtesy of Michele Wambaugh.

Traditional Agroforestry in the High Islands of Micronesia

Traditional subsistence agriculturalists of the higher and larger islands of Micronesia developed a wide range of agricultural technologies and systems for the production of food under different environmental conditions and in different locations. These included permanent systems of wetland taro cultivation, mixed tree gardening, intermittent tree gardening (shifting cultivation), home or kitchen gardening, and open grassland cultivation. The use of any one of these systems did not preclude the use of others. Most islanders produced food using all five systems and, in addition, exploited the reefs and ocean for marine products. Most attention will be given here to mixed tree gardening, an elaborate and refined indigenous agroforestry that is the predominant rural land use on many Micronesian high islands. The other four systems will be briefly discussed, particularly with regard to their relationship to trees.

Areas in hectares of land classes in Micronesian high islands. (Total agroforest includes agroforest/tree gardens and coconut plantations; Truk data are for the high islands of Moen, Dublon, Fefan, and Eten only)

	Belau	Kosrae	Pohnpei	Truk	Yap
Forest	28,093	7,066	19,683	986	3,882
Secondary forest and vegetation	594	1,272	1,843	252	553
Agroforest/Tree gardens	187	2,585	11,741	2,378	2,379
Coconut plantation	743	124			
Total agroforest	930	2,585	11,865	2,378	2,379
Non-forest	8,285	263	2,102	554	2,743
Total area	38,832	11,186	35,493	4,170	9,557
(percentage agroforest)	(0.48)	(23)	(33)	(57)	(25)

(*Continued*)

(Continued)

Wetland Taro Cultivation

The production of *Colocasia esculenta* and *Cyrtosperma chamissonis* taros in essentially permanent patches has long been practiced on all the high islands of Micronesia, often by means of sophisticated systems of water management that minimized soil erosion and stagnation of water. Favoured areas for the wetland cultivation of taro are alluvial bottom lands and the lowland swamps and marshes located inland of the mangroves, generally within agroforests. The taro patch may act as a filter to minimize sedimentation of the lagoons and ocean. When a patch under fallow is selected for replanting with *Colocasia esculenta*, it is cleared of vegetation and drained, the soil is dug up, and various leaves, twigs, and sea grasses are added as a mulch or green manure, with *Wedelia biflora*, *Carica papaya*, and *Macaranga* sp. recognized as good manuring species. After mulching, the patch is worked to produce a fertile muck of desired consistency and planted with cormels or corm tops. Harvesting occurs six months to a year after planting, and the patch may be replanted to provide a continuous supply. If the yield or quality declines, the patch is allowed to lie fallow for a number of years.

The cultivation of *Cyrtosperma chamissonis* requires less labour and attention than is needed for *Colocasia*. Green manures are not added to the soil to improve its fertility for cultivation of *Cyrtosperma chamissionis*, and cultivation methods include the periodic removal of fallen vegetation and debris in order to maintain the flow of water through the system. *Cyrtosperma* is more shade-tolerant than *Colocasia* and thus more compatible with tree culture, and an integral part of the agroforestry system of Yap. *Cyrtosperma* is the preferred aroid in Yap and Truk; *Colocasia* remains the preferred food in Pohnpei, where not all of the trees were cut back during garden clearance, with some, such as *Hibiscus tiliaceus*, being left standing to provide shade for the young plants.

(*Continued*)

(Continued)

Mixed Tree Gardening

A very well developed form of traditional agroforestry in Micronesia is the tree garden or agroforest. Except for the initial planting and care while the trees are young, this agroforestry system requires little energy input, but provides Micronesians with an abundant supply of different tree crops and products from marginal lands over a long time period, while maintaining a permanent canopy cover. Except in Palau, such agroforests cover considerable areas of high-island Micronesia.

The composition and structure of these agroforest gardens varies from place to place. In coastal areas, they tend to be relatively simple, consisting of few species, dominated by coconuts. In Truk, breadfruit is a dominant species of mixed tree gardening. In Guam, breadfruit, coconuts, and *Cycas circinalis* were harvested from mixed tree gardens. In Palau, these mixed forests were called chereomel and consisted of a wide range of food, fruit, and timber trees, including coconuts, mango, breadfruit, *Terminalia catappa*, and *Inocarpus fagifer*. The forests serve as a habitat for feral and domesticated animals, provide traditional medicines and other culturally valued products, and are or were a source for canoe hulls, building materials, and firewood.

The likely development process of tree gardens was the planting of trees for food and other uses around homesteads and in the drained areas created by the excavation of taro patches and the construction of drained paths between homes and villages. These home and path-side tree gardens became confluent and today make up a significant proportion of the island's terrestrial vegetation and contain about 50 native and introduced tree species, such as coconut, breadfruit, betel-nut (*Areca catechu*), *Inocarpus fagifer*, many varieties of banana and citrus species, *Pangium edule*, papaya, cacao, and guava. Of species that have long been cultivated on Yap, there are numerous named varieties — for example, 21 named coconut varieties and 28 named breadfruit varieties. Tree gardens are reported to be largely self-fertilizing. Owners maintain their tree gardens by selective pruning and removal of undesirable trees; occasionally

(*Continued*)

(Continued)

desired trees are planted. Once established, such tree gardens require little maintenance.

Traditional agroforestry system still supports most of the island's 28,000 people despite increasing emphasis on the cash economy.

The general characteristics of the Pohnpeian agroforestry system are an extensive, permanent overstory of breadfruit, coconut, and forest trees above fruit and multipurpose trees, and an understory of shrubs, root crops, and herbaceous plants. Although sharing many characteristics of home gardens, the system is best classified as a "multistory" tree garden because it is not limited to the area immediately surrounding the house compound but extends throughout the landscape. Breadfruit and yams (*Dioscorea* spp.) are the major staples, and complement each other in seasonality. *Hibiscus tiliaceus* is the premier "multipurpose" tree, because it is used for firewood and light construction, poles or whole trees are used for yam trellises, its leaves are used as green manure in soil pits dug for yams, its bark for rope, and its inner bark for best fibers for straining mashed *Piper methysticum* roots for drinking.

Raynor and Fownes (1991) carried out intensive sampling in 54 landholdings on Pohnpei and enumerated 102 different species, 26 of which were upper canopy species, 39 were sub-canopy, and the remainder were understorey. A few species found on nearly every holding constituted the typical Pohnpeian agroforest: coconut, breadfruit, *Cananga odorata*, mango, *Musa* spp., *Hibiscus tiliaceus*, *Morinda citrifolia*, *Alocasia macrorriza*, *Dioscorea alata*, and *Piper methysticum*. As on Yap and elsewhere, several of the important species have many recognized cultivars, a diversity that is an important component of the aggregate biological diversity of Micronesian agroforestry systems (Raynor and Fownes 1991).

Because they could detect no patterns in species composition dependent on elevation, soil type, or particular region, Raynor and Fownes concluded that "... the Pohnpeian agroforest is a managed landscape, despite its superficial resemblance to forest. Species presence

(*Continued*)

(Continued)

or absence was apparently more strongly controlled by farmers' decisions than ecophysiological constraints."

Intermittent Tree Gardening (Shifting Cultivation)

Intermittent tree gardening (i.e., shifting cultivation, or swidden) is practiced in secondary forest fallows on all high islands of Micronesia. Structurally and functionally, this system is little different from shifting cultivation systems described for the other parts of the Pacific, except that in Kosrae burning was not used in garden clearing (Wilson 1968). In contrast to mixed tree gardening, intermittent tree gardening is not a permanently productive form of land use. When the gardens, cleared from forest, are no longer harvested after one to two years of production, the site returns to fallow, succeeding through stages to a forest of spontaneous secondary species, here and there enriched with useful species such as breadfruit planted during the garden phase. The major crop in such gardens is yams of the genus *Dios corea* — six species of which and 34 locally recognized varieties have been recorded from Yap.

Home Gardens and Lanchos

Throughout high-island Micronesia, home gardens are a common feature of most households. In Guam and the Northern Marianas a variation of the home garden is known as the lancho. These provide villagers with a ready source of food, fruit, spices, herbs, and, in some cases, medicinal plants. In urban areas, these gardens are, in the main, supplementary to a wage income (Falanruw, 1999).

Out of an extensive pool of fruit-trees, the most commonly found are varieties of citrus, coconuts, breadfruit, and bananas. Ornamental trees and shrubs, some of which have ritual or ceremonial significance, are other components of kitchen gardens. *Hibiscus* hybrids, *Cordyline fruticosa*, and *Codiaeum varigatum* are as significant in

(*Continued*)

(Continued)

the Micronesian high islands as they are in Melanesian and Polynesian societies. The latter two species were sometimes not planted in Palauan house gardens because of their association with death and the supernatural. In Guam, the "pickle" tree, or bilimbi (*Averrhoa bilimbi*), carambola, mango, coconuts, soursop (*Annona muricata*), perennial chilli (*Capsicum frutescens*), annatto (*Bixa orellana*), *Citrus* spp., *Jatropha integerrima*, *Cycas circinalis*, *Plumeria rubra*, *P. serratifolia*, *Araucaria excelsa*, and *Dracaena marginata* are found in many home gardens. In Palau, the betel pepper vine (*Piper* betle) was zealously guarded against theft. Other useful plants were *Areca catechu*, *Citrus* mitts, and *Muntingia calabura*.

For Guam and the Northern Marianas, there is little documentation of traditional subsistence cultivation. Prior to European contact, the indigenous Chamorros were mainly dependent on the ocean; root crop agriculture was rudimentary and supplemented by hunting for fruit bats, birds, and land crabs. However, under Spanish rule, and by the end of the nineteenth century, subsistence agriculture on the ranch (lancho) became accepted as the Chamorro way of life. Today, most ranchos are located in southern Guam and consist of a simply built cooking and sleeping house surrounded by food trees, chickens, pigs, and garden. Relatively few ranchos are cultivated without the use of fertilizers or pesticides, and not all of the production is consumed at home. In the Northern Marianas, ranchos are more difficult to find because of the impacts of economic development, division of family lands, foodstamp programmes, and population increases. During the Japanese administration of the Northern Marianas, traditional subsistence agriculture was replaced by the development of sugar-cane plantation agriculture.

Traditional Open-Canopy (Ked) Agriculture

The ked area lying in the interior region of Babeldaob, Palau, is characterized by exposed and eroded oxisols and ultisols, and a degraded

(*Continued*)

(Continued)

vegetation. Ked also refers to the fertile grassland areas in Palau, Yap, and Pohnpei, which are used for traditional subsistence agriculture.

Ked agriculture involves burning the grass, turning the friable soil, and hoeing ridges along the contour to reduce erosion. Mulching (with sea grass and *Cymbopogon citratus*) and ditching are other features of this open-canopy agriculture. Sweet potatoes, *Colocasia* taro, *Manihot esculenta*, and pineapple are the most commonly planted crops. *Cymbopogon citratus* (lemon grass) is planted as a border and to prevent soil erosion. With crop rotation to reduce species-specific insect predation, the ked garden can be cultivated for many years without fallowing.

On Yap, crops grown in open areas include sweet potatoes, cassava, and vegetables. The making of sweet-potato gardens involves the construction or reconditioning of drained garden beds by cutting or burning the vegetation on the site, piling the debris on the garden area, and sometimes adding additional mulch, including washed up sea grass. Ditches are dug or deepened around the garden bed, and the soil piled on top of the mulch, a process that drains the garden bed, suppresses weeds, and provides a fertile soil-mulch-soil sandwich. Open-canopy gardening on Yap today is reported to be causing a retreat of forest and an exhaustion of soils under secondary vegetation because of careless and too-frequent burning and too-short fallow periods between gardening.

The ditching, mulching, and other garden preparation activities, and the relatively long period of cultivation of ked agriculture, are very similar to the intensive, semi-permanent forms of cultivation found in Papua New Guinea grassland ecosystems.

Micronesian high islands, agricultural mounds and terraces, with or without stone facing, attest to the intensive and long-term cultivation of food crops. In Pohnpei, for example, bananas, coconuts, *Piper methysticum*, and *Alocasia macrorrhiza* are grown in earthen mounds and hillside terraces (Hunter-Anderson 1987).

(*Continued*)

(Continued)

Sustainability of Micronesian High-Island Agroforestry

It is clear that in the high islands of Micronesia, subsistence agroforestry has a long history and has successfully incorporated many recently introduced cultigens into the continuing evolution of traditional systems. High-island peoples developed both permanent and impermanent forms of agroforestry that provided food and materials from a large number of cultivated species and varieties. Raynor and Fownes (1991, 164) specifically address the question of the sustainability of the Pohnpeian agroforestry system, noting that Pohnpei is thought to have been settled for at least 2,500 years, and supported a pre-European-contact population as large as 50,000, compared with 30,000 in 1990. Twenty-two of the 54 farms surveyed by Raynor and Fownes were reported to be over 100 years old, but farmers claimed that many of them had been farmed as long as people had been there. The presence of domesticated trees and abandoned garden beds in today's forests on Pohnpei and Yap, and probably on other Micronesian high islands, demonstrates that the conversion of forest to garden and back to forest has a long history (Falanruw 1990, 102; Raynor and Fownes 1991, 164). The future sustainability of high island agroforestry probably depends less on ecological than on human factors, including not only knowledge of the systems and attention to their careful management but also the desire to maintain them in the face of food imports and the tendency of young people to seek wage employment in town or to emigrate (Raynor and Fownes 1991, 164). For the present, traditional agroforestry and subsistence agriculture remain important land uses in many high islands of Micronesia but are hardly practiced at all any longer in Guam and the Northern territories.

9

GMOs and the Law

The courts have not been consistent with respect to the planting of GM crops. The GM crops that are allowed to be planted vary from country to country, or even within different parts of a country. For instance, in May 2007, the Supreme Court of India cautiously allowed field trials of GM crops to proceed further. At the same time, a Federal judge in the U.S. made a final ruling that the U.S. Department of Agriculture's (USDA's) 2005 approval of Monsanto's genetically engineered (GE) "Roundup Ready" alfalfa was illegal. The Judge called on the USDA to ban any further planting of the GE seed until it conducts a complete Environmental Impact Statement (EIS) on the GE crop. Surely, countries can benefit from each other's experiences in some respects, in spite of some obvious differences.

Supreme Court Allows Field Trials of GM Crops in India

The Supreme Court of India in New Delhi recently permitted field trials of genetically modified (GM) crops approved by the Genetic Engineering Approval Committee (GEAC) in 2006 under certain conditions. The Judges ordered that the GM crop fields should be at least 200 meters away from fields with normal crops. One scientist should be in charge of the trials and make sure that non-GM crop fields are

(*Continued*)

(Continued)

not contaminated by pollen flow from GM crops. The GEAC should define a protocol for ensuring that contamination by GM crops does not exceed 0.01%. The GEAC is directed by the court to produce data relating to the toxicity and allergenicity of GM crops under trials. The court's ruling was in response to a writ petition filed by Aruna Rodrigues, PV Satheesh and others, calling for a moratorium on GM crops on health and environmental hazards.

The Supreme court's order delighted the biotech industry, which had been hoping for approvals for fresh GM crop field trials.

Crop biotech update. *Seed Quest*, May 11, 2005.

Judge Halts Planting of GE "Roundup Ready" Alfalfa

In a decision handed down in February 2007, a Federal Court has ruled, for the first time ever, that the U.S. Department of Agriculture failed to abide by federal environmental laws when it approved a genetically engineered crop without conducting a full Environment Impact Statement (EIS). A Federal judge in San Francisco made a final ruling that the U.S. Department of Agriculture's (USDA's) 2005 approval of Monsanto's genetically engineered (GE) "Roundup Ready" alfalfa was illegal. The Judge called on the USDA to ban any further planting of the GE seed until it conducts a complete Environmental Impact Statement (EIS) on the GE crop. Judge Charles Breyer in the Federal Northern District of California affirmed his preliminary ruling, which echoed the Center for Food Safety's arguments in their lawsuit against USDA, that the crop could harm the environment and contaminate natural alfalfa. The ruling requires Forage Genetics to provide the locations of all existing Roundup Ready alfalfa plots to the USDA within 30 days. The Judge ordered the USDA to make the location of these plots "publicly available as soon as practicable" so that growers of organic and conventional alfalfa "can test their own crops to determine if there has been contamination."

The Center for Food Safety in Washington D.C. first initiated the legal action in February 2006, representing itself and the following

(*Continued*)

(Continued)

coplaintiffs in the suit: Western Organization of Resource Councils, National Family Farm Coalition, Sierra Club, Beyond Pesticides, The Cornucopia Institute, Dakota Resource Council, Trask Family Seeds, and Geertson Seed Farms (www.centerforfoodsafety.org).

Judge Breyer's decision is consistent with his earlier ruling in February, when he found that the USDA failed to address concerns that Roundup Ready alfalfa will contaminate conventional and organic alfalfa. In calling for a permanent injunction, Judge Breyer noted that contamination of natural and organic alfalfa by the GE variety has already occurred, and noted that "Such contamination is irreparable environmental harm. The contamination cannot be undone."

- "This ruling is good news for organic farmers and most conventional farmers across the country," said Andrew Kimbrell, Executive Director of the Center. "This crop represents a very real threat to their crops and their livelihood. This ruling is a turning point in the regulation of biotech crops in this country," Kimbrell concluded.
- The permanent injunction ordered today by Judge Breyer follows his ruling last month finding that the USDA violated national environmental laws by approving GE alfalfa without a full Environmental Impact Statement. Monsanto and Forage Genetics, the developers of the GE alfalfa seed, failed to convince the Judge that their interests outweighed the public interest in food safety, freedom to farm natural crops, and environmental protection. In fact, Judge Breyer specifically noted that Monsanto's fear of lost sales "does not outweigh the potential irreparable damage to the environment."
- Judge Breyer found that the USDA failed to address the problem of Roundup-resistant "superweeds", which could follow commercial planting of GE alfalfa. Commenting on the agency's refusal to assess this risk, the Judge stated, "Finally, the court rejects the defendants' assertion that allowing an expansion in the Roundup

(*Continued*)

(Continued)

Ready alfalfa market is in the public interest because it promotes the use of less toxic herbicides. The record reflects that organic and most conventional forage alfalfa is grown without the use of any herbicides. In any event, a finding that increasing the use of Roundup is in the public interest is premature in light of APHIS's failure to analyze the potential for the development of Roundup-resistant weeds."

Reaction to the Judge's ruling was immediate and jubilant by various environmental groups.

- "As a consumer of organic foods, I'm relieved to know that a U.S. District Court judge understands the regulatory role the USDA plays, even though the agency itself seems to have forgotten," said Dean Hulse, an organic food consumer from Fargo and past chair of Dakota Resource Council. "Judge Breyer's ruling forces the USDA to do its job — that is, to conduct the research necessary to determine the effects of Roundup Ready alfalfa on the environment."
- "I'm hopeful that Judge Breyer's precedent-setting ruling will induce a rebirth of values at the USDA, in particular, and federal regulatory agencies generally," added Hulse. "The USDA's role with respect to regulating transgenic crops should be that of watch dog, not lap dog."
- Organic alfalfa seed producer Blaine Schmaltz, Rugby ND, said the ruling helps farmers in a time of uncertainty. "The judge's order to make public the location of Roundup Ready alfalfa fields is a critical part of the decision," said Schmalz. "It allows GM-free and organic producers like me to make sound planting decisions."
- "This ruling protects the ability of farmers producing organic meat and milk to obtain non-GMO alfalfa seed to grow feed for their animals and preserve the organic integrity of their products," said Jim Munsch, a certified organic livestock producer from Coon Valley, Wisconsin, who represents The Cornucopia Institute, one of the plaintiffs. "This is precedent-setting. For the first time the

(*Continued*)

(Continued)

courts have intervened on a USDA ruling to ensure that proper environmental evaluation and consideration for the livelihood of family farmers are accounted for and balance the desires of large companies," Munsch added.

- "This landmark decision curtails a genetically engineered crop that, among other serious environmental problems, increases farmers' dependency on toxic weed killers that hurt farmers, food consumers, and the environment," said Jay Feldman, executive director of Beyond Pesticides.
- "Today's final ruling reaffirms what Sierra Club has been saying all along: the government needs to look before it leaps and must comprehensively examine through an EIS how genetically engineered alfalfa could impact the environment before approving its widespread use," said Neil Carman of the Sierra Club's genetic engineering committee. "Conducting an EIS is plain common sense."
- "This is a huge victory for family farmers in the livestock and dairy industry," said Bill Wenzel, National Director of Farmer to Farmer Campaign on Genetic Engineering. "It is unfortunate that it took lengthy and expensive litigation to achieve what should have been apparent to the bureaucrats at the USDA — that nobody but Monsanto benefits from the commercialization of GE alfalfa."

U.S. alfalfa exports in total nearly $480 million per year, with about 75% headed to Japan. The court disagreed with the USDA's assertion that exports to Japan would not be harmed by deregulation of GM alfalfa. However, in the early part of 2006, Japan did endorse biotech alfalfa through its own environmental approval process, and it can be exported to Japan. Canada also submitted and approved biotech alfalfa through its own regulatory process.

Meanwhile, despite a separate ruling that the USDA was wrong to allow a field trial of engineered bentgrass without an environmental assessment — a decision that caused the agency to temporarily suspend all field trial approvals — the USDA has quietly started allowing field trials again. The USDA recently approved new trials of loblolly pine, wheat,

(*Continued*)

(Continued)

alfalfa, and canola without requiring environmental assessments even though, as in the bentgrass case, some of these species have wild relatives with which they could hybridize growing near the chosen sites.

U.S. Department of Agriculture, Biotechnology Report. Washington, D.C., February 2007.

EU Council Sets New Standards for GM Contamination

In a report published on 13 June 2007, the EU-27's agriculture ministers decided to set new organic food production and labelling standards from 2009, but the environmental groups reacted that the rules are lax and will allow widespread contamination of organic products by genetically modified organisms. Despite the opposition of Belgium, Greece, Italy, and Hungary, the Council adopted, on 12 June 2007, a controversial regulation on organic production and labelling, which is expected to benefit both farmers and consumers by creating an EU organic logo for all products containing at least 95% organic ingredients.

Although the text bans the use of genetically modified organisms (GMOs) in organic food, it allows for products containing up to 0.9% of "adventitious or technically unavoidable" GMO content to be labelled and sold as organic. This is considered too high by the European Parliament and environmental groups who had called for this accidental contamination threshold to be set at 0.1% — the lowest level at which genetically modified organisms can be technically detected — saying that any threshold higher than this would make it too difficult for organic farmers to keep their crops free from "genetic pollution."

Environmental groups Greenpeace, Friends of the Earth and the European Environmental Bureau criticised the decision. Friends of the Earth Europe (Helen Holder) issued a statement: "Now that the EU has declared traces of genetic contamination in organic crops

(*Continued*)

(Continued)

acceptable, organic farmers will find it increasingly difficult to keep their crops GM-free. The EU must urgently introduce cross-border legislation to protect organic and conventional farmers from genetic pollution."

The European Parliament and environmental groups had called for the threshold of contamination of organic food to be 0.1%, which is the lowest level at which genetically modified organisms can be technically detected.

It is feared that the contamination in some countries outside of Europe, where GMOs are grown without any controls, is now affecting the choice of European consumers. As the success of organic farming shows, people are ready to pay for higher quality food free of GMOs.

Marco Contiero, policy Officer at Greenpeace EU Unit, explained: "The lax attitude towards contamination taken by the European Commission and some member states disregards the preferences of European consumers and may put the whole organic sector at risk. In practice, low levels of genetically modified material could start slipping into all organic food."

The European Commission is now reviewing measures for the containment of commercially grown GM crops — called "coexistence".

Mauro Albrizio from EEB said: "Organic agriculture is a vibrant sector, creating jobs and protecting the environment. The 0.9% threshold should not relax the necessity for stringent anti-contamination measures. If the EU is committed to preserving and supporting the organic farming sector, then strict co-existence measures are a necessity, protecting conventional and organic farming from genetic contamination, with stiff penalties for GMO farmers and biotech companies if contamination does occur."

Agriculture Commissioner Mariann Fischer Boel said that the new regulation would "help consumers to recognise organic products throughout the EU more easily and give them assurances of precisely what they are buying."

(Continued)

(Continued)

The use of the new EU logo will be compulsory as of 1 January 2009, but national or private logos, often reflecting more stringent standards desired by some member states, will also be authorised.

Green NGOs stress that the new standards do not lessen the necessity for stringent anti-contamination measures. According to Mauro Albrizio from the European Environmental Bureau: "If the EU is committed to preserving and supporting the organic farming sector, then strict co-existence measures are a necessity, protecting conventional and organic farming from genetic contamination, with stiff penalties for GMO farmers and biotech companies if contamination does occur."

In 2008, the Commission will review national 'co-existence' rules aimed at containing commercially-grown GM crops, and further assess the need for an EU-wide law.

Accidental GMO content permitted. www.euractive.com. June 13, 2007.

2007 Food Democracy Legislation Tracker

The legislation tracker and accompanying maps provide up-to-date information on state legislation and regulation that impacts local sustainable farming systems and community decision-making. The types of activity we monitor include:

1. Preemption: Bills and regulations intended to remove local community decision-making over food and farming systems.
2. GMOs: Bills and regulations addressing the risks of genetically modified food crops and the inadequacy of federal regulations.
3. Local Food: Bills and regulations supporting sustainable farming systems, farmers, local economies, and the regional distribution of foods.

State	Year	Bill	Type	Language	Date/Link
Alabama	2007	*none*			
Alaska	2005	SB25	GMO	This bill amends the Alaska Food, Drug, and Cosmetic Act, AS 17.20.040 by adding genetically modified fish or fish product to the list of misbranded food, unless conspicuously labeled or identified as such.	PASSED May 2005
Arizona	2005	SB1282	Preemption	Amended: The regulation and use of seeds are of statewide concern. The regulation of seeds pursuant to this article and their use is not subject to further regulation by a county, city, town, or other political subdivision of this state.	PASSED 4/22/05
Arkansas	2007	State Plant Board	GMO	***Regulation:*** The Arkansas State Plant Board banned Cheniere and Clearfield 131 rice varieties for the 2007 and 2008 growing seasons and mandated genetic testing of all seed stocks.	PASSED 3/2/07
California	2007	CA Rice Comm.	GMO	***Regulation:*** The California Rice Commission voted to support a moratorium "on the field testing of all genetically modified (GM) rice cultivars in the State of California for the 2007 crop, and for future crops, until such time as research protocol and safeguards are acceptable to the California Rice Commission."	PASSED 3/14/07

(Continued)

(Continued)

State	Year	Bill	Type	Language	Date/Link
	2007	AB541	GMO	Establishes the right of farmers and landowners to compensation for economic losses due to genetic contamination of their crops. Protects farmers from being sued by a GE manufacturer if their crop is contaminated by that company's GE product. Establishes a county-level GE crop notification process so that farmers can trace contamination to the GE manufacturer. Protects the food supply by prohibiting the open-field cultivation of genetically engineered food crops used to produce drugs and biologics such as hormones and antibiotics.	Held over until 2008
	2007	SB63	GMO	Requires every livestock producer who sells or transfers any cloned animal or its progeny to disclose to the buyer or transferee that the animal is cloned or is the progeny of a cloned animal. It would also require food for human consumption that contains any product from a cloned animal or its progeny to be labeled to indicate that the food includes the product of a cloned animal or its progeny.	Activity

(Continued)

(Continued)

State	Year	Bill	Type	Language	Date/Link
	2006	SB1056	Preemption	This bill would state that these provisions of law relating to nursery stock and seed are of statewide concern and occupy the entire field of regulation regarding registration, labeling, sale, storage, transportation, distribution, notification of use, and use of nursery stock and seeds to the exclusion of local regulations.	*Session ended; bill not passed*
Colorado	2007	*none*			
Connecticut	2007	HHB 6582	GMO	To protect farmers and innocent landowners from liability and damage due to trespass by genetically modified organisms onto their property.	*Session ended; bill not passed*
Delaware	2007	*none*			
Florida	2005	HB1717	Preemption	In order to ensure uniform health and safety standards, the adoption of standards and fines in the subject areas of paragraphs (a)-(n) is expressly preempted to the state and the Department of Agriculture and Consumer Services. Any local government enforcing the subject areas of paragraphs (a)-(n) must use the standards and fines set forth in the pertinent statutes or any rules adopted by the department pursuant to those statutes.	PASSED 5/6/05

(Continued)

(Continued)

State	Year	Bill	Type	Language	Date/Link
Georgia	2005	SB87	Preemption	No county, municipal corporation, consolidated government, or other political subdivision of this state shall adopt or continue in effect any ordinance, rule, regulation, or resolution regulating the labeling, packaging, sale, storage, transportation, distribution, notification of use, or use of seeds.	PASSED 2/18/05
Hawaii	2007	SB179	GMO	Provides for procedures to protect nongenetically engineered plant life from their genetically engineered counterparts. Establishes the liability of manufacturers and growers of genetically engineered plants, plant parts, or seeds.	*Session ended; bill not passed*
	2007	SB717	GMO	Regulates production of pharmaceutical crops and industrial crops. Imposes civil penalties for violations.	*Session ended; bill not passed*
	2007	SB1290	GMO	Genetically modified organisms notification and certification. Any person that proposes to research, test, propagate, cultivate, grow, or produce any genetically modified organism shall: (1) Notify the department of the person's intent to conduct the activity; and (2) Obtain certification of approval from the department prior to conducting the activity.	*Session ended; bill not passed*

(Continued)

(Continued)

State	Year	Bill	Type	Language	Date/Link
	2007	HB704	GMO	Provides a 10-year moratorium on testing, propagating, cultivating, growing, and raising genetically engineered taro.	*Died in committee*
	2007	HB1577	GMO	Temporarily prohibits the growing of genetically modified coffee. Allows research on genetically modified coffee in environmentally secure facilities.	*Session ended; bill not passed*
Idaho	2005	HB38	Preemption	This chapter and its provisions are of statewide concern and occupy the whole field of regulation regarding the registration, labeling, sale, storage, transportation, distribution, notification of use, and use of seeds to the exclusion of all local ordinances, rules or regulations. Except as otherwise specifically provided in this chapter, no ordinance, rule or regulation of any political subdivision may prohibit or in any way attempt to regulate any matter relating to the registration, labeling, sale, storage, transportation, distribution, notification of use, or use of seeds.	PASSED 3/23/05

(Continued)

(Continued)

State	Year	Bill	Type	Language	Date/Link
Illinois	2007	HB1300	Local Food	Provides that the Task Force shall develop a plan for expanding and supporting a State local and organic food system and for assessing and overcoming obstacles to an increase in locally grown food and local organic food production.	Activity
Indiana	2005	HB1302	Preemption	Regulation of seeds. Prohibits a political subdivision from regulating the storage and use of seeds unless the political subdivision is granted a waiver by the state seed commissioner.	PASSED 3/25/05
Iowa	2005	HF642	Preemption	It is the intent of the general assembly in enacting this Act to accomplish uniformity in oversight and regulation of seed used in agriculture. It is not intended that this Act preclude a local governmental entity from pursuing governmental activities not in conflict with this Act.	PASSED 4/6/05
Kansas	2005	HB2341	Preemption	No local authority shall enact or enforce any law, ordinance, rule, regulation or resolution in conflict with, in addition to, or supplemental to, the provisions of the Kansas seed law unless expressly authorized by law to do so. Any law, ordinance, rule, regulation or resolution in conflict with, in addition to, or supplemental to, the provisions of the Kansas seed law is hereby declared to be invalid and of no effect.	PASSED 4/1/05

(*Continued*)

(Continued)

State	Year	Bill	Type	Language	Date/Link
Kentucky	2007	*session ended*			
Louisiana	2007	*none*			
Maine	2007	LD1650	GMO	Creates a right of action as and damages for a private nuisance against a manufacturer of a genetically engineered plant part, seed or plant that cross-contaminates a person's land and limits the liability of knowing and unknowing users and possessors of a genetically engineered plant part, seed or plant.	*Session ended; bill not passed*
	2007	LD1684	Preemption	A municipality or political subdivision may not enact a law or ordinance that unreasonably restricts farm structures or farm practices within an agriculture protection area unless the law or ordinance bears a direct relationship to public health or safety.	*Session ended; bill not passed*
Maryland	2007	*session ended*			
Massachusetts	2007	*none*			

(Continued)

(Continued)

State	Year	Bill	Type	Language	Date/Link
Michigan	2006	SB777	Preemption	Local unit of government shall not adopt, maintain, or enforce an ordinance that prohibits or regulates the labeling, sale, storage, transportation, distribution, use, or planting of agricultural, vegetable, flower, or forest tree seeds.	PASSED 5/5/06
Minnesota	2007	HF1663	GMO	Environmental Quality Board shall notify interested parties if a permit to release genetically engineered wild rice is issued anywhere in the United States. The board shall adopt rules that require an environmental impact statement and otherwise comply with chapter 116D and rules adopted under it for a proposed release and a permit for a release of genetically engineered wild rice. The board may place conditions on the permit and may deny, modify, suspend, or revoke the permit.	*Session ended; bill not passed*
	2007	HF2006	Local Food	A bill creating a pilot incentive grant program to encourage school districts to purchase locally grown foods.	*Session ended; bill not passed*
Mississippi	2007	*session ended*			

(Continued)

(Continued)

State	Year	Bill	Type	Language	Date/Link
Missouri	2007	SB570	Preemption	Bill specifies the establishment of a joint committee to conduct a comprehensive study of concentrated animal feeding operations in Missouri, to include the economic impact of such operations, issues pertaining to the regulation of such operations at the local and state levels of government, and any other issue the committee deems relevant.	*Session ended; bill not passed*
	2007	SB364	Preemption	Local Preemption: This act provides that state laws and regulations shall preempt local laws regarding agricultural operations and provides that any farm or farming-related activity that is in compliance with all applicable state and federal laws shall also therefore be in compliance with any local law or ordinance.	*Issue moved to SB570 – see above*
	2006	SB1009	Preemption	The provisions of sections 266.011 to 266.111 are of statewide concern and occupy the entire field of regulation regarding the registration, labeling, sale, storage, transportation, distribution, notification of use, use, and planting of seeds and other propagules to the exclusion of all local ordinances or regulations.	*Session ended; bill not passed*
	2006	HB1842	Preemption	Specifies that no ordinance of any political subdivision will regulate seeds or other propagules and the state will not enforce any rule more restrictive than federal rules regarding seed technologies.	*Session ended; bill not passed*

(*Continued*)

(Continued)

State	Year	Bill	Type	Language	Date/Link
Montana	2007	HB381	Local Food	Whenever a public agency seeks to procure food for human consumption, the specifications for the food may, at the discretion of the public agency procuring the food, include a requirement that the food be produced or processed in Montana.	*Bill not passed*
	2007	HB432	GMO	Liability for GMO's that escape property boundaries	*Bill not passed*
Nebraska	2007	LB516	Local Food	Authorizes a study with respect to corporate farming and agricultural production. It is the intent of the Legislature to foster and enhance legal, social, and economic conditions in Nebraska consistent with and which advance those state interests that structure, develop, and increase agricultural production.	PASSED 5/24/07
	2006	LB834	Preemption	Political subdivisions shall not prohibit or in any other manner regulate any matter relating to the registration, labeling, sale, storage, transportation, distribution, notification of use, use, planting, or cultivation of seeds.	*Session ended; bill not passed*
Nevada	2007	*none*			
New Hampshire	2007	*none*			

(Continued)

(Continued)

State	Year	Bill	Type	Language	Date/Link
New Jersey	2007	A385	Local Food	Local Sourcing: An Act establishing a task force in the Department of Agriculture to develop place-based food markets.	Activity
New Mexico	2007	SB90	Local Food	Local Sourcing: New Mexico department of agriculture to purchase New Mexico-grown fresh fruits and vegetables to administer to school lunch programs statewide.	*Session ended; bill not passed*
New York	2007	A03990	GMO	Requires persons who sell or distribute genetically engineered plants, planting stock or seeds to provide written instructions to purchasers or growers of such stock.	*Session ended; bill not passed*
	2007	A01403	GMO	Establishes an affirmative defense for causes of actions related to violation of a patent, trademark or other intellectual property right on grounds that a party possessed or used seeds or plants that contained genetically engineered or genetically modified organisms without entering into an agreement or paying fees to the manufacturer or licensed distributor of such products.	*Session ended; bill not passed*

(Continued)

(Continued)

State	Year	Bill	Type	Language	Date/Link
North Carolina	2006	H671	Preemption	§106 423.10. Regulation of genetically modified and genetically engineered plants. Except as provided in Article 15 of Chapter 113A of the General Statutes, the Board of Agriculture shall have the sole authority to regulate genetically engineered and genetically modified plants.	*Session ended; bill not passed*
North Dakota	2005	SB2277	Preemption	A political subdivision, including a home rule city or county, may not adopt or continue in effect any ordinance, resolution, initiative, or home rule charter regarding the registration, labeling, distribution, sale, handling, use, application, transportation, or disposal of seed. This section does not apply to city zoning ordinances.	PASSED 3/16/05
Ohio	2005	HB66	Preemption	Operating Budget Bill Amended with the following clause: Prohibits political subdivisions from regulating or enacting legislation relating to the registration, packaging, labeling, sale, storage, distribution, use or application of fertilizer and from regulating or enacting legislation relating to the registration, sale, storage, transportation, distribution, notification of use, use, or planting of seed.	PASSED 6/30/05

(*Continued*)

(Continued)

State	Year	Bill	Type	Language	Date/Link
Oklahoma	2005	HB1471	Preemption	An Act relating to agriculture; providing for certain jurisdiction for seeds; preempting local jurisdiction for seeds; providing for codification; and providing an effective date.	PASSED 4/18/05
Oregon	2007	SB234	GMO	Authorizes Director of Agriculture and appointee of Director of Human Services to enter into memoranda of understanding or other intergovernmental agreements to further collaboration between state and federal agencies, and to increase state input, regarding biopharmaceutical crop issues and requirements of specific interest to state.	PASSED 6/25/07
Pennsylvania	2004	HB2387	Preemption	Statewide jurisdiction and preemption. This chapter and its provisions are of Statewide concern and occupy the whole field of regulation regarding the registration, labeling, sale, storage, transportation, distribution, notification of use, and use of seeds to the exclusion of all local regulations. Except as otherwise specifically provided in this act, no ordinance or regulation of any political subdivision or home rule municipality may prohibit or in any way attempt to regulate any matter relating to the registration, labeling, sale, storage, transportation, distribution, notification of use or use of seeds, if any of these ordinances, laws or regulations are in conflict with this chapter.	PASSED 11/29/04

(Continued)

(Continued)

State	Year	Bill	Type	Language	Date/Link
Rhode Island	2007	*none*			
South Carolina	2007	*none*			
South Dakota	2005	SB152	Preemption	Enjoin the enforcement of any ordinance, regulation, or action of any political subdivision of this state which is in conflict with any provision of this chapter or any rule promulgated pursuant to § 38-12A-20.	PASSED 2/25/05
Tennessee	2007	*None*			
Texas	2005	HB2313	Preemption	Sec. 71.153. LOCAL REGULATION. (a) A political subdivision may not adopt an ordinance or rule that restricts the planting, sale, or distribution of noxious or invasive plant species.	PASSED 6/17/05
Utah	2007	*session ended*			
Vermont	2007	H0133	Local Food	Local Sourcing: Direct state agencies and institutions to procure Vermont farm and food products whenever available at a cost of no more than eight percent above like products produced outside the state.	*Session ended; bill not passed*

(*Continued*)

(Continued)

State	Year	Bill	Type	Language	Date/Link
Virginia	2007	SB594	Preemption	Prohibits localities from regulating the registration, packaging, labeling, sale, storage, distribution, use, or application of fertilizer.	*Session ended; bill not passed*
Washington	2007	HB1888	GMO	Any grower or processor of a Brassica seed crop may petition the Department of Agriculture Director to request establishment of a Brassica seed production district. The Director may adopt rules to establish a district including, but not limited to, production districts and agreements; a centralized notification process for growers intending to plant a crop within a district; isolation distances; exclusion of crops; and control of volunteer and weed plants within a district.	PASSED 4/21/07
West Virginia	2007	SB605	GMO	State Board of Education to require foods to be free from genetically modified components.	*Session ended; bill not passed*

(Continued)

(Continued)

State	Year	Bill	Type	Language	Date/Link
	2005	SB580	Preemption	Statewide jurisdiction and preemption. — This chapter and its provisions are of statewide concern and occupy the whole field of regulation regarding the registration, labeling, sale, storage, transportation, distribution, notification of use, and use of seeds to the exclusion of all local regulations. Except as otherwise specifically provided in this article, no ordinance or regulation of any political subdivision or home rule municipality may prohibit or in any way attempt to regulate any matter relating to the registration, labeling, sale, storage, transportation, distribution, notification of use or use of seeds, if any of these ordinances, laws or rules are in conflict with this chapter.	PASSED 4/16/05
Wisconsin	2007	SB89	Local Food	Requires DATCP to conduct a program to increase awareness and consumption of locally produced foods and related products and to increase the production and improve the distribution of foods and related products for local consumption.	Activity
Wyoming	2007	*none*			

Australian Govt Agency Accused of Rigging GM Survey

A Biotechnology Australia survey found 73 per cent of consumers accept GM food compared to 46 per cent in 2005. However, it has been accused of manipulating the results of a survey showing more people are ready to accept genetically modified food (July 2007) (http://www.biotechnology.gov.au). In its response, the Gene Ethics Network stated that the survey questions were loaded to promote GM crops as environmentally friendly.

"People no longer see it as something novel and scary and dangerous and unknown and the other one is that people see GM crops as potentially solving bigger problems in their minds which are drought and climate change and pandemics and all those sorts of things, so biotechnology crops, GM crops are no longer percieved to be such a problem, more now a solution," he said.

Public support for genetically modified food crops rose dramatically to 73 per cent in 2007, up from 46 per cent in 2005, due to perceptions about the role they can play in countering drought and pollution.

Australian Industry Minister, Ian Macfarlane, said a *Biotechnology Australia* report, released today, found a major change in public attitudes towards biotechnology in all areas.

"When asked if GM crops should be grown in their state, 50 per cent of respondents from all states said Yes, with a further 30 per cent approving as long as they were strongly regulated," Mr Macfarlane said.

"This marks a significant change in public attitudes and coincides with an increased confidence in science across society. The perceived benefits from biotechnology are increasing while the perceived risks are declining.

"Changes in attitudes have been influenced by the public's increased familiarity with gene technology and a perception that GM crops could be used to counter major environmental concerns."

The study looked at public concerns about biotechnology applications and sought to understand what benefits people wanted from the technology.

(*Continued*)

(Continued)

The highest values placed on biotechnology applications were:

- cleaning up pollution (97 per cent support);
- developing environmentally-friendly vehicle fuels (97 per cent);
- recycling water more effectively (96 per cent);
- helping address climate change (91 per cent); and
- combating salinity (90 per cent).

The survey was developed with input from industry, research organisations and non-government organisations and the full study can be found at www.biotechnology.gov.au/reports.

ABC Rural News, Australia, July 2007.

10

Human Rights and Ethical Issues

Biodiversity loss can (and should) be viewed as an ethical and moral problem for humanity. The implications for food and economic security have already been addressed in earlier chapters. Leading international organizations, as they are constituted at present, are not equipped to address the needs of the marginalized and disadvantaged populations of the world. Biodiversity loss and biopiracy are clearly tied to patents and intellectual property rights (IPRs), among other factors. While the commercial and developmental aspects of IPRs have received much attention over the years, their implications for human rights issues have been ignored. Destruction of habitats, erosion of biodiversity, appropriation of plant life resources, and neglect or deliberate suppression of traditional knowledge have all impacted deeply on the human rights of large sections of humanity.

Genetic Erosion

The Canada-based, nonprofit biotech organization Rural Advancement Foundation International (RAFI) analyzed old United States Department of Agriculture (USDA) lists of common fruits and vegetables, and found that 97% of the varieties had become extinct over a period of 80 years (Fowler and Mooney 1990).[a]

U.S. Vegetable Varieties Lost (1903–1983)

Vegetable	Total 1903 varieties	Varieties lost (%)
Artichoke	34	94.1
Asparagus	46	97.8
Beans		
Runner bean	14	92.9
Lima bean	96	91.7
Garden bean	578	94.5
Beets	288	94.1
Mangel beet	178	98.3
Broccoli	34	100.0
Brussels sprouts	35	88.6
Burnet	1	100.0

(*Continued*)

[a] In 1984, Cary Fowler, Hope Shand, and Pat Mooney cofounded Rural Advancement Foundation International (RAFI) in Canada, whose name was changed to ETC Group (pronounced "etcetera" group) in 2001. ETC Group is a small international civil society organization (CSO) addressing the impact of new technologies on rural communities. ETC has offices in Canada, the United States, and Mexico; and works closely with CSO partners around the world. ETC Group is dedicated to the conservation and sustainable advancement of cultural and ecological diversity and human rights. To this end, ETC Group supports socially responsible developments of technologies useful to the poor and marginalized, and it addresses international governance issues and corporate power.

ETC Group works in partnership with CSOs for cooperative and sustainable self-reliance within disadvantaged societies, by providing information and analysis of socioeconomic and technological trends and alternatives. This work requires joint action in community, regional, and global fora. ETC Group's strengths lie in the research and analysis of technological information (particularly, but not exclusively, plant genetic resources, biotechnologies, and in general biological diversity), and in the development of strategic options related to the socioeconomic ramifications of new technologies.

(Continued)

Vegetable	Total 1903 varieties	Varieties lost (%)
Cabbage	544	94.9
Cardoon	6	83.3
Carrot	287	92.7
Cauliflower	158	94.3
Celeriac	25	88.0
Celery	164	98.2
Chervil	8	100.0
Chicory	17	82.4
Chives	1	0.0
Chufas	2	100.0
Collards	28	82.1
Corn		
Field corn	434	90.8
Pop corn	48	100.0
Sweet corn	307	96.1
Corn salad	21	95.2
Cress	39	94.9
Cucumber	285	94.4
Pickling cucumber	10	80.0
Dandelion	25	100.0
Eggplant	97	90.7
Endive	64	93.7
Horseradish	1	100.0
Kale	124	92.7
Kohlrabi	55	94.5
Leek	39	87.2
Lettuce	497	92.8
Martynia	4	100.0
Muskmelon	338	92.0
Mustard	44	88.6
Okra	38	89.5
Onion	357	94.1
Orach	5	80.0
Parsley	82	85.4
Parsnip	75	93.3
Pea	408	93.9
Peanut	31	93.5

(Continued)

(Continued)

Vegetable	**Total 1903 varieties**	**Varieties lost (%)**
Peppers	126	89.7
Radish	463	94.2
Rampion	1	100.0
Rhubarb	35	97.1
Roquette	1	100.0
Rutabaga	168	97.0
Salsify	29	93.1
Spanish salsify	1	100.0
Skirrel	1	100.0
Sorrel	10	00.0
Spinach	109	93.6
Malabar spinach	1	0.0
Squash	341	88.3
Sunflower	14	92.9
Swiss chard	23	95.7
Tomato	408	80.6
Husk tomato	17	88.2
Turnip	237	89.9
Watercress	2	0.0
Watermelon	223	91.0

Source: Thrupp (2000).

U.S. Fruit Varieties Lost (1804–1904)

	Total varieties	**% lost**
Apple varieties	7098	86.2 (6121)
Pear varieties	2683	87.7 (2354)

Source: Thrupp (2000).

Ethical Issues

Ethical issues are involved in almost every aspect of biotechnology. I have discussed this subject in detail in my previous books (Dronamraju 1998; Dronamraju 2002). Transfer of technology, especially when it involves developing countries, requires careful consideration of ethical and moral dilemmas. The biological, social, and economic impacts of new technologies on a population and the intellectual property rights involved are (or should be) of great interest to all individuals in today's society. In particular, such knowledge should be made available to the indigenous communities of all developing countries because they are impacted directly by these new developments. To be more specific, these communities have been the targets of exploitation and biopiracy, which have adversely impacted on their lifestyle. Similar ethical problems arise when biotechnology is exported to developing countries which lack the technical resources or knowledge to monitor the consequences.

Ethical and Moral Framework

The ethical standards of the Western world towards nature were strongly influenced by the Aristotlean approach, which implies that animals and plants are created solely for the use of human beings. This homocentric approach has led not only to the unlimited exploitation of this planet's biodiversity, causing many species to become extinct, but also to extreme cruelty and irreverence towards all life. This is in complete contrast to the beliefs of several Eastern religions including Hinduism, Buddhism, Jainism, and others which advocate nonviolence and kindness towards all living beings.

The idea that humans are the ultimate masters of the world has also led to the denigration of animals and plants as lacking rational thought or self-awareness and as devoid of intrinsic value of their own. These unfortunate and erroneous ideas have been encouraged and practiced in close association with Christianity for two millennia. It is not surprising that the best scientists are those who have rejected religion in its traditional form. Most of them are either atheists like Albert Einstein and Robert Oppenheimer, or agnostics like J.B.S. Haldane.

As modern research is increasingly making clear, several animal species possess self-awareness and intelligence that was barely imaginable a few years ago. Indeed, both animals and plants are now found to possess faculties which are not unlike those of humans. For instance, some chimpanzees can not only recognize and understand many human commands, but also make up their own sentences and converse with their human friends. They are capable of expressing their desires, feelings, and emotions. They possess intuitive and reasoning faculties just like humans. Other animal species are also found to possess unique faculties which were not known before. Recent research on certain plants has shown that when a part of the plant is attacked by an invader, the affected cells of the plant send an "SOS", a cry for help and to warn other parts of the plant.

So, it is clear that from a purely moral and ethical perspective, humans have unjustifiably disrespected the intrinsic rights of other living species on the planet. This attitude has, in turn, led to various natural calamities such as erosion, floods, depletion of biodiversity, and extinction of rare species of animals and plants.

Ethics and Genetic Engineering

New biology, including genetic engineering and biotechnology, has given rise to a host of ethical issues:

1. *Ecocentric concerns*. The ecological balance between humans and other species is perceived to be so delicate that any imbalance, such as genetic contamination through transgenic intervention, is viewed with much concern. Ecological ethic implies that the earth is a single ethical system, the ethical norm being the well-being of the comprehensive community as a whole, not just of human society.
2. *Health concerns*. These are centered around the direct toxic effects of transgenic products and the indirect effects of a genetically contaminated world where the long-existing boundaries between the species disappear. Much concern is also expressed about gene targeting, introduction of a gene, and the control of gene expression in a new host. There are a great number of unanswered questions because the technology is hardly precise. There is potential for carriers such as viruses

and bacteria as well as DNA segments to escape from target cells and infect nontarget organs and tissues, leading to cancer. Recombination between plant-produced viral genes and homologous genes of introduced viruses may produce new virulent strains, which could infect a wider range of hosts.

3. *Biodiversity concerns*. Biodiversity loss is also regarded as an ethical problem. This aspect has already been discussed in detail. From an ethical and moral point of view, any technology that results in loss of life and loss of biodiversity is unacceptable.
4. *Economic concerns*. The high cost of new technology has already caused many suicides in the Warangal District of Andhra Pradesh in India. This poses both ethical and social dilemmas for many farmers. Many developing countries do not have appropriate safeguards and regulatory resources to screen for possible risks arising from new technologies. There are various ethical concerns regarding the export of technologies to communities which are economically and technologically backward.
5. *Religious concerns*. Many religions believe in the sanctity of life as a unique gift from God. They object to its alteration by means of genetic engineering and its patenting as a commercial entity. This is a problem in religious ethics.

Ethics of Patenting

Indian Court Ruling in Novartis Case Protects India as the 'Pharmacy of the Developing World'

The landmark decision by the High Court in Chennai to uphold India's Patents Act in the face of the challenge by Swiss pharmaceutical company Novartis is a major victory for patients' access to affordable medicines in developing countries (Statement issued by the international medical humanitarian organization Doctors Without Borders/Médecins Sans Frontières (MSF), August 2007).

(*Continued*)

(Continued)

Novartis took the Indian government to court over its 2005 Patents Act because it wanted a more extensive granting of patent protection for its products than offered by the law. Novartis claimed that India's Patents Act did not meet rules set down by the World Trade Organization and was in violation of the Indian constitution. Apparently all of Novartis's claims have been rejected by the High Court today.

India only began giving patents on medicines to comply with WTO rules, but it designed its law with safeguards so that patents can only be granted for real innovations. This means that companies seeking a patent for modifications to a molecule already invented, in order to extend ever further their monopolies on existing drugs, would be unsuccessful in India. It is this aspect of the law that Novartis was seeking to have removed. A ruling in favor of the company would have drastically restricted the production of affordable medicines in India that are crucial for the treatment of diseases throughout the developing world.

Developing country governments and international agencies like UNICEF and the Clinton Foundation rely heavily on importing affordable drugs from India, and 84% of the antiretrovirals that MSF prescribes to its patients worldwide come from Indian generic companies. India must therefore be allowed to remain the "pharmacy of the developing world."

Nearly half a million people worldwide voiced their concern about the impact Novartis's case could have on access to medicines in the developing world. Among them were the Indian Health Minister Anbumani Ramadoss, Archbishop Desmond Tutu, Global Fund Director Michel Kazatchkine, members from the European Parliament and the US Congress, former Swiss President Ruth Dreifuss, former UN Special Envoy for AIDS in Africa Stephen Lewis, German Development Minister Heidemarie Wieczorek-Zeul, Norwegian Development Minister Erik Solheim, as well as authors John Le Carré and Naomi Klein. An F petition urging Novartis to drop the case gathered over 420,000 signatures.

Medecins sans Frontiers. Indian court ruling in Novartis case. August 6, 2007.

Monsanto Patents Asserted Against American Farmers Rejected by Patent Office

In July 2007, the United States Patent and Trademark Office rejected four key Monsanto patents related to genetically modified crops that PUBPAT (Public Patent Foundation) challenged last year because the agricultural giant is using them to harass, intimidate, sue — and in some cases literally bankrupt — American farmers. In its Office Actions rejecting each of the patents, the USPTO held that evidence submitted by PUBPAT, in addition to other prior art located by the Patent Office's Examiners, showed that Monsanto was not entitled to any of the patents.

Monsanto has filed dozens of patent infringement lawsuits asserting the four challenged patents against American farmers who saved seed from one year's crop to replant the following year, something farmers have done for centuries. Many of these farmers are unable to hire adequate representation to defend themselves in court.

One study found that, "Monsanto has used heavy-handed investigations and ruthless prosecutions that have fundamentally changed the way many American farmers farm. The result has been nothing less than an assault on the foundations of farming practices and traditions that have endured for centuries in this country and millennia around the world, including one of the oldest, the right to save and replant crop seed." The lawsuits filed by Monsanto against American farmers include Monsanto Company v. Mitchell Scruggs, *et al.*, 459 F.3d 1328 (Fed. Cir. 2006), Monsanto Company v. Kem Ralph individually, *et al.*, 382 F.3d 1374 (Fed. Cir. 2004) and Monsanto Company v. Homan McFarling, 363 F.3d 1336 (Fed. Cir. 2004).

Although Monsanto has the opportunity to respond to the Patent Office's rejections of the patents (U.S. Patents Nos. 5,164,316, 5,196,525, 5,322,938 and 5,352,605), third party requests for re-examination, like the ones filed by PUBPAT against the four Monsanto patents, are successful in having the reviewed patents either changed or completely revoked more than two-thirds of the time.

Papua New Guinea

It is the "nature of the beast", one might say, when powerful multinational companies appropriate large sections of the forest in Brazil, India, Borneo, or elsewhere with the sole purpose of financial gain, against which the indigenous communities are helpless and vulnerable. Many years ago, Fowler and Mooney (1990) cited a specific example from Papua New Guinea. Between the Eia and Mambara rivers in Papua New Guinea lies the land where the Binandere tribe has lived for centuries in a "stone age" fashion. They have survived and flourished in harmony with the rainforest. However, their survival was severely threatened in the 1970s by an Amerian-owned timber company, Parsons & Whitmore, which arrived on the scene to cut down the forest for lumber and other products.

The tribe protested. They refused the financial compensation that was offered by the company. In an emotional interview with a Swedish documentary television crew through an interpreter, the Binandere chief, Kipling Jiregari, explained the tribe's response.

> My name is Kipling Jiregari and I shall speak of the land and of the forest. When God created the world and Papua New Guinea, he created our country. . . . Here we all . . . tended our gardens. The land gives us our food and everything we need.
>
> Money has no future. Money disappears. Only Man and the land remain, when all else has disappeared. Therefore I have stopped the Company. No company shall destroy our land.
>
> If the Company's men come here again, we will kill them and eat them up. They shall never touch our wood, where our food and medicine is. The wood is our skin, and without his skin, Man dies. . . . The woods, the land, the fish, the swamp, the birds — everything we will protect — not only for our own sake, but also for coming generations.

The company was stopped from cutting down the forest and the tribe was saved, at least for a while.

Another community, living nearby, was not so lucky. JANT, a subsidiary of the huge Japanese multinational corporation Honshu Paper, clear-cut the tropical forest, turning the entire forest into wood chips. Their motto appears to be, "The only good tree is a stump", the same motto adopted by the Montana loggers before them. It only takes a few seconds to fell one of the giant trees. As it falls to the ground, the tree pulls down smaller surrounding trees connected to it by vines in the canopy. Machines with giant revolving blades reduce the trees into millions of chips in a few minutes. The machines have destroyed the forest. Birds and wild game have left. Both Japan and the US have remained the major importers of tropical hardwoods.

Africa

There has been a great deal of pressure on African governments in recent years to buy GM seeds and adopt GM crops. However, a Michigan State University study showed that it may take up to 15 years to develop GM crops, create regulatory frameworks, field test and deliver GM cultivars to smallholder farmers in Africa. Their research is based on seven African GM case studies including the spectacular failure of Kenya's GM sweet potato project and the wildly premature acclaim of "success" of the Bt cotton smallholder in South Africa.

Bans and Restrictions Imposed by African Countries on GM Imports

- Algeria introduced a ban on the import, distribution, commercialisation and use of GM plant material in December 2000.
- Angola introduced a ban on imports of unmilled GM food aid in April 2004.
- Benin has taken measures to prevent imports of GM food aid, with a moratorium on import of GMOs until national legislation comes into force.

(Continued)

(Continued)

- Lesotho has permitted the distribution of non-milled GM food aid, with a public warning that the grain should be consumed and not used for cultivation.
- Malawi has had a ban on importing unmilled GM crops since 2002.
- Mozambique's government is prepared to accept GM food aid provided that maize is milled prior to distribution.
- Namibian government rejected GM maize in 2002 and received wheat for food aid instead.
- Nigeria's government prepared to accept GM food aid provided maize is milled prior to distribution.
- Sudan banned the import of GM food aid during May 2003, but issued a series of temporary waivers under pressure by the US.
- Swaziland permitted the distribution of non-milled GM food aid, with a public warning that the grain should be consumed and not used for cultivation.
- Zambia refused to accept GM grain donated as food aid in 2002.

GM crops in Africa? No thanks! Institute of Science in Society, London, April 10, 2005.

Biodiversity destruction by the United States in Vietnam

Indochina or Vietnam occupies an area that is only slightly larger than Texas. That country absorbed twice the tonnage of munitions that the U.S. used in World War II. Twenty-six billion pounds fell on Indochina — the equivalent of 450 Hiroshima bombs, 142 pounds for every acre of land, 584 pounds for every person (Fowler and Mooney 1990, p. 103).

The typical 500-pound bomb (108 of which could be carried by a B-52) left a crater 15 feet deep and 30 feet across. There are twenty-six million of these craters in Vietnam, covering 423,000 acres. In addition to the "wasting" of large areas of the country, shrapnel inflicted additional damage on a large part of the vegetation. During

(*Continued*)

(Continued)

the final years of the war, the U.S. dropped a highly destructive bomb capable of completely clearing out a 1.3-hectare area for use as a helicopter landing pad. The "casualty" zone of the bomb was a staggering 49 hectares.

About 2% of the land area of South Vietnam was cleared by "Rome ploughs" (a huge chain pulled by two tractors or bulldozers), including a 1500-acre area which was carved out in the shape of the emblem of the U.S. 1st infantry Division, 25 miles from Saigon.

We must also remember the heavy damage inflicted on the vegetation by the chemicals dropped by the U.S. Approximately 14 million pounds of 2,4,5-T were used by the U.S. Department of Defense. This amount is adequate to obliterate the vegetation on ten million acres if used properly. Fifty-five million kilograms of herbicides were used. They were used to destroy not only tropical forests but also rubber plantations and rice fields. Many areas were sprayed repeatedly. They have destroyed enormous areas of biodiversity, known varieties of rice and other crops but also several wild species, including some waiting to be discovered!

Dr. David Ehrenfeld of the Department of Horticulture and Forestry at Rutgers University characterized the biodiversity destruction as follows: "To say that every ecosystem in Vietnam, major and minor, has been seriously altered or wrecked beyond hope of repair, is to make a safe and conservative statement."

(Fowler and Mooney 1990, 102–103)

References

Note: In many pages, I have cited from the proceedings of various UN bodies such as General Assembly, FAO, WHO, UNDP, etc. These are well-known agencies and the proceedings are all available on the Internet; moreover, many libraries do not carry these documents. Therefore, I have not cited these references here.

Abdul Salam, M. and Sreekumar, D. (1991) Kerala homegardens: a traditional agroforestry system from India. *Agroforestry Today* **3**: 10–15.

Behrens, M.R., Mutlu, N., Chakraborty, S., Dumitru, R., Jiang, W.Z., Lavallee, B.J., Herman, P.L., Clemente, T.E. and Weeks, D.P. (2007) Dicamba resistance: enlarging and preserving biotechnology-based weed management strategies. *Science* **316**: 1185–1188.

Bickham, J.W., Sandhu, S., Hebert, P.D., Chikhi, L. and Athwal, R. (2000) Effects of chemical contaminants on genetic diversity in natural populations: implications for biomonitoring and ecotoxicology. *Mutat Res* **463**: 33–51.

Biello, D. (2007) Getting the bugs out of genetically modified crops. Are crops genetically altered to resist insects really better for the environment? *Sci Am*, June 7, 2007.

Brooks, T.M., Mittermeier, R.A., Mittermeier, C.G., da Fonseca, G.A.B., Rylands, A.B., Konstant, W.R., Flick, P., Pilgrim, J., Oldfield, S., Magin, G. and Hilton-Taylor, C. (2002) Habitat loss and extinction in the hotspots of biodiversity. *Conserv Biol* **16**: 909–923.

Chakrabarty, A.M. (1981) Microbial genetic engineering and environmental pollution. In: Keenberg, M. (ed.), *Genetic Engineering*, Vol. IV, Seattle: Battelle Seminars and Studies Program, pp. 211–219.

Chapin, M. (1988) *Defending Kuna Yala: PEMASKY, The Study Project for the Management of the Wildlands of Kuna Yala, Panama – A Case Study for Shifting the Power: Decentralization and Biodiversity Conservation.* Biodiversity Support Program, Washington, D.C.: U.S. Agency for International Development.

Correa, C.M. (2001) Review of the TRIPS Agreement: fostering the transfer of technology to developing countries. Penang, Malaysia: Third World Network.

Crichton, M. (2007) Who owns your genes? Patenting life. *International Herald Tribune*, Paris, February 13, 2007.

Dalen, L., Gotherstrom, A. and Angerbjorn, A. (2004) Identifying species from pieces of faeces. *Conserv Genet* **5**(1): 109–111.

Dronamraju, K.R. (1998) *Biological and Social Issues in Biotechnology Sharing*. Aldershot (UK): Ashgate Publishing Co.

Dronamraju, K.R. (2002) *Biological Wealth and Other Essays*. London: World Scientific.

Ecosystems and Human Well-being: Biodiversity Synthesis. (2005) Millennium Ecosystem Assessment Report. Washington, D.C.: World Resources Institute.

Ermakova, I. (2006) Influence of genetically modified soya on the birth-weight and survival of rat pups. In: *Proceedings of the Conference "Epigenetics, Transgenic Plants and Risk Assessment 2006"*, pp. 41–48.

Ewers, R.M. (2006) Interaction effects between economic development and forest cover determine deforestation rates. *Glob Environ Change* **16**: 161–169.

Falanruw, M.C. (1999) Ecological rationality of the traditional homegarden system in the Chao Phraya Basin, Thailand. *Agroforestry Syst* **46**: 3–23.

Fowler, C. and Mooney, P. (1990) *Shattering: Food, Politics and the Loss of Genetic Diversity*. Tucson: The University of Arizona Press.

Gari, J. and Villarreal, M. (2003) Agrobiodiversity and indigenous knowledge in mitigation of the consequences of HIV/AIDS. Paper presented at the HIV/AIDS Mitigation Workshop. Pretoria, Italy: FAO, HIV/AIDS Programme, May 2003.

Gari, J.A. (2004) Plant diversity, sustainable rural livelihoods and the HIV/AIDS crisis. UNDP SE Asia HIV and Development Programme.

Gouse, M., Praay, C., Kirsten, J. and Shimmelpfenning, D. (2005) A GM subsistence crop in Africa: the case of Bt white maize in South Africa. *Int J Biotechnol* **7**: 84–94.

Gullison, R.E., Frumhoff, P.C., Canadell, J.G., Field, C.B., Nepstad, D.C., Hayhoe, K., Avissar, R., Curran, L.M., Friedlingstein, P., Jones, C.D. and Nobre, C. (2007) Tropical forests and climate policy. *Science* **316**: 985–986.

Gupta, V.S.R. and Watson, S. (2004) Ecological impact of GM cotton on soil biodiversity. Consultancy Report, Australian Government Department of the Enviornment, Water, Heritage and the Arts. Canberra, Australia: CSIRO.

Harshberger, J.W. (1896) The purpose of ethnobotany. *Am Antiq* **17**: 73–81.

Hellmich, R.L., Siegfried, B.D., Sears, M.K., Stanley-Horn, D.E., Mattila, H.R., Spencer, T., Bidne, K.G., Daniels, M.J. and Lewis, L.C. (2001) Monarch larvae sensitivity to *Bacillus thuringiensis* purified proteins and pollen. *Proc Natl Acad Sci USA* **98**: 11925–11930.

Hileman, B. (2007) U.S. Patent Office rejects four patents on Monsanto crops. *Chem Eng News* **85**: 15.

Ho, M.-W. (2007) *Genetic Engineering: Dream or Nightmare?* Penang, Malaysia: Third World Network.

Hunter-Anderson, R. (1987) Indigenous fresh water management technologies of Trak, Ponape and Kosrae, Eastern Caroline Islands, and of Cuam, Mariana Islands, Micronesia. Technical Report 65, Water and Energy Research Institute of the Western Pacific, University of Guam, Mangilao, Guam.

IFATPC. (2004) GM technology: assessing the issues confronting developing countries. International Food and Agriculture Trade Policy Council. Washington, D.C., USA.

Jaramillo, C., Rueda, M.J. and Mora, G. (2006) Cenozoic plant diversity in the Neotropics. *Science* **311**: 1893–1896.

Kate, K. ten. (1995) Biodiversity prospecting partnerships. The role of providers, collectors and users. *Biotechnol Dev Monit* **25**: 16–21.

Kimbrell, A. and Newman, N. (2007) *Your Right to Know: Genetic Engineering and the Secret Changes in Your Food*. Washington, D.C.: Center for Food Safety.

King, S.R. (1996) Conservation and tropical medicinal plant research. In: Balick, M.J., Elisabetsky, E. and Laird, S.A. (eds.), *Medicinal Resources of the Tropical Forest*, New York: Columbia University Press, pp. 63–74.

Kloppenburg, J. (1988) *First the Seed: The Political Economy of Plant Biotechnology, 1942–2000*. Cambridge (UK): Cambridge University Press.

Kurien, V. (2002) No more barriers in analyzing Ayurveda. *The Hindu Business Line*, Chennai, India, October 14, 2002.

Laurance, W.F., Nascimento, H.E., Laurance, S.G., Andrade, A.C., Fearnside, P.M., Ribeiro, J.E. and Capretz, R.L. (2006) Rain forest fragmentation and the proliferation of successional trees. *Ecology* **87**(2): 469–482.

Laurance, W.F. and Peres, C.A. (eds.) (2006) *Emerging Threats to Tropical Forests*. Chicago: University of Chicago Press.

Maheshwari, J.K. (1988) Ethnobotanical research and documentation. *Acta Univ Uppsala* **28**: 207–217.

Maheshwari, J.K. (1990) Tribal ecosystem: an overview. In: Doria, R.S. *et al.* (eds.), *Man Development and Environment*, New Delhi: Ashish Publishing House.

Marvier, M., McCreedy, C., Regetz, J. and Kareiva, P. (2007) A meta-analysis of effects of Bt cotton and maize on nontarget invertebrates. *Science* **316**: 1475–1477.

Mead, A. and Ratuva, S. (2007) *Pacific Genes and Life Patents*. Tokyo: United Nations University.

Mgbeoji, I. (2006) *Global Biopiracy: Patents, Plants and Indigenous Knowledge*. Vancouver: UBC Press.

Morse, S., Bennett, R. and Ismael, Y. (2005) Comparing the performance of official and unofficial genetically modified cotton in India. *Ag Bio Forum* **8**: 1–6.

Myers, N., Mittermeier, R.A., Mittermeier, C.G., da Fonseca, G.A. and Kent, J. (2000) Biodiversity hotspots for conservation priorities. *Nature* **403**: 853–858.

Ortiz-García, S., Ezcurra, E., Schoel, B., Acevedo, F., Soberón, J. and Snow, A.A. Absence of detectable transgenes in local landraces of maize in Oaxaca, Mexico (2003–2004). *Proc Natl Acad Sci USA*, August 10, 2005 [published online before print].

Packer, K. and Webster, A. (1996) Patenting culture in science. *Sci Technol Human Values* **21**: 427–453.

Quist, D. and Chapela, I. (2001) Transgenic DNA introgressed into traditional maize landraces in Oaxaca, Mexico. *Nature* **414**: 514–543.

RAFI Report on Corporate Control of Food, Farming and Health. (1999) RAFI International 3rd Annual Report, Winnipeg, Canada.

Raney, T. (2006) Economic impact of transgenic crops in developing countries. *Curr Opin Biotechnol* **17**: 1–5.

Raynor, W.C. and Fownes, J.H. (1991) Indigenous agroforestry of Pohnpei. I. Plant species and cultivars. *Agroforestry Syst* **16**: 139–157, 159–165.

Roht-Arriaza, N. (1996) Of seeds and shamans: the appropriation of the scientific and technical knowledge of indigenous and local communities. *Mich J Int Law* **17**: 940–945.

Sahai, S. (2000) Farmers' rights and food security. *Econ Polit Wkly*, March 11, 2000.

Sahai, S. (2001) Plant variety protection and farmers' rights law. *Econ Polit Wkly*, September 1, 2001.

Sahai, S. (2003) India's Plant Variety Protection and Farmers' Rights Act, 2001. *Curr Sci* **84**(3): 407–412.

Sears, M.K., Hellmich, R.L., Stanley-Horn, D.E., Oberhauser, K.S., Pleasants, J.M., Mattila, H.R., Siegfried B.D. and Dively, G.P. (2001) *Proc Natl Acad Sci USA* **98**: 11937–11942, 2001.

Sen, N. (2001) TKDL — a safeguard for Indian traditional knowledge. *Curr Sci* **82**: 1070–1071.

Seralini, G.E. (2005) CRIIGEN Report on MON863 GM Maize produced by Monsanto CIE. University of Caen, France, June 2005.

Seralini, G.-E. (2007) New analysis of a rat feeding study with a genetically modified maize reveals signs of hepatorenal toxicity. *Arch Environ Contam Toxicol* **52**: 596–602.

Shiva, V. (2001) *Protect or Plunder? Understanding Intellectual Property Rights*. London: Zed Books.

Shiva, V. (2007) Opinion: commodification and privatization of the planet. DW-World-De, *Deutsche Welle*, Germany, Bonn.

Singh, R.P., Chari, M.S., Raheja, A.K. and Kraus, W. (eds.) (1996) *Neem and Environment*. New Delhi: Oxford & IBH Publishing Ltd.

Sloan, S. (2007) Fewer people may not mean more forest for Latin American forest frontiers. *Biotropica* **39**(4): 443–446.

Swaminathan, M.S. (ed.) (1996) *Agrobiodiversity and Farmers' Rights*. Delhi: Konark Publishers Pvt Ltd.

Swaminathan, M.S. (1999) *I Predict: A Century of Hope: Harmony with Nature and Freedom from Hunger*. Madras: East West Books Ltd.

Swaminathan, M.S. and Jana, S. (eds.) (1992) *Biodiversity: Implications for Global Food Security*. Madras: Macmillan India Ltd.

Thrupp, L.A. (2000) Linking agricultural biodiversity and food security: the valuable role of agrobiodiversity for sustainable agriculture. *Int Aff* **76**(2): 283–297.

U.S. National Research Council. (1992) Report of an Ad Hoc Panel of the Board of Science and Technology for International Development. Washington, DC: National Research Council, National Academy of Sciences.

Wilson, W.S. (1968) Land, activity and social organization of Lelu, Kusaie. PhD thesis, University of Pennsylvania, Philadelphia.

World Trade Organization (WTO). (1994) TRIPS Agreement. Geneva: WTO.

Wright, S.J. and Muller-Landau, H.C. (2006) The future of tropical forest species. *Biotropica* **38**(3): 287–301.

Appendix

Indian Peasants Institute

Acharya N G Ranga Memorial Lecture

Hyderabad, 9 June 2001

"Science and Our Agricultural Future"

by

MS Swaminathan*
UNESCO Chair in Ecotechnology
M S Swaminathan Research Foundation, Chennai

I. Introduction

For over 60 years, Prof N G Ranga was the soul of Indian agriculture. Among the political leaders of the preindependence era, he was foremost in fighting for farmers' rights. Like Gandhiji, he was of the firm conviction that India's destiny is in rural upliftment and that if Indian agriculture goes wrong nothing else will have a chance to go

* Reproduced with the kind permission of Prof. M.S. Swaminathan.

right. He also believed that the scientific transformation of Indian agriculture is a must, if the economic well-being of small farm families is to be improved. Above all, he was a strong advocate of according a high priority to building a sustainable food security system for the country. In a lecture devoted to his memory, I would like to speak about the role of science in our agricultural future.

II. A Biovision for Indian Agriculture

The Roman farmer Varro (1st century B.C.) is reported to have stated, "Agriculture is a science which teaches us what crops should be planted in each kind of soil, and what operations are to be carried out, in order that the land may produce the highest yields in perpetuity" (Prof G T Scarascia Mugnozza, personal communication). Realising this goal will call for continuous improvements in technology without associated ecological or social harm. In the Presidential Address to the Agricultural Sciences Section of the Indian Science Congress, held at Varanasi in January 1968, I gave the following description of the implications of unsustainable agriculture (Swaminathan 1968; Swaminathan 1993).

> Exploitive agriculture offers great dangers if carried out with only an immediate profit or production motive. The emerging exploitive farming community in India should become aware of this. Intensive cultivation of land without conservation of soil fertility and soil structure would lead, ultimately, to the springing up of deserts. Irrigation without arrangements for drainage would result in soils getting alkaline or saline. Indiscriminate use of pesticides, fungicides, and herbicides could cause adverse changes in biological balance as well as lead to an increase in the incidence of cancer and other diseases, through the toxic residues present in the grains or other edible parts. Unscientific tapping of underground water will lead to the rapid exhaustion of this wonderful capital resource left to us through ages of natural farming. The rapid replacement of numerous locally adapted varieties with one or two high-yielding strains in large contiguous areas would

> result in the spread of serious diseases capable of wiping out entire crops, as happened prior to the Irish potato famine of 1854 and the Bengal rice famine in 1942. Therefore the initiation of exploitive agriculture without a proper understanding of the various consequences of every one of the changes introduced into traditional agriculture, and without first building up a proper scientific and training base to sustain it, may only lead us, in the long run, into an era of agricultural disaster rather than one of agricultural prosperity.

Since then, there has been extensive research on the development of gene deployment strategies to match the physiologic races of pathogens, integrated pest and nutrient management systems, and other forms of environment-friendly technologies. What is the role of transgenic crops in such a quest for sustainable advances in productivity?

There is now considerable controversy surrounding the methods of assessing risks and benefits in relation to transgenic or genetically modified (GM) crops. It is becoming increasingly clear that scientific data alone cannot allay ecological, economic, health, and ideological apprehensions. Hence there is need to adopt a "hasten slowly" attitude in relation to the spread of GM crops, particularly those intended for human consumption. There are still several uncertainties around allergenicity; mechanisms of allergenicity are yet to be fully understood. Hasten we must in terms of scientific research, but slowly in terms of covering large areas with transgenic crops until the various doubts are cleared. The Cartagena Protocol on Biosafety represents the first step in developing internationally agreed guidelines for undertaking risk-benefit analyses in a manner which inspires public confidence.

There are many potentially valuable applications of GM technologies in tropical agriculture, where biotic and abiotic pressures are high (Swaminathan 1982). The transition from Mendelian to molecular breeding represents a shift from generalised to precision breeding. Precision farming is an important component of sustainable agriculture. What role can precision breeding play in taking the ecofarming movement forward?

The following are a few of the major scientific issues requiring particular attention.

- **Soil Health Care:** Maintenance of soil health requires attention to the physical, chemical, microbiological, and erodability characteristics of the soil.
- **Water Quality:** The quality of irrigation water, with particular reference to salt concentration, is important in relation to crop growth.
- **Plant Health Care:** Steps will have to be taken to protect crops from the triple alliance of weeds, pests, and pathogens. The pest pressure is particularly high in tropical and subtropical agriculture, since crops and alternate hosts are available in the field throughout the year.
- **Genetic Homogeneity:** Experience has shown that genetic homogeneity enhances genetic vulnerability to pests and diseases. Monoculture of transgenic crop varieties over large areas will enhance prospects for both the breakdown of resistance and the outbreak of pest epidemics.
- **Abiotic Stresses:** With intensive agriculture, problems of salinisation, water logging, and pollution are increasing in intensity. Bioremediation techniques will hence become increasingly important. Droughts, floods, cyclones, and other natural calamities pose additional threats to crop security. The consequences of potential changes in climate as a result of global warming are yet to be understood fully, but it is clear that anticipatory research should be initiated to meet potential adverse changes in temperature, precipitation, sea level, and ultraviolet B radiation.
- **Postharvest Management:** Uniform ripening, uniform skin colour, processing and keeping quality, and capacity to withstand transportation over long distances are all becoming important in the market, particularly in vegetables, fruits, and flowers. Globalisation of trade is opening up new markets for agricultural produce, but markets are also getting to be very choosy in terms of the quality of the produce.

Thus, there will be a need for genetic material which can help to reduce or eliminate the dependence on market-purchased chemicals on the one hand, and enhance adaptation to market preferences on the other. Research on bioremediation techniques will have to be stepped up to clean up problems arising from soil and water pollution. It is not surprising that the very first patent given to any living organism was for a microorganism developed via genetic engineering by Dr Ananda Chakrabarty for cleaning up pollution caused by oil spills (Chakrabarty 1981).

Blending of Recombinant DNA Technology and Organic Farming Methods

For every problem, there is a solution. Methods are being developed to find substitutes for antibiotic markers in recombinant DNA experiments. Also, techniques which help to be as precise as possible in the transfer of alien DNA are being standardised. It is likely that most of the current biosafety and environmental concerns associated with GM crops will be satisfactorily addressed scientifically during the next few years, so that precision breeding will become an important component of an economically and ecologically efficient precision farming system. The following examples from the work and approach of scientists at the M S Swaminathan Research Foundation (MSSRF) will help to illustrate the power of blending traditional practices with frontier technologies.

A. *Prebreeding and Participatory Breeding*

An integrated prebreeding procedure leading to the production of novel genetic combinations and designer genotypes, and participatory breeding involving the development of location-specific varieties jointly with farming families, would help to combine genetic efficiency and diversity in a mutually reinforcing manner. This will help to avoid the danger inherent in spreading single genotypes over large areas. Also, sustainable agriculture needs for its sustenance location-specific varieties.

At MSSRF, a group of scientists headed by Dr Ajay Parida have been working during the past 8 years on the transfer of salinity-tolerance genes from mangrove tree species to annual crops. This programme was initiated with support from the Department of Biotechnology, Government of India, to prepare genetic material which will prove to be useful if sea levels rise, thereby safeguarding coastal agriculture.

One such gene — betaine aldehyde dehydrogenase (*BADH*), cloned from a highly salt-tolerant mangrove species, *Avicennia marina* — is currently being evaluated in transgenic tobacco and *Brassica* systems for its efficacy. *BADH* converts betaine aldehyde to glycine betaine. Glycine betaine is an effective compatible solute, and its accumulation confers salinity tolerance in plants. The transgenic tobacco and *Brassica*, overexpressing the *BADH* from *Avicennia*, conferred salinity tolerance up to 250 mM NaCl. Work on isolation of a gene that can convert the ubiquitous choline into betaine aldehyde is being actively pursued. Other genes relating to stress resistance isolated and characterised from the mangrove species include catalase (*CAT*), superoxide dismutase (*SOD*), glyoxalase and sodium hydrogen antiporter. These genes are being evaluated for their expression in transgenic systems. Also, transformation work is in progress in rice and *Vigna* (Ajay Parida, personal communication).

Once transgenic plants with the desired salinity tolerance are developed, they will be used, after obtaining the necessary clearance from the regulatory authority, in breeding programmes undertaken jointly with farm families. The aim will be to transfer to numerous locally adopted varieties the salinity-tolerance characteristic in coastal and in inland areas where salinity is a problem.

B. *Bioremediation: Sequestration of Salt*

Salinity is responsible for major crop losses, particularly in semi-arid and irrigated agriculture. High salinity in soil may result from excessive irrigation or excessive application of chemical fertilizer. Usually, sulfates, chlorides, and bicarbonates of Na^{+}, Mg^{2+}, and

Ca^{2+} contribute to the salinity of soil. Sodium (Na^+) is the predominant soluble cation in most saline soil water, particularly in coastal areas. An alternate approach for practising crop cultivation in saline environments is the amelioration of soil salinity in agricultural habitats. Conventionally, it is done by the addition of gypsum followed by leaching out of excess salts by flooding. A biological approach to solving this problem will be preferable. *Anabaena torulosa,* a blue-green alga, was found to grow and enrich the nitrogen status of moderately saline "Kharland" soils. *A. torulosa* was found not to intracellularly accumulate Na^+ but rather entrap the cation in its extracellular mucopolysaccharide sheath, thereby reducing the availability of this deleterious cation to the crops. Research on the biological sequestration of salt from the soil may be rewarding.

Under a collaborative research programme between the Bhabha Atomic Research Centre and MSSRF, a salt-tolerant culture of *A. torulosa* along with *AL31* were given for testing in the southern coastal region. Several field trials have been conducted so far. The trials have shown that *A. torulosa* established very well along with local *Ananaena* sp. in the field condition. Enhanced nitrogenase activity was observed in the field after transplantation. Up to 64% salt sequestration was observed when 1000 mL of innoculum was added (Sudha Nair, personal communication).

Biotechnological Applications in Organic Farming

MSSRF scientists are integrating a wide variety of biotechnological applications in improving the productivity, profitability, stability, and sustainability of major cropping systems. Among the techniques of particular value are vermiculture, biopesticides, biofertilisers including stem-nodulating green manure crops, azolla, blue-green algae, and improved rhizobial strains. Such biopesticides and biofertilisers are best produced by village-level self-help groups. In fact, there are good opportunities for gainful employment in the area of producing such biological software for sustainable agriculture.

Transgenic Crops: The Way Ahead

The era of precision breeding opened up by advances in genomics and genetic engineering has become an ally in the movement for environmentally sustainable advances in agriculture, a phenomenon I have christened as "evergreen revolution" (see Swaminathan 2000). Knowledge is a continuum. The 20th century was marked by spectacular advances in crop productivity triggered by Mendelian breeding. The 21st century will witness even more spectacular progress from an intelligent integration of Mendelian and molecular breeding. The enormous power which transgenic technology has conferred on humankind imposes an ethical obligation, which should be discharged by developing a transparent and multistakeholder method of risk-benefit analysis, capable of inspiring public confidence and trust.

At the same time, the tendency to decry all advances in the breeding of transgenic crops will not be in the interest of sustainable food and nutrition security. India's population exceeds a billion, and there is no option in the future except to produce more crops per unit of land and per every drop of water. In my Coromandel Lecture titled "Agriculture on Spaceship Earth" delivered on 26 February, 1973, I mentioned that we are fortunately in a position to build a positive policy of economic ecology based on a series of do's rather than don't's (Swaminathan 2001).

Getting the best out of the new genetics for farm families will be possible only if the principles of economic ecology as well as a "do" philosophy underpin science and public policies.

III. Sustainable Food Security

The concept of food security has been undergoing an evolutionary change during the last 50 years. In the 1950s, food security was considered essentially in terms of production. It was assumed that adequate production will assure adequate availability of food in the market as well as in the household. In the 1970s, it became clear that availability alone does not lead to food security, since those who lack purchasing power will not be able to have access to a balanced diet; purchasing power in turn is related to jobs or livelihood opportunities. More recently, it is becoming evident that even if availability and access are satisfactory, the biological absorption of food in the body is related to the consumption of clean drinking water as well as environmental hygiene, primary health care, and primary education. Finally, even if physical and economic access to food is assured, ecological factors will determine the long-term sustainability of food security systems. Based on the above considerations, the M S Swaminathan Research Foundation and the United Nations World Food Programme (WRP) have recently brought out a *Food Insecurity Atlas of Rural India* (Vepa *et al.* 2001).

The *Food Insecurity Atlas* of MSSRF and WFP reveals that every State in the country has its strengths and weaknesses in relation to the five major dimensions considered in the analyses. These are: **availability of food**, which is a function of production; **access** to food, which is related to purchasing power; **absorption** of food in the body, which is determined by the availability of safe drinking water, environmental hygiene, primary health care, and primary education; **vulnerability** to transient hunger, which is related to natural and manmade calamities and disasters; and **sustainability** of production, which is influenced by the extent of attention given to the ecological foundations essential for sustained advances in production. The *Atlas* reveals that nonfood factors like livelihood and income-earning opportunities, health care facilities, education, sanitation, and environmental hygiene are as important for food security at the level of every individual, as factors relating to the availability of food grains in the market and access to clean drinking water.

The *Atlas* provides an opportunity for State Governments to draw up food security balance sheets based on strengths and weaknesses, and to identify the "hot spots" with reference to endemic and transient hunger as well as to open (i.e. protein-calorie undernutrition) and hidden (i.e. micronutrient deficiencies) hunger.

While we should give the highest priority to improving food consumption and equitable distribution, we should not decelerate our efforts in improving agricultural production through yield improvement, higher factor productivity, and better postharvest management. Agricultural production, factor productivity, and investment in irrigation and postharvest and rural infrastructure are all declining in India. Prices of many agricultural commodities have collapsed, and Indian farm families are in deep economic and psychological distress. This trend, if not arrested immediately, will lead to social chaos, since agriculture (crop and animal husbandry, fisheries, forestry, agroprocessing, and small-scale agribusiness) is not just a food-producing machine, but is the backbone of the livelihood security system for nearly 700 million children, women, and men in the country. There is no time to relax on the food production front just because the major problem today is in the area of marketing and distribution. At the same time, we should not continue to remain silent spectators to the coexistence of mountains of grains and millions of hungry.

IV. From a Green to an Evergreen Revolution

The first 60 years of the 20th century were marked by a sense of despair and frustration regarding India's capability to achieve a balance between human numbers and the production of foodgrains and other agricultural commodities. In 1968, this mood of despair and diffidence gave way to one of optimism and self-confidence in relation to our agricultural potential and our farmers' ability to adapt and adopt new technologies, a phenomenon which was christened in that year as "Green Revolution". This agricultural transformation helped to strengthen national sovereignty in many areas, including the capacity to remain nonaligned in foreign policy.

Our agriculture is now at a crossroads. On the one hand, our national capability in frontier areas of science and technology — as for example in biotechnology, information, communication and space technologies, nuclear and renewable energy technologies, and management science — has opened up uncommon opportunities for achieving an "evergreen revolution", i.e. sustainable advances in crop productivity per units of land, water, and time without associated ecological harm. Thus, sustainable food security will have to be defined as "physical, economic, social, and ecological access to balanced diets and safe drinking water, so as to enable every individual to lead a productive and healthy life in perpetuity". A life cycle approach will have to be followed in the case of nutrition, ranging from *in utero* to old age. Achieving such a form of food security will require synergy between technology and public policy.

There are, on the other hand, both internal and external threats to our agricultural progress. The most important among the internal threats is damage to the ecological foundations essential for sustained agricultural advance, like land, water, forests, and biodiversity. The other major internal weakness is the mismatch between production and postharvest technologies, and the consequent need for the Government of India to undertake "trade relief" operations on lines similar to those of cyclone, flood, and drought relief. External threats include the unequal trade bargain inherent in the WTO agreement of 1994; the rapid expansion of proprietary science; and potential adverse changes in temperature, precipitation, sea level, and ultraviolet B radiation.

The global threats to the agricultural destiny of developing countries can be overcome only if industrialized countries, particularly the United States of America, are willing to take the following steps.

- Ensure that the Kyoto protocol relating to the Climate Convention is implemented.
- Extend adequate support to public good research at the national and international levels, thereby fostering a new social contract between science and society (Swaminathan 1999).
- Revise the Agreement on Agriculture of the World Trade Organisation in a manner such that trade becomes a powerful tool for poverty eradication. At present, there is no level playing field between the produce emanating from factory farming and farmers' farming (i.e. small-scale producers).

V. Sustainable Advances in Agricultural Productivity

The smaller the farm, the greater the need for marketable surplus to ensure cash income. Fortunately, the gap between potential and actual yields is high in most farming systems. Even in the case of rice and wheat, the present average yield is just 40% of what can be achieved even with technologies currently on the shelf. Therefore, a massive effort should be made to launch a productivity revolution in farming. An integrated approach is necessary to remove the technological, infrastructure, and social and policy constraints responsible for the productivity gap and, in some cases, productivity decline. Reducing the cost of production through ecotechnologies and improving income through efficient production and postharvest technologies will help to enhance opportunities for both skilled employment and farm income. Precision farming methods which can help to enhance income and yield per drop of water and per units of land and time need to be standardised, demonstrated, and popularised speedily, if a reduction in the cost of production is to be achieved without reduction in yield.

As an immediate measure for strengthening food security at the level of individuals and households, there is no better option than initiating a systematic effort in each agroclimatic zone to identify and remove the constraints responsible for the prevailing yield gaps. This is true not only of crop plants, but also of livestock and fisheries. The local Panchayati Raj institutions or other forms of local bodies should be fully involved both in identifying constraints that limit production and in removing them. The following are some of the other steps needed.

Land and Water Care

Some of the measures needed to conserve land for agricultural purposes are:

- Arresting land degradation and the loss of biological potential of the soil (desertification)

- Promoting land and water use on the basis of agroecological, meteorological, and marketing factors
- Restoring degraded and wasted land through agroforestry and other appropriate methods of restoration ecology
- Launching community-centred water harvesting, conservation, and use programmes to ensure the efficient harvest of rainwater and the sustainable use of ground water. The adoption of conjunctive water use practices ensures integrated use of river, rain, ground, treated sewage, and sea water in such a manner that every drop yields more crop and income.
- Introducing public policies to prevent the diversion of prime farm land for nonfarm uses and the unsustainable exploitation of ground water.

There are currently several Central and State Government programmes dealing with wasteland development, rainwater harvesting, watershed development, command area management, shallow tube well construction, social forestry and agroforestry, and prevention of damage to hydrologic cycles in hilly areas. There is an urgent need for convergence and synergy among these programmes so that land and water conservation and use can be dealt with in a scientific and holistic manner. Land and water management problems (the term "management" is used to denote concurrent attention to conservation, sustainable use, and equitable sharing of benefits) are multidimensional, so unidimensional approaches through numerous independent schemes implemented by separate departments of Central and State governments will only result in inefficient and ineffective use of financial and technical resources.

1. Schedule 11 of Constitution amendment 73, relating to Panchayati Raj institutions, entrusts to Panchayats/local bodies responsibilities for the management of land, water, and common property resources. If these bodies, in which a third of the members are women, are legally, technically, and financially enabled to discharge the functions listed in Schedule 11, a beginning can be made in fostering land and water management in

an environmentally and socially sustainable manner. There are some legislative, administrative, and legal hurdles in the way of decentralised planning. Hence, it is important to address these issues immediately.

2. The existing State Land Use Boards should be revitalised and reorganised in such a manner that they can give proactive advice to farm families on land use during the southwest and northeast monsoon periods based on the following factors:

    - Farming systems (crop, livestock, fish, and agroforestry) which will be most efficient under the given soil, water, and climatic conditions
    - Short- and medium-range weather forecasts (the country has developed considerable capability in this area)
    - Projected market demand (both home and external markets)
    - Cost of production, risks involved, and expected return
    - Potential for on-farm and nonfarm livelihood generation, so as to maximize income and employment per units of land and water.

If such advice is given at least several weeks before the sowing season, a proper match can be achieved between production and potential market demand. Uneconomic market interventions can then be avoided. The agroecological potential of every village can be utilised in an ecologically and economically optimum manner. The "Blue Box" of the Agreement on Agriculture of the World Trade Organisation provides for expenditure on achieving a balance between demand and supply in farm products. Seed banks of alternative crops will have to be established at the local level.

The reorganised Land Use Boards should also be able to develop contingency cropping patterns to suit different rainfall and water availability patterns. Thanks to the long-range weather data available by the Meteorological Department, it is now possible to develop computer simulation models of likely deviations in monsoon behaviour. These can be used for formulating land use advice based on GIS maps, which also take into consideration the moisture-holding capacity of soils, physiological efficiency of crops, home needs, and market demand.

If such steps are taken, we can promote land use based on considerations of both ecological sustainability and economic efficiency. Since land use decisions are also water use decisions, land and water care and use are best dealt with in a simultaneous and interactive manner. For example, if the ongoing technology missions in crops like oilseeds, pulses, maize, and cotton are linked to the watershed development and dry farming programmes, these missions will become more effective.

VI. Reorganisation of Extension Services

Advances in information technology also provide opportunities for farm graduates to establish computer-aided and internet-connected Rural Knowledge Centres. These centres should help to convert generic information into location-specific information. The present extension service has outlived its utility. It can be replaced over time by farmer-owned and farmer-operated knowledge centres. A virtual college linking such village knowledge centres to agricultural universities and research institutions can be established, so that farm women and men are able to get up-to-date and authentic technical advice. Nearly a million farm graduates (both men and women) can be involved in establishing and operating such Rural Knowledge Centres based on modern information and communication technology. Such centres can also operate local community radio stations. Such a restructuring and retooling of extension services will help to provide demand-driven environment and farming systems as well as specific advice to farmers. They will trigger a knowledge revolution in agriculture, and will lead to an efficient and ecosensitive precision farming movement. This great opportunity for achieving a transition from unskilled to skilled work and for designing a new extension service for the new economy should not be missed.

VII. Linking Food and Ecological Security

The provisional census 2001 figures reveal that our population is higher by nearly 20 million than expected and that the sex ratio continues to be adverse to women (933 women per 1000 males). The sex ratio is even more adverse in the 0–6-year age group, suggesting the possibility of increasing female foeticide. Fortunately, female literacy is improving and Madhya Pradesh has shown the way for achieving a quantum leap in both male and female literacy through its education guarantee programme. While we must relentlessly pursue the goals of literacy and health for all and gender justice and equity, we should take advantage of a rare and unique opportunity in the history of independent India provided by the growing grain stocks in Government godowns to leapfrog in our efforts to realize Gandhiji's vision of a hunger-free India.

Current government stocks of wheat, rice, and other grains exceed 45 million tonnes. The Government may have to purchase another 15 to 20 million tonnes of wheat and rice during the next few months. A considerable proportion of these stocks remains in gunny bags and temporary storage structures. The Government of India has announced a scheme for the construction of large numbers of rural godowns. Severe drought in several parts of Rajasthan, Gujarat, Madhya Pradesh, and other States is compounding the problems of poverty-induced endemic hunger and drinking water scarcity.

The time is therefore opportune to launch an imaginative "Community Grain Bank" movement. On average, one tonne of wheat or rice supports the food needs of five individuals in our country. Community Grain Banks each with 200 tonnes of wheat or rice or other locally acceptable staples like *ragi, jowar, bajra,* and *maize* could be established, to begin with, in "hunger hot spot" villages. Remote areas with poor communication such as the desert areas of Rajasthan as well as hilly, tribal, and drought-affected areas can be given priority in starting the Community Grain Bank movement. As many as 25,000 Grain Banks can be established during the next few months if the Government of India will immediately approve the release of five million tonnes of grain for this purpose. Because large quantities of Government stocks are in gunny bags, it is easy to move them to the Community Grain Banks, where they can be stored using the low-cost technology standardised by the Food Corporation of India. It will be sad if the Government sits on over 60 million tonnes of food grains, allowing some of it to rot, rather than take them to places where, in Gandhiji's words, "God is Bread".

Based on the experience of the initial 25,000 village-level grain banks, another 25,000 can be established later this year, thus using 10 million tonnes of the surplus stock in a socially meaningful manner. Let the first year of the new millennium be a year of decisive action in our resolve to provide every individual in the country an opportunity for a productive and healthy life.

The Community Grain Banks can be sustained with locally procured grains, wherever feasible. They should be linked to the rural godowns scheme. The Banks could function under the overall umbrella of the Gram Sabha, and can be operated by local self-help groups of women and men. This will ensure their relevance to local conditions in addition to involving low transaction costs. The Community Grain Banks could be used for initiating at the local level food for work, food for nutrition (i.e. distribution of food among pregnant and nursing mothers, infants and old, and infirm persons), waste land and watershed development, ecological restoration of common property resources, and community water banks (see MS Swaminathan, *Sunday Hindu*, 15 October 2000). They can also be the vehicles for operating the targeted public distribution, Antyodaya Anna Yojana, and other Central and State Government schemes. Thus, the Community Grain Banks can become instruments of ecorestoration, water harvesting, and hunger elimination.

We should link conservation, cultivation, and consumption in a mutually reinforcing manner. For this, it will be useful to foster the establishment of community gene, seed, water, and grain banks in every village.

VIII. Increasing Production and Productivity

Future agricultural production programmes will have to be based on a three-pronged strategy designed to foster an evergreen revolution, which leads to increased production without associated ecological and social harm. The following are the four major elements of this strategy for producing more in an environment-friendly manner:

a. Defending the gains already made: This will call for conservation and enhancement of soil and water resources as well as forests and biodiversity through an integrated package of government regulation, education, and social mobilisation (through Panchayats and local bodies). The traditional "green revolution" areas are in urgent need of such an integrated natural resource management strategy so that the pattern of present production does not erode future prospects. The Punjab, which is India's granary today, will become food insecure in 15 to 20 years from now if the current unsustainable land and water use practices continue. Defending the gains already achieved will also need stepping up maintenance research for ensuring that new strains of pests and pathogens do not cause crop losses. Special steps are needed to prevent the introduction of invasive alien species, which are coming into the country along with imported food and agricultural commodities. These invasive alien species, like new and aggressive weeds, nematodes, etc., can cause incalculable harm to the future of Indian agriculture.

Conservation and enhancement of land and water resources are important. Water harvesting, watershed development, and economic and efficient water use can help to enhance productivity and income considerably. Conjunctive use of different water sources should become the rule rather than the exception. Unless there is equity in water sharing, there will be no cooperation in water saving. Therefore, equitable methods of water sharing should be promoted. Where water is scarce, high-value but low-water-requiring crops should be grown. **In this context, the organization of Pulses and Oilseed Villages should become a national movement.**

This should be a major aim of the Pulses and Oilseeds Technology Missions. Solving internal shortages of pulses and oilseeds through imports will only add to the economic woes of dry land farming communities. Pulses and oilseeds are important income-earning and soil-enriching crops in dry land areas. Various estimates of land degradation exist. The Ministry of Rural Development has also published a Wasteland Atlas of India. The following kinds of soil degradation have been quantified:

Wind erosion	19.7 million ha.
Salinisation	4.1 million ha.
Water logging	3.1 million ha.
Water erosion	69.6 million ha.
Soil fertility decline	13.7 million ha.

Thus, there are vast opportunities for launching Wasteland Development Enterprises by local self-help groups through the following strategy.

- Identify the precise nature of soil degradation and develop scientific restoration measures.
- Based on agro-ecological conditions, choose tree species which can help to initiate suitable enterprises. For example, a plant pesticide model of wasteland development could involve the planting of neem and melia.
- Appropriate species can be chosen and planted, depending on soil and water conditions, for undertaking the preparation of furniture, doors, windows, etc. or paper, fibre, or fruit packaging industries.

The aim is to add value to wasteland development through an integrated strategy of restoration and commercialisation. Such a twin approach will impart greater momentum to wasteland reclamation, particularly in periurban areas.

b. Extending the gains to rain-fed and semiarid hill and island areas, which have so far been bypassed by yield enhancement

technologies: Regional imbalances in agricultural development are growing, based largely on the availability of assured irrigation on the one hand and assured and remunerative marketing opportunities on the other. North Bihar is an exception, where water is plentiful but agricultural growth is slow. Eastern India has a large untapped yield reservoir and, by and large, falls under the "green but no green revolution" category. West Bengal has made impressive progress since the 1990s; while more recently Assam has started making progress, thanks to a large shallow-tube well programme designed to tap ground water during rabi and summer (*boro*) seasons. The introduction of ecoregional technology missions aimed to provide appropriate packages of technology, technoinfrastructure, services, and input and output pricing and marketing policies will help to include the excluded in agricultural progress.

Technologies for elevating and stabilising yields are available for semiarid and dry farming areas, as a result of the work done by the Indian Council of Agricultural Research institutions, state agricultural universities, and the International Crop Research Institute for Semi Arid Tropics. Agroforestry and animal husbandry are extremely important in arid and semiarid regions. Livestock and livelihood are closely linked in such areas. A major effort in water conservation and management and land use planning is needed in all areas which have been bypassed by scientific agriculture. Attention to horticulture, with particular emphasis on postharvest technology, will help to optimise income and employment from every drop of water. Both livestock and tree farming will provide opportunities for downstream employment. Therefore, the emphasis should be on farming systems that can optimise the benefits of natural resources in a sustainable manner, and not merely on cropping systems. Also, in coastal areas, there is need for a massive programme of coastal system research and development, involving capture and culture fisheries, coastal forestry and agroforestry, and integrated crop and animal husbandry. Coastal forestry and agroforestry can provide much of the fuel wood needs of inland areas. As in the crop sector, there is a vast untapped production reservoir in the fisheries sector. With the coming into force of the UN

Convention on the Law of the Sea, India's exclusive economic zone in the oceans around the country extends to over two million square kilometers.

Dry farming areas are also ideal for the cultivation of low-water-requiring but high-value pulses and oilseeds. The Pulses and Oilseeds Technology Missions should be revitalised and linked to water harvesting and watershed development during the Tenth Plan (2002–2007). As emphasised earlier, taking the easy option of importing large quantities of pulses and oilseeds forecloses the great opportunity for improving the economic well-being of farm families in dry farming areas through improving the production and productivity of pulses and oilseeds.

c. **Making new gains through farming system intensification, diversification, and value addition:** During the past decade, discoveries in information and biological technologies have contributed in unanticipated ways to fundamental changes in the global economy and to unprecedented economic growth, particularly in industrialised countries. There are also growing bonds of partnerships between universities and industries. In the USA, for example, industrial research parks surround leading research universities. We have more than 40 agricultural, animal science, fisheries, and rural universities, in addition to numerous agricultural and forestry research institutions. Universities and research institutions should serve as the engines of growth in a knowledge-based economy. They should also address, through their research and training agendas, the great challenges that confront our country in terms of poverty and the lack of basic human needs.

Detailed agroclimatic and soil maps are available for the country. Watershed and wasteland atlases are also available. We have considerable capacity in remote sensing and GIS mapping. These should be used for developing improved farming systems, which can provide more income and jobs. Value addition to primary products should be done at the village itself. Integrated crop-livestock-fish production systems should be fostered. Opportunities for nonfarm employment will then improve.

d. Institutional support: Higher production can be sustained only if there are opportunities for assured and remunerative marketing. A major challenge relates to reducing the cost of production by improving productivity. This will call for appropriate institutional structures which can help to provide key centralised services to small and marginal farm families, and to provide them with the power of scale in ecofarming (i.e. integrated pest management, scientific water management, integrated nutrient supply, precision farming, etc.) as well as in marketing. The role of the Small Farmers Agribusiness Consortium (SFAC) which was established for this purpose should be reviewed, and appropriate institutional structures owned and controlled by farm families should be promoted. Federations of self-help groups, farmer-controlled cooperatives and corporate business entities, and other socially relevant institutional structures should be promoted. Without enhanced efficiency in the production and postharvest phases of agriculture, Indian farmers will not be able to face the challenge of globalisation in terms of cost competitiveness, quality of produce, and consistency of supply.

It should not be forgotten that but for the existence of a very capable and professionally run National Dairy Development Board as well as a dairy farmers' cooperative movement, we would not have been able to achieve the top position in milk production in the world. User-controlled and user-driven institutional structures characterised by low transaction costs are essential to provide the needed assistance in postharvest technology, like drying, storage, processing, and marketing.

Adequate food availability is necessary for both stabilising prices and ensuring the operation of an effective public distribution system. There is therefore no time to relax on the food production front. There is particularly an urgent need for greater investment in irrigation, power supply, rural roads, cold storages, godowns, and food processing units. By extending the benefits of technological transformation and institutional reform to more areas and farming systems, India can become a leader in world agriculture.

IX. Policies for Improving Economic Access to Food

As early as 1856, Col. Baird Smith, who investigated the causes of a serious famine in Northwest India, wrote: "Indian famines are famines of work, and not of food. Where there is work, there is money. Where there is money, there is food." This situation is as relevant today as it was 150 years ago. Food security in India is best described in million-person-years of jobs and livelihoods rather than in million tonnes of food grains.

Agriculture — comprising crop and animal husbandry, inland and marine fisheries, forestry and agroforestry, agroprocessing, and agribusiness — constitutes the backbone of the livelihood security system of India, particularly in rural areas. Our agriculture is still "farmers' farming" and not "factory farming", as in industrialised countries. This is our great strength, since the health of plants and animals and other hazards associated with factory farming are now becoming evident. **Therefore, jobs/livelihoods for Indians must be the bottom line of all our economic and development policies.** Unfortunately, modern industry is not labour absorbing, and usually enhances its efficiency by downsizing of staff to improve output per person. The "new economy" based on information technology and knowledge industries is also by itself not employment intensive, but could lead to new employment if intelligently used.

Farming is the largest private sector enterprise in India. Nearly 58.9% of workers depend upon agriculture for their income and livelihoods. Now that the share of agriculture in total workers is declining, we must think of increasing their productivity and alleviating rural poverty. Rural poverty is greater than urban poverty. Nearly 50% of the rural population, belonging to labour families, are engaged in unskilled low-wage work in several parts of the country.

New Economy and New Employment Opportunities

Opportunities for new employment include the production of ecofoods; biological software for sustainable agriculture like biofertilizers, biopesticides, and vermiculture; bioprocessing; health foods; herbal medicines; recycling of solid and liquid wastes; and agriculture and agroprocessing machinery. In the new knowledge-based economy, good ecology will be fundamental to good business.

Thanks to both the ongoing technological revolution — particularly in molecular genetics, information, and space applications — and the spread of democratic systems of governance at the grassroots level, it is now possible to work towards achieving a substantial reduction in chronic and hidden hunger by the year 2007, which marks the 60th anniversary of India's independence. What is important to ensure is that the means adopted to achieve this end are not at the expense of the prospects for sustainable food security for the generations yet to be born.

The best tribute we can pay to the life and work of Acharya NG Ranga is to work with determination for achieving the goal of a hunger-free India. This will call for happy farming families as well as sustainable farming and equitable trading systems. Trade should be not only free, but also fair. Above all, farmers' farming should be safeguarded against the onslaught of factory farming, since farmers' farming is based not just on the economics of money, but on the economics of livelihood security and human dignity.

References

1. Chakrabarty AM (1981). Microorganisms having multiple compatible degradative energy-generating plasmids and preparation thereof. US Patent & Trademark Office, March 31, 1981.
2. Provisional Population Total Paper I of 2001. Registrar General and Census Commissioner, India.
3. NSS, 55th Round (2000).
4. Swaminathan MS (1968). The age of algeny, genetic destruction of yield barriers and agricultural transformation. Presidential Address, Agricultural Science Section, Fifty-fifth Indian Science Congress, January 1968. Proceedings of Indian Science Congress, Varanasi, India.
5. Swaminathan MS (1982). Biotechnology research and third world agriculture. *Science* **218**:967–972.

6. Swaminathan MS (ed.) (1993). *Wheat Revolution: A Dialogue.* Macmillan India Ltd, Madras, India.
7. Swaminathan MS (1999). Science in response to basic human needs. *Curr Sci* **77**(3):341–353. Keynote Address delivered on 21 June 1999 at the World Conference on Science, Budapest, Hungary.
8. Swaminathan MS (2000). An evergreen revolution. *Biologist* **47**(2):85–89.
9. Swaminathan MS (2001). Platform for a common present and future for humankind: introduction. *Coromandel Lectures,* Coromandel Fertilizers Ltd, Secunderabad, Andhra Pradesh, India. Pp. vii–xxxx.
10. Vepa SS, Bhavani RV, *et al.* (2001). *Food Insecurity Atlas of Rural India.* MS Swaminathan Research Foundation and the World Food Programme, Chennai, India.

Index